Reliability and Maintenance Modeling with Optimization

Reliability and maintenance modeling with optimization is the most fundamental and interdisciplinary research area that can be applied to every technical and management field. *Reliability and Maintenance Modeling with Optimization: Advances and Applications* aims at providing the most recent advances and achievements in reliability and maintenance.

The text discusses replacement, repair, and inspection, offers estimation and statistical tests, covers accelerated life testing, explores warranty analysis manufacturing, and includes service reliability.

The targeted readers are researchers interested in reliability and maintenance engineering, and the book can serve as supplemental reading in professional seminars for engineers, designers, project managers, and graduate students.

Advanced Research in Reliability and System Assurance Engineering

Series Editor: Mangey Ram, Professor, Graphic Era University, Uttarakhand, India

Modeling and Simulation Based Analysis in Reliability Engineering
Edited by Mangey Ram

Reliability Engineering
Theory and Applications
Edited by Ilia Vonta and Mangey Ram

System Reliability Management
Solutions and Technologies
Edited by Adarsh Anand and Mangey Ram

Reliability Engineering
Methods and Applications
Edited by Mangey Ram

Reliability Management and Engineering
Challenges and Future Trends
Edited by Harish Garg and Mangey Ram

Applied Systems Analysis
Science and Art of Solving Real-Life Problems
F. P. Tarasenko

Stochastic Models in Reliability Engineering
Lirong Cui, Ilia Frenkel, and Anatoly Lisnianski

Predictive Analytics
Modeling and Optimization
Vijay Kumar and Mangey Ram

Design of Mechanical Systems Based on Statistics
A Guide to Improving Product Reliability
Seong-woo Woo

Social Networks
Modeling and Analysis
Niyati Aggrawal and Adarsh Anand

Operations Research
Methods, Techniques, and Advancements
Edited by Amit Kumar and Mangey Ram

Statistical Modeling of Reliability Structures and Industrial Processes
Edited by Ioannis S. Triantafyllou and Mangey Ram

Industrial Reliability and Safety Engineering
Applications and Practices
Edited by Dilbagh Panchal, Mangey Ram, Prasenjit Chatterjee, and Anish Kumar Sachdeva

Reliability and Maintenance Modeling with Optimization
Advances and Applications
Edited by Mitsutaka Kimura, Satoshi Mizutani, Mitsuhiro Imaizumi, and Kodo Ito

For more information about this series, please visit: https://www.routledge.com/Advanced-Research-in-Reliability-and-System-Assurance-Engineering/book-series/CRCARRSAE

Reliability and Maintenance Modeling with Optimization

Advances and Applications

Edited by
Mitsutaka Kimura
Satoshi Mizutani
Mitsuhiro Imaizumi
Kodo Ito

CRC Press
Taylor & Francis Group
Boca Raton London New York

CRC Press is an imprint of the
Taylor & Francis Group, an **informa** business

First edition published 2023
by CRC Press
6000 Broken Sound Parkway NW, Suite 300, Boca Raton, FL 33487-2742

and by CRC Press
4 Park Square, Milton Park, Abingdon, Oxon, OX14 4RN

CRC Press is an imprint of Taylor & Francis Group, LLC

Library of Congress Cataloging-in-Publication Data

Names: Kimura, Mitsutaka, editor.
Title: Reliability and maintenance modeling with optimization : advances and applications / Mitsutaka Kimura, Aichi University, Japan, Satoshi Mizutani, Aichi Institute of Technology, Japan, Mitsuhiro Imaizumi, Aichi Gakusen University, Japan, Kodo Ito, Tottori University, Japan.
Description: Boca Raton : CRC Press, 2023. | Series: Advanced research in reliability and system assurance engineering | Includes bibliographical references and index.
Identifiers: LCCN 2022045317 (print) | LCCN 2022045318 (ebook) | ISBN 9780367558055 (hardback) | ISBN 9780367558062 (paperback) | ISBN 9781003095231 (ebook)
Subjects: LCSH: Reliability (Engineering)
Classification: LCC TA169 .R435 2023 (print) | LCC TA169 (ebook) | DDC 620/.00452--dc23/eng/20221202
LC record available at https://lccn.loc.gov/2022045317
LC ebook record available at https://lccn.loc.gov/2022045318

ISBN: 978-0-367-55805-5 (hbk)
ISBN: 978-0-367-55806-2 (pbk)
ISBN: 978-1-003-09523-1 (ebk)

DOI: 10.1201/9781003095231

Typeset in CMR10
by KnowledgeWorks Global Ltd.

Publisher's note: This book has been prepared from camera-ready copy provided by the authors.

Contents

SECTION I Stochastic Maintenance Policies

SECTION III Warranty Analysis Manufacturing

SECTION IV Software Reliability and Testing

SECTION V Maintenance Optimization and Applications

Preface

Reliability and Maintenance Modeling with Optimization: Advances and Applications is based on original research and survey articles from the latest research results. This widely covers mechanical engineering, electrical and systems engineering, computer science, management science, operations research and nuclear engineering. Reliability and maintenance modeling with optimization is the most fundamental but interdisciplinary research area. This can be applied to every technical and management field, because reliability and safety are the highest prioritized issues in our daily life. As the recent technical development rapidly grows, we often encounter technical and economic problems with reliability and maintenance. So we need to update the state-of-the-art knowledge of reliability and maintenance techniques, in which the key issue is optimization, to realize highly dependable systems and social platforms.

Since the failure and degradation phenomena are essentially uncertain, the techniques employed are based on probability and statistics. More specifically, several advanced mathematical techniques, such as Markov and non-Markov modeling, statistical inference and optimization algorithms, are useful to overcome the underlying technical problems. Also, the recent development of machine learning techniques enables us to provide more comprehensive solutions for practical problems by utilizing big data. Since reliability and maintenance modeling is an interdisciplinary research area, the latest research results should be summarized often and updated for applications for the real world in the comprehensive way. In this book, we aim at providing recent progress in reliability and maintenance modeling with optimization to academia and industries.

Professor Toshio Nakagawa has greatly contributed to this book, as shown in the impressive reference list of Chapter 1. He received B.S.E. and M.S. degrees from Nagoya Institute of Technology in 1965 and 1967, respectively; and a Doctor Engineering degree from Kyoto University in 1977. He worked as a Research Associate at Syracuse University for two years from 1972 to 1973. He is now an Honorary Professor at Aichi Institute of Technology, Japan. His research interests are optimization problems in operations research and management science, and also analysis of stochastic and computer systems in reliability and maintenance theory. He has published 7 books from Springer, and more than 300 papers in research journals.

Professor Hisahi Yamamoto has also greatly contributed to this book, as shown in the impressive reference list of Chapter 8. He received his B.S., M.S., and Ph.D. degrees in industrial engineering from Tokyo Institute of Technology in 1981, 1983, and 1996, respectively. His main research interests include optimizations based on reliability engineering for fault-tolerant systems, facility layout problems, and worker allocation optimization problems. He is a professor at Tokyo Metropolitan University, Japan.

Professor Lirong Cui has greatly contributed to this book, as shown in the impressive reference list of Chapter 4. He received his bachelor's degree from Tiangong University in 1983, master's degree in Science from the Institute of System Sciences, Chinese Academy of Sciences in 1986, and his PhD degree in Probability and Statistics from the University of Wales, UK, in 1994, respectively. He worked in the China Aerospace industry from 1986 to 1999 and at Beijing Institute of Technology from 2003 to 2021. From 2000 to 2002, he was a Research Fellow at the National University of Singapore. In May of 2021, he joined Qingdao University and took a part-time job in Southern University of Science and Technology.

This book covers the following topics:

- Stochastic Maintenance Policies
- Reliability Modeling & Application
- Warranty Analysis Manufacturing
- Software Reliability and Testing
- Maintenance Optimization and Applications

Mitsutaka KIMURA,
Satoshi MIZUTANI,
Mitsuhiro IMAIZUMI,
Kodo ITO

Contributors

Tomoaki Akiba
Chiba Institute of Technology
Japan

Richard Arnold
Victoria University of Wellington
New Zealand

Jiajia Cai
Nanjing University of Aeronautics and Astronautics
China

Mingchih Chen
Fu Jen Catholic University
Taiwan

Stefanka Chukova
Victoria University of Wellington
New Zealand

Lirong Cui
Qingdao University
China

Tadashi Dohi
Hiroshima University
Japan

Lance Fiondella
University of Massachusetts Dartmouth
USA

Takaji Fujiwara
SRATECH Laboratory, Inc.
Japan

Yu Hayakawa
Waseda University
Japan

Yoshiyuki Higuchi
Fukushima University
Japan

Mitsuhiro Imaizumi
Aichi Gakusen University
Japan

Shinji Inoue
Kansai University
Japan

Kodo Ito
Tottori University
Japan

Fengming Kang
Beijing Institute of Technology
China

Jau-Chuan Ke
National Taichung University of Science and Technology
Taiwan

Mitsutaka Kimura
Aichi University
Japan

Tzu-Hsin Liu
Chaoyang University of Technology
Taiwan

Sarah Marshall
Auckland University of Technology
New Zealand

Satoshi Mizutani
Aichi Institute of Technology
Japan

Vidhyashree Nagaraju
University of Tulsa
USA

Toshio Nakagawa
Aichi Institute of Technology
Japan

Syouji Nakamura
Kinjo Gakuin University
Japan

Taishin Nakamura
Tokai University
Japan

Fumio Ohi
Nagoya Institute of Technology
Japan

Hiroyuki Okamura
Hiroshima University
Japan

Cunhua Qian
Nanjing Tech University
China

Sun-Keun Seo
Dong-A University
South Korea

Jingyuan Shen
Nanjing University of Science and Technology
China

Shey-Heui Sheu[1,2]
[1]Asia University, Taiwan
[2]China Medical University Hospital, China Medical University
Taiwan

Wei-Teng Sheu
National Taiwan University of Science and Technology
Taiwan

Hironobu Sone
IBM Japan, Ltd.
Japan

Yoshinobu Tamura
Yamaguchi University
Japan

Jinting Wang
Central University of Finance and Economics
China

Hisashi Yamamoto
Tokyo Metropolitan University
Japan

Akihiro Yamane
Tottori University
Japan

Shigeru Yamada
Tottori University
Japan

Nan Yang
Beijing Jiaotong University
China

Won Young Yun
Pusan National University
South Korea

Zhe-George Zhang
Western Washington University
USA

Xufeng Zhao
Nanjing University of Aeronautics and Astronautics
China

Junjun Zheng
Ritsumeikan University
Japan

Lei Zhou
Yamaguchi University
Japan

Section I

Stochastic Maintenance Policies

1 Nine Memorial Research Works

Toshio Nakagawa
Aichi Institute of Technology, Japan

CONTENTS

1.1 INTRODUCTION

I have mainly studied reliability and maintenance theory since 1971 with S. Osaki [1]. Looking back to the past half century, I think back to the memorial nine papers written with my colleagues together and by myself [2]. In this chapter, I introduce the selected nine theoretical topics which are deeply impressed on my mind, and also have been cited greatly by other papers and books as follows: (1) Two-unit Standby System: First-passage distributions, renewal functions and transition probabilities of a two-unit standby system are obtained, using Markov renewal processes with nonregeneration points. (2) Imperfect Preventive Maintenance (PM): Three imperfect PM models are introduced; the unit becomes as good as new with probability q, and its age becomes x units younger and reduces to at at each PM. (3) Discrete Weibull Distribution: A discrete Weibull distribution is defined as $1 - q^{k^{\alpha}}$. (4) Definition of Minimal Repair: Minimal repair explained by the sentence of "the failure rate remains undisturbed by any failures" is defined mathematically and some theorems are shown. (5) Shock and Damage Models: A new policy where the unit is replaced at damage Z is proposed, and replacement policies with time T, shock N and damage Z are considered. (6) Finite Interval: Two finite interval models of imperfect PM and inspection policy are introduced and their optimal policies to minimize the total expected cost are derived. (7) Replacement First, Last, Overtime and Middle: Four replacement first, last, overtime and middle policies are proposed, and their expected cost rates are obtained. (8) Random K-out-of-n System: Reliability function of general redundant systems based on a K-out-of-n system is given and its mean time to failure (MTTF) is shown to be computed easily when the failure time is exponential. (9) Asymptotic Calculations: Using an exponential distribution, the mean time to jth failure and MTTF of redundant systems, and optimal checking times are derived approximately when the failure time has a Weibull distribution.

I believe that such studies would provide a good stimulus to young researchers and a useful resource to practical engineers and managers. Hereafter, I aim constantly to make further interesting studies and to write good papers and books, which might be cited by others and would be applied practically to actual fields.

1.2 TWO-UNIT STANDBY SYSTEM

Two-unit systems are the most fundamental redundant model in reliability theory and have been used widely in various ways and fields. Many researchers over the past 60 years have analyzed such systems [3, 4, p. ix] which were extensively summarized [5].

As a typical example of two-unit systems, we consider a two-unit standby redundant system which consists of two identical and independent units [5, 6]: An operating unit has a general failure distribution $F(t)$ with finite mean $1/\lambda$, and a failed unit has a general repair distribution $G(t)$ with finite mean $1/\mu$. If an operating unit fails and the other unit is in standby, the failed unit undergoes repair immediately and the standby unit takes over its operation immediately. However, an operating unit fails while the other is under repair, and the failed unit has to wait for repair until a repairman is free, which means system failure. Two units are used alternately for its operation as described above.

To analyze the system, we define the following four system states:

State -1: One unit begins to operate and the other is in standby.

State 0: One unit is operating and the other unit is in standby.

State 1: One unit is operating and the other unit is under repair.

State 2: One unit is under repair and the other unit is waiting for repair.

The system states defined above form a Markov renewal process [7, p. 123] in Fig 1.1: An epoch at which the system makes a transition into State 1 is a regeneration point, however, the epochs at which the system makes a transition into State j ($j = 0, 2$) are not regeneration points except both failure and repair times are exponential.

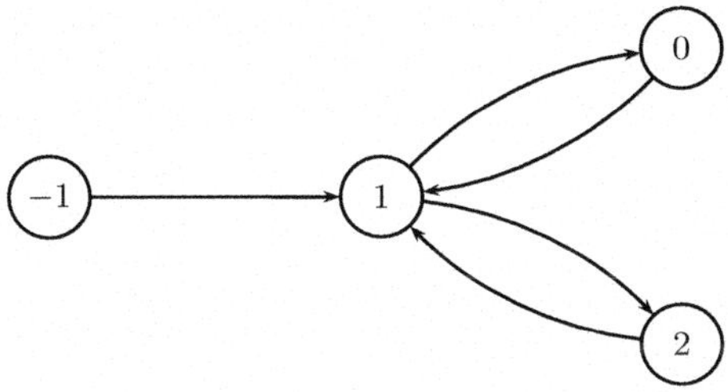

Figure 1.1 Transition diagram of a two-unit standby system.

Define a mass function, i.e., one step transition distribution $Q_{ij}(t)$ from State i $(i = -1, 1)$ to State j $(j = 0, 1, 2)$ by the probability that after making a transition into State i, the system next makes a transition into State j in time t. Then, the following mass functions are obtained:

$$Q_{-1\,1}(t) = F(t), \quad Q_{1\,0}(t) = \int_0^t \overline{F}(u)\mathrm{d}G(u), \quad Q_{1\,2}(t) = \int_0^t \overline{G}(u)\mathrm{d}F(u), \tag{1.1}$$

where $\overline{F}(t) \equiv 1 - F(t)$ and $\overline{G}(t) \equiv 1 - G(t)$.

We cannot define the mass functions $Q_{0\,1}(t)$ and $Q_{2\,1}(t)$ because the epochs for States 0 and 2 are not regeneration points. Then, we thought of defining new mass functions $Q_{1\,1}^{(0)}(t)$ and $Q_{1\,1}^{(2)}(t)$ which are the probability that after making a transition into State 1, the system next makes a transition into State 0 and State 2 and returns to State 1 in time t, respectively, which are given by

$$Q_{1\,1}^{(0)}(t) = \int_0^t G(u)\mathrm{d}F(u), \qquad Q_{1\,1}^{(2)}(t) = \int_0^t F(u)\mathrm{d}G(u). \tag{1.2}$$

Using the above mass functions, we obtain the following first-passage distributions, renewal functions and transition probabilities when the system starts from State -1 at time 0: Letting $H_{ij}(t)$ denote the first-passage distributions from State i $(i = -1, 1)$ to State j $(j = 0, 1, 2)$,

$$H_{1\,0}(t) = Q_{1\,0}(t) + Q_{1\,1}^{(2)}(t) * H_{1\,0}(t), \qquad H_{1\,2}(t) = Q_{1\,2}(t) + Q_{1\,1}^{(0)}(t) * H_{1\,2}(t),$$
$$H_{-1\,j}(t) = Q_{-1\,1}(t) * H_{1\,j}(t) \qquad (j = 0, 2), \tag{1.3}$$

where the asterisk mark represents the Stieltjes convolution, i.e., $A(t) * B(t) \equiv \int_0^t B(t-u)\mathrm{d}A(u)$. Thus, taking LS (Laplace Stieltjes) transforms of (1.3), LS transforms of the first-passage distributions from State -1 are

$$h_{-1\,0}(s) = \frac{q_{-1\,1}(s)q_{1\,0}(s)}{1 - q_{1\,1}^{(2)}(s)}, \qquad h_{-1\,2}(s) = \frac{q_{-1\,1}(s)q_{1\,2}(s)}{1 - q_{1\,1}^{(0)}(s)}, \tag{1.4}$$

where $\varphi(s) \equiv \int_0^\infty e^{-st} d\Phi(t)$ for $\mathrm{Re}(s) > 0$ and a general function $\Phi(t)$, and their mean times are

$$l_{-10} = \frac{1}{\lambda} + \frac{1}{\mu \int_0^\infty \overline{F}(t)dG(t)}, \qquad l_{-12} = \frac{1}{\lambda}\left[1 + \frac{1}{\int_0^\infty \overline{G}(t)dF(t)}\right]. \tag{1.5}$$

Similarly, the renewal functions $M_{ij}(t)$ from State i ($i = -1, 1$) to State j ($j = 0, 1, 2$) which are the expected numbers of occurrences of State j in $(0, t)$ starting from State -1 are

$$m_{-11}(s) = \frac{q_{-11}(s)}{1 - h_{11}(s)}, \qquad m_{-1j}(s) = \frac{q_{-11}(s)q_{1,j}(s)}{1 - h_{11}(s)} \quad (j = 0, 2), \tag{1.6}$$

where

$$H_{11}(t) = Q_{11}^{(0)}(t) + Q_{11}^{(2)}(t),$$
$$h_{11}(s) = \int_0^\infty e^{-st} G(t)dF(t) + \int_0^\infty e^{-st} F(t)dG(t).$$

The transition probabilities that the system is in State j ($j = 0, 1, 2$) at time t starting from State -1 are

$$P_{10}(t) = Q_{10}(t) - Q_{11}^{(0)}(t) + H_{11}(t) * P_{10}(t),$$
$$P_{11}(t) = 1 - Q_{10}(t) - Q_{12}(t) + H_{11}(t) * P_{11}(t),$$
$$P_{12}(t) = Q_{12}(t) - Q_{11}^{(2)}(t) + H_{11}(t) * P_{12}(t),$$
$$P_{-10}(t) = 1 - Q_{-11}(t) - Q_{-11}(t) * P_{10}(t), \qquad P_{-1j}(t) = Q_{-11}(t) * P_{1j}(t).$$

Thus, LS transforms of $P_{-1j}(t)$ ($j = 0, 1, 2$) are

$$p_{-10}(s) = 1 - q_{-11}(s) + \frac{q_{-11}(s)[q_{10}(s) - q_{11}^{(0)}(s)]}{1 - h_{11}(s)},$$

$$p_{-11}(s) = \frac{q_{-11}(s)[1 - q_{10}(s) - q_{12}(s)]}{1 - h_{11}(s)}, \quad p_{-12}(s) = \frac{q_{-11}(s)[q_{12}(s) - q_{11}^{(2)}(s)]}{1 - h_{11}(s)}, \tag{1.7}$$

and limiting probabilities $P_j \equiv \lim_{t\to\infty} P_{ij}(t)$ are

$$P_0 = 1 - \frac{1}{\mu l_{11}}, \qquad P_1 = -1 + \frac{1/\lambda + 1/\mu}{l_{11}}, \qquad P_2 = 1 - \frac{1}{\lambda l_{11}}, \tag{1.8}$$

where

$$l_{11} \equiv \int_0^\infty t dH_{11}(t) = \frac{1}{\lambda} + \frac{1}{\mu} - \frac{1}{\int_0^\infty \overline{F}(t)\overline{G}(t)dt}.$$

Even if there exist some nonregeneration points in stochastic models of objective redundant systems, we could derive the three reliability measures, using the above techniques [7, p. 140, 8, p. 30]. Such techniques were used effectively in analysis of computer systems [9].

1.3 IMPERFECT PM (PREVENTIVE MAINTENANCE)

It has been well known that reliability is very important for transportation such as cars, trains, ships and airplanes. Japanese National Railways did Wakagaeri Koji (Rejuvenation Work) for inspecting and maintaining railways and others from 1976 to 1982, and several trains of Sinkansen (Super-express Tokaido) stopped during their maintenance time. At that time, we were studying PM (Preventive Maintenance) policies of redundant systems [10] and thought an idea that such maintenances might be called imperfect PM and formulated its reliability models [8, p. 171, 11]. At the same time, imperfect repair models were considered and generalized [12, 13].

We introduce the following classical three imperfect PM models with minimal repair at failures [8, p. 176]:

(1) MODEL A PROBABILITY

Consider the periodic PM policy for a one-unit system which should operate for an infinite time span:

1. An operating unit is maintained preventively at times kT $(k = 1, 2, \dots)$ and undergoes only minimal repair at failures.
2. The failure rate $h(t)$ remains undisturbed by any minimal repair.
3. The unit after PM has the same failure rate as before PM with probability p $(0 \le p < 1)$ and becomes as good as new with probability $q \equiv 1 - p$.
4. Cost of each minimal repair is c_1 and cost of each PM is c_2.

The expected cost rate is

$$C_A(T; p) = \frac{1}{T}\left[c_1 q^2 \sum_{j=1}^{\infty} p^{j-1} \int_0^{jT} h(t)\mathrm{d}t + c_2\right], \tag{1.9}$$

and optimal T^* to minimize $C_A(T; p)$ satisfies

$$\sum_{j=1}^{\infty} p^{j-1} \int_0^{jT} t\mathrm{d}h(t) = \frac{c_2}{c_1 q^2}. \tag{1.10}$$

(2) MODEL B AGE

3. The age becomes x units younger at each PM, where x $(0 \le x \le T)$ is constant, and the unit is replaced if it operates for the time interval NT $(N = 1, 2, \dots)$ for a specified T $(0 < T < \infty)$.
4. Cost of each minimal repair is c_1, cost of each PM is c_2, and cost of replacement at time NT is c_3 with $c_3 > c_2$.

1, 2. Same as Model A.

The expected cost rate is

$$C_B(N;T,x) = \frac{1}{NT}\left[c_1 \sum_{j=0}^{N-1} \int_{j(T-x)}^{T+j(T-x)} h(t)\mathrm{d}t + (N-1)c_2 + c_3\right], \quad (1.11)$$

and optimal N^* to minimize $C_B(N;T,x)$ satisfies

$$\sum_{j=0}^{N-1} \int_0^T \{h[t+N(T-x)] - h[t+j(T-x)]\}\mathrm{d}t \geq \frac{c_3 - c_2}{c_1}. \quad (1.12)$$

(3) MODEL C RATE

3. The age after PM reduces to at $(0 < a \leq 1)$ when it was t before PM, i.e., the age becomes $t(1-a)$ units of time younger at each PM, and the unit is replaced at time NT $(N = 1, 2, \ldots)$.

1,2,4. Same as Model B.

The expected cost rate is

$$C_C(N;T,a) = \frac{1}{NT}\left[c_1 \sum_{j=0}^{N-1} \int_{A_jT}^{(A_j+1)T} h(t)\mathrm{d}t + (N-1)c_2 + c_3\right], \quad (1.13)$$

where $A_j \equiv a + a^2 + \cdots + a^j$ $(j = 1, 2, \ldots)$ and $A_0 \equiv 0$, and optimal N^* to minimize $C_C(N;T,a)$ satisfies

$$\sum_{j=0}^{N-1} \left[\int_{A_NT}^{(A_N+1)T} h(t)\,\mathrm{d}t - \int_{A_jT}^{(A_j+1)T} h(t)\mathrm{d}t\right] \geq \frac{c_3 - c_2}{c_1}. \quad (1.14)$$

Similar imperfect sequential PM models were summarized [14, 15].

1.4 DISCRETE WEIBULL DISTRIBUTION

It has been well known that the geometric and the negative binomial distributions of a discrete distribution correspond to an exponential distribution and a gamma distribution of a continuous distribution, respectively. We had a doubt that "What is a discrete distribution corresponding to a Weibull distribution?", because a Weibull distribution is the most important distribution in reliability theory and applications.

It is easily noted that when a continuous exponential survival function is $\overline{F}(t) = \mathrm{e}^{-\lambda t}$, replacing $\mathrm{e}^{-\lambda}$ with q and t with k $(k = 0, 1, 2, \ldots)$, we have a geometric survival distribution q^k. In a similar way, from the survival function

$\overline{F}(t) = \exp[-(\lambda t)^\alpha]$ of a Weibull distribution, we define the following discrete Weibull survival function [8, p. 17, 16]:

$$\overline{P}(k; q, \alpha) \equiv (q)^{k^\alpha} \qquad \text{for } \alpha > 0,\ 0 < q < 1, \tag{1.15}$$

the probability function p_k, the failure rate h_k, and the mean μ are

$$p_k = (q)^{k^\alpha} - (q)^{(k+1)^\alpha}, \quad h_k = 1 - (q)^{(k+1)^\alpha - k^\alpha}, \quad \mu = \sum_{k=1}^{\infty} (q)^{k^\alpha}. \tag{1.16}$$

The failure rate h_k increases (decreases) with k for $\alpha > 1$ ($\alpha < 1$) and agrees with a geometric distribution for $\alpha = 1$.

Since 1975, a discrete Weibull distribution has been used in many research areas and practical fields [17] and has been generalized theoretically. This has been cited by more than 450 papers and books, which increases a little with every year. Furthermore, referring to the above method, a survey of discrete distributions used in reliability models was presented [18] and other discrete distributions have been proposed [19].

1.5 DEFINITION OF MINIMAL REPAIR

Replacement with minimal repair has been defined that the unit is replaced periodically at planned times kT ($k = 1, 2, \dots$) and only minimal repair after each failure is made so that the failure rate remains undisturbed by any repair of failures between successive replacements [20, p. 96].

Suppose that the unit begins to operate at time 0. If the unit fails then it undergoes minimal repair and begins to operate again. Let denote by $0 \equiv Y_0 \le Y_1 \le \cdots \le Y_n \le \dots$ the successive failure times and $X_n \equiv Y_n - Y_{n-1}$. Under these assumptions, minimal repair at failures was defined theoretically by M. Kowada as follows [8, p. 96, 21].

Definition 1.5.1 *Let $F(t) \equiv \Pr\{X_1 \le t\}$ for $t \ge 0$. A unit undergoes minimal repair if and only if*

$$\Pr\{X_n \le x | X_1 + X_2 + \cdots + X_{n-1} = t\} = \frac{F(t+x) - F(t)}{\overline{F}(t)} \qquad (n = 2, 3, \dots) \tag{1.17}$$

for $x > 0$, $t \ge 0$ such that $F(t) < 1$.

The function $[F(t+x) - F(t)]/\overline{F}(t)$ is called the failure rate and represents the probability that the unit with age t fails in $[t, t+x]$. This means that the failure rate remains undisturbed by any minimal repair of failures, i.e., the unit after each minimal repair has the same failure rate as before failure. When $F(t)$ has a density function $f(t) \equiv \mathrm{d}F(t)/\mathrm{d}t$, $h(t) \equiv f(t)/\overline{F}(t)$ is called the instantaneous failure rate and has the same monotone property as $[F(t+x) -$

$F(t)]/\overline{F}(t)$. Furthermore, $H(t) \equiv \int_0^t h(u)\mathrm{d}u$ is called the cumulative hazard function which satisfies $\overline{F}(t) = \mathrm{e}^{-H(t)}$, where we note that $H(t)$ represents the expected number of minimal repairs during $[0, t]$.

Using the above definition of minimal repair, we get the following theorems:

Theorem 1.5.1 *Letting* $G_n(x) \equiv \Pr\{Y_n \leq x\}$ *and* $F_n(x) \equiv \Pr\{X_n \leq x\}$ *(*$n = 2, 3, \dots$*),*

$$G_n(x) = \sum_{j=n}^{\infty} \frac{H(x)^j}{j!} \mathrm{e}^{-H(x)}, \qquad F_n(x) = \int_0^{\infty} F(t+x) \frac{H(t)^{n-2}}{(n-2)!} h(t)\mathrm{d}t.$$

Theorem 1.5.2 *If the failure rate* $h(t)$ *increases with* t *to* $h(\infty)$ *then* $E\{X_n\} = \int_0^{\infty} [H(x)^{n-1}/(n-1)!]\mathrm{e}^{-H(x)}\mathrm{d}x$*, which is the mean time between failures, decreases with* n *(*$n = 1, 2, \dots$*) to* $1/h(\infty)$ *as* $n \to \infty$*.*

Theorem 1.5.3 *If the failure rate* $h(t)$ *increases with* t *then*

$$\frac{\int_0^T [H(t)^n/n!] f(t)\mathrm{d}t}{\int_0^T [H(t)^n/n!] \overline{F}(t)\mathrm{d}t} \qquad (n = 0, 1, 2, \dots)$$

increases with n *to* $h(T)$ *as* $n \to \infty$ *for any* $T > 0$*.*

Let $G(t)$ represent any distribution with failure rate $r(t) \equiv g(t)/\overline{G}(t)$ of $G(t)$, where $g(t)$ is a density function of $G(t)$.

Theorem 1.5.4 *If both* $h(t)$ *and* $r(t)$ *increase then*

$$\frac{\int_0^T [H(t)^{n-1}/(n-1)!]\overline{G}(t) f(t)\mathrm{d}t}{\int_0^T [H(t)^n/n!]\overline{G}(t)\overline{F}(t)\mathrm{d}t}$$

increases with n *and converges to* $h(T)$ *and* $r(T)$ *as* $n \to \infty$ *for any* $T > 0$*.*

From Theorems 1.5.3 and 1.5.4, for any function $\phi(t)$ which is continuous and $\phi(t) \neq 0$ for any $t > 0$, if $h(t)$ increases then

$$\frac{\int_0^T H(t)^n \phi(t) f(t)\mathrm{d}t}{\int_0^T H(t)^n \phi(t) \overline{F}(t)\mathrm{d}t}$$

increases with n to $h(T)$ as $n \to \infty$ for any $T > 0$.

Using the above theorems, several modified and extended periodic replacements with minimal repair have been proposed and their optimal policies have been obtained [22, 23] which will be shown in Section 1.8.

1.6 SHOCK AND DAMAGE MODEL

Consider a standard cumulative damage model [24, 25, 26]: Successive shocks occur at time intervals X_j $(j = 1, 2, \dots)$, and each shock causes some damage with amount W_j to the unit, where $W_0 \equiv 0$. When the total damage due to shocks is additive and $N(t)$ denotes the total number of shocks up to time t, the total damage $Z(t)$ at time t is

$$Z(t) \equiv \sum_{j=0}^{N(t)} W_j.$$

Letting $F(t) \equiv \Pr\{X_j \leq t\}$ with mean $1/\lambda$ and $G(x) \equiv \Pr\{W_j \leq x\}$ with mean $1/\mu$ for any j, the distribution of $Z(t)$ is

$$\Pr\{Z(t) \leq x\} = \sum_{j=0}^{\infty} G^{(j)}(x)[F^{(j)}(t) - F^{(j+1)}(t)]. \tag{1.18}$$

Suppose that the unit fails when the total damage has exceeded a failure level K $(0 < K < \infty)$, and it is replaced preventively at time T $(0 \leq T < \infty)$. Then, the total expected cost rate is

$$C(T) = \frac{c_K - (c_K - c_T)\sum_{j=0}^{\infty}[F^{(j)}(T) - F^{(j+1)}(T)]G^{(j)}(K)}{\sum_{j=0}^{\infty} G^{(j)}(K)\int_0^T [F^{(j)}(t) - F^{(j+1)}(t)]\mathrm{d}t}, \tag{1.19}$$

where c_K =replacement cost at failure K and c_T = replacement cost at time T with $c_T < c_K$. Optimal replacement policies which minimize the expected cost rates were discussed by Taylor [27], Feldman [28] and Zuckerman [29].

We now propose that the unit is replaced preventively at damage Z $(0 \leq Z \leq K)$ [30]. Then, the expected cost rate is

$$C(Z) = \frac{c_K - (c_K - c_Z)[G(K) - \int_0^Z \overline{G}(K - x)\mathrm{d}M_G(x)]}{[1 + M_G(Z)]/\lambda}, \tag{1.20}$$

where c_Z = replacement cost at damage Z with $c_Z < c_K$, and optimal Z^* to minimize $C(Z)$ satisfies

$$\int_{K-Z}^{K} [1 + M_G(K - x)]\mathrm{d}G(x) = \frac{c_Z}{c_K - c_Z}, \tag{1.21}$$

where $M_G(x) \equiv \sum_{j=1}^{\infty} G^{(j)}(x)$. The left-hand side of (1.21) increases strictly with Z from 0 to $M_G(K)$. Thus, if $M_G(K) > c_Z/(c_K - c_Z)$, then there exists a finite and unique Z^* $(0 < Z^* < K)$ which satisfies (1.21). Such discussions for deriving optimal policies are much simpler and easier than those for deriving optimal T^* in (1.19).

Furthermore, when the unit is replaced preventively at time T $(0 \le T < \infty)$, shock N $(N = 1, 2, \dots)$ and damage Z $(0 \le Z \le K)$, whichever occurs first, the expected cost rate is [31, p. 42]

$$C_F(T,N,Z) = \frac{\begin{array}{l} c_K - (c_K - c_T)\sum_{j=0}^{N-1}[F^{(j)}(T) - F^{(j+1)}(T)]G^{(j)}(Z) \\ -(c_K - c_N)F^{(N)}(T)G^{(N)}(Z) \\ -(c_K - c_Z)\sum_{j=0}^{N-1}F^{(j+1)}(T)\int_0^Z [G(K-x) - G(Z-x)]\mathrm{d}G^{(j)}(x) \end{array}}{\sum_{j=0}^{N-1}G^{(j)}(Z)\int_0^T [F^{(j)}(t) - F^{(j+1)}(t)]\mathrm{d}t}, \quad (1.22)$$

and when it is preventively replaced at time T, shock N and damage Z, whichever occurs last, the expected cost rate is

$$C_L(T,N,Z) = \frac{\begin{array}{l} c_K - (c_K - c_T)\sum_{j=N}^{\infty}[F^{(j)}(T) - F^{(j+1)}(T)][G^{(j)}(K) - G^{(j)}(Z)] \\ -(c_K - c_N)[1 - F^{(N)}(T)][G^{(N)}(K) - G^{(N)}(Z)] \\ -(c_K - c_Z)\sum_{j=N}^{\infty}[1 - F^{(j+1)}(T)]\int_0^Z [G(K-x) - G(Z-x)]\mathrm{d}G^{(j)}(x) \end{array}}{\begin{array}{l} \sum_{j=N}^{\infty}G^{(j)}(Z)\int_T^{\infty}[F^{(j)}(t) - F^{(j+1)}(t)]\mathrm{d}t \\ +\sum_{j=0}^{N-1}G^{(j)}(K)\int_T^{\infty}[F^{(j)}(t) - F^{(j+1)}(t)]\mathrm{d}t \\ +\sum_{j=0}^{\infty}G^{(j)}(K)\int_0^T [F^{(j)}(t) - F^{(j+1)}(t)]\mathrm{d}t \end{array}}. \quad (1.23)$$

Optimal policies which minimize $C_F(T,N,Z)$ and $C_L(T,N,Z)$ were discussed extensively, and their modified and extended models were proposed [32, 33]. Furthermore, it was shown that when $c_T = c_N = c_Z$, the policy with damage Z is the best one among three. Using shock and damage models, garbage collection [31, p. 131] and backup [33, p. 183, 34] policies were proposed and their optimal policies were discussed.

1.7 FINITE INTERVAL

It has been supposed in most maintenance policies that the unit has to be operating for an infinite interval [8, 20]. We introduce the following two maintenance policies in which the unit has to be operating for a finite interval S $(0 < S < \infty)$:

(1) Imperfect PM [14, 35, p. 60]
PM is done at sequential times $0 \equiv T_0 < T_1 < \cdots < T_N \equiv S$ and the unit is replaced at time $T_N \equiv S$. The failure rate in the kth PM becomes $b_k h(x)$ when it was $h(x)$ in the $(k-1)$th PM, where $1 = b_0 < b_1 \le b_2 \le \cdots \le b_{N-1}$ and $B_k \equiv \Pi_{j=0}^{k-1} b_j$ $(k = 1, 2, \dots, N)$. Then, the total expected cost until replacement is

$$C(N) = c_1 \sum_{k=1}^{N} B_k \int_0^{T_k - T_{k-1}} h(t)\mathrm{d}t + (N-1)c_2 + c_3, \quad (1.24)$$

where c_1 = cost of minimal repair, c_2 =cost of PM and c_3 = cost of replacement.

Optimal T_k^* $(k = 1, 2, \ldots, N-1)$ to minimize $C(N)$ satisfy

$$h(T_k - T_{k-1}) = b_k h(T_{k+1} - T_k). \tag{1.25}$$

Thus, we compute T_k which satisfy (1.25), and substituting them for (1.24), we obtain the total expected cost $C(N)$. Next, comparing $C(N)$ for all $N \geq 1$, we can get the optimal PM number N^* and times T_k^* $(k = 1, 2, \ldots, N^*)$.

(2) Inspection Policy [35, p. 66, 36]

The unit is checked at successive times $0 < T_1 < T_2 < \cdots < T_N \equiv S$ to detect failures. Then, the total expected cost until replacement or time S is

$$C(\mathbf{T}_k) = \sum_{k=0}^{N-1} \int_{T_k}^{T_{k+1}} [c_1(k+1) + c_2(T_{k+1} - t)]\mathrm{d}F(t) + c_1 N\overline{F}(T_N) + c_3, \tag{1.26}$$

where c_1 = cost of one check, c_2 = loss cost per unit of time for the time elapsed between failure and its detection at the next checking time, and c_3 = cost of replacement.

Optimal T_k^* to minimize $C(\mathbf{T}_k)$ satisfy

$$\begin{aligned} T_2 - T_1 &= \frac{F(T_1)}{f(T_1)} - \frac{c_1}{c_2}, \\ T_{k+1} - T_k &= \frac{F(T_k) - F(T_{k-1})}{f(T_k)} - \frac{c_1}{c_2} \qquad (k = 2, \ldots, N-1), \\ S - T_{N-1} &= \frac{F(T_{N-1}) - F(T_{N-2})}{f(T_{N-1})} - \frac{c_1}{c_2}. \end{aligned} \tag{1.27}$$

Solving the above simultaneous equations, we can compute T_k^* numerically.

Table 1.1 presents sequential checking times T_k and the expected cost

$$\widetilde{C}(N) \equiv C(N) + c_2 \int_0^S \overline{F}(t)\mathrm{d}t - c_3,$$

when $S = 100$, $c_1/c_2 = 2$, $\overline{F}(t) = 1 - \mathrm{e}^{-\lambda t^2}$, and $S = \int_0^\infty \mathrm{e}^{-\lambda t^2}\mathrm{d}t = \sqrt{\pi/\lambda}/2$, which is the mean failure time of the unit. Comparing $\widetilde{C}(N)$ for $N = 1, 2, \ldots, 8$, optimal checking number is $N^* = 4$ and optimal checking times are 44.1, 66.0, 84.0, 100.

When S is a random variable with a distribution $G(t) \equiv \Pr\{S \leq t\}$, the total expected cost in (1.26) is

$$C(\mathbf{T}_k) = \sum_{k=0}^{\infty} \overline{F}(T_k)\left[c_1\overline{G}(T_k) + c_2\int_{T_k}^{T_{k+1}} \overline{G}(t)\mathrm{d}t\right] + c_3 - c_1 - c_2\int_0^\infty \overline{F}(t)\overline{G}(t)\mathrm{d}t. \tag{1.28}$$

Table 1.1
Checking time T_k and expected cost $\widetilde{C}(N)$ for $N = 1, 2, \ldots, 8$.

N	1	2	3	4	5	6	7	8
T_1	100	64.14	50.9	44.1	40.3	38.1	36.8	36.3
T_2		100	77.1	66.0	60.0	56.2	54.3	53.3
T_3			100	84.0	75.4	70.5	67.8	66.6
T_4				100	88.6	82.3	78.9	77.3
T_5					100	91.1	87.9	85.9
T_6						100	94.9	92.5
T_7							100	97.2
T_8								100
$\widetilde{C}(N)/c_2$	102.00	93.55	91.52	91.16	91.47	92.11	92.91	93.79

Similar discussions of obtaining optimal policies to minimize $C(\mathbf{T}_k)$ were made [37, p. 93] and optimal policies to minimize the expected cost rate were derived [38]. Furthermore, this was applied to checkpoint models [35, p. 123, 39].

1.8 REPLACEMENT FIRST, LAST, OVERTIME AND MIDDLE

Suppose that the unit works for a job with random working times [40, 41]: When the unit undergoes minimal repair at failures in Section 1.5, we consider the preventive replacement policies in which it is replaced at a planned time T, a number N of working cycles and a number K of failures: It is assumed that the unit has a failure distribution $F(t)$ with mean $\mu \equiv \int_0^\infty \overline{F}(t)\mathrm{d}t$, and failure rate $h(t) \equiv f(t)/\overline{F}(t)$, where $f(t) \equiv \mathrm{d}F(t)/\mathrm{d}t$, $H(t) \equiv \int_0^t h(u)\mathrm{d}u$, $P_k(t) = \sum_{j=k}^{\infty}[H(t)^j/j!]\mathrm{e}^{-H(t)}$ $(k = 0, 1, 2, \ldots)$. In addition, the unit operates for random cycles Y_n $(n = 1, 2, \ldots)$ with distribution $G(t) \equiv \Pr\{Y_n \leq t\}$, mean $1/\theta$, and $G^{(n)}(t)$ $(n = 1, 2, \ldots)$ which is the n-fold Stieltjes convolution of $G(t)$ with itself and $G^{(0)}(t) \equiv 1$ for $t \geq 0$.

Combining T, N and K appropriately, we propose the following five replacement models, where c_i = replacement cost at i $(i = T, N, K)$ and c_M = cost of minimal repair [37, p. 53, 42, p. 33].

(1) Replacement First

When the unit is replaced at time T, work N or failure K, whichever occurs first, the expected cost rate is

$$\frac{\begin{array}{l} c_T + (c_N - c_T)\int_0^T \overline{P}_K(t)\mathrm{d}G^{(N)}(t) \\ \quad + (c_K - c_T)\int_0^T [1 - G^{(N)}(t)]\mathrm{d}P_K(t) + c_M\int_0^T [1 - G^{(N)}(t)]\overline{P}_K(t)h(t)\mathrm{d}t \end{array}}{\int_0^T [1 - G^{(N)}(t)]\overline{P}_K(t)\mathrm{d}t}. \tag{1.29}$$

(2) Replacement Last

When the unit is replaced at time T, work N or failure K, whichever occurs last, the expected cost rate is

$$\frac{\begin{array}{l} c_T + (c_N - c_T)\int_T^\infty P_K(t)\mathrm{d}G^{(N)}(t) + (c_K - c_T)\int_T^\infty G^{(N)}(t)\mathrm{d}P_K(t) \\ \quad + c_M\{H(T) + \int_T^\infty [1 - G^{(N)}(t)P_K(t)]h(t)\mathrm{d}t\} \end{array}}{T + \int_T^\infty [1 - G^{(N)}(t)P_K(t)]\mathrm{d}t}. \tag{1.30}$$

(3) Replacement Middle

When the unit is replaced at time T, work N or failure K, whichever occurs in middle, the expected cost rate is

$$\frac{\begin{array}{l} c_T + (c_N - c_T)[\int_T^\infty \overline{P}_K(t)\mathrm{d}G^{(N)}(t) + \int_0^T P_K(t)\mathrm{d}G^{(N)}(t)] \\ \quad + (c_K - c_T)\{\int_T^\infty [1 - G^{(N)}(t)]\mathrm{d}P_K(t) + \int_0^T G^{(N)}(t)\mathrm{d}P_K(t)\} \\ \quad + c_M\{\int_0^T [1 - G^{(N)}(t)P_K(t)]h(t)\mathrm{d}t + \int_T^\infty [1 - G^{(N)}(t)]\overline{P}_K(t)h(t)\mathrm{d}t\} \end{array}}{\int_0^T [1 - G^{(N)}(t)P_K(t)]\mathrm{d}t + \int_T^\infty [1 - G^{(N)}(t)]\overline{P}_K(t)\mathrm{d}t}. \tag{1.31}$$

(4) Modified Replacement Middle

Let t_N and t_K be the respective occurrence times of work N and failure K. When the unit is replaced at time $\mathrm{Max}\{t_N, t_K\}$ before time T or $\mathrm{Min}\{t_N, t_K\}$ after time T, whichever occurs first, the expected cost rate is

$$\frac{\begin{array}{l} c_N + (c_N - c_K)\{G^{(N)}(T)\overline{P}_K(T) + \int_0^T G^{(N)}(t)\mathrm{d}P_K(t) \\ \quad + \int_T^\infty [1 - G^{(N)}(t)]\mathrm{d}P_K(t)\} + c_M\{\int_0^T [1 - P_K(t)G^{(N)}(t)]h(t)\mathrm{d}t \\ \quad + \int_T^\infty \overline{P}_K(t)[1 - G^{(N)}(t)]h(t)\mathrm{d}t + G^{(N)}(T)\int_T^\infty \overline{P}_K(t)h(t)\mathrm{d}t\} \end{array}}{\begin{array}{l} \int_0^T [1 - P_K(t)G^{(N)}(t)]\mathrm{d}t + \int_T^\infty [1 - G^{(N)}(t)]\overline{P}_K(t)\mathrm{d}t \\ \quad + P_K(T)\int_T^\infty [1 - G^{(N)}(t)]\mathrm{d}t + G^{(N)}(T)\int_T^\infty \overline{P}_K(t)\mathrm{d}t \end{array}}. \tag{1.32}$$

It is shown that this policy is better than replacement middle.

(5) Replacement Overtime T

When the unit is replaced at the first working time over a planned time T or work N, whichever occurs first, the expected cost rate is

$$\frac{c_{OT} + (c_N - c_{OT})G^{(N)}(T) + c_M\sum_{n=0}^{N-1}\int_0^T [\int_0^\infty \overline{G}(u)h(t+u)\mathrm{d}u]\mathrm{d}G^{(n)}(t)}{(1/\theta)\sum_{n=0}^{N-1} G^{(n)}(T)}, \tag{1.33}$$

and when it is replaced at the first failure over time T or failure K, whichever occurs first, the expected cost rate is

$$\frac{c_{OT} + (c_K - c_{OT})P_K(T) + c_M \sum_{k=0}^{K-1} P_k(T)}{\int_0^T \overline{P}_K(t)\mathrm{d}t + \overline{P}_K(T)\int_T^\infty \mathrm{e}^{-H(t)+H(T)}\mathrm{d}t}, \tag{1.34}$$

where c_{OT} = replacement cost over time T. Optimal policies to minimize the expected cost rates have been discussed and their comparative results have been shown [43, 44].

1.9 RANDOM K-OUT-OF-n SYSTEM

Consider a K-out-of-n ($K = 0, 1, 2, \ldots, n; n = 1, 2, \ldots$) system which can operate if and only if at least K units of the total n units are operable [20, p. 216, 35, p. 12]. When each unit has an identical failure distribution $F(t)$, reliability of the system at time t is

$$R_{n,K}(t) = \sum_{i=K}^{n} \binom{n}{i} \overline{F}(t)^i F(t)^{n-i}. \tag{1.35}$$

Furthermore, when K is a random variable with a probability $p_i \equiv \Pr\{K = i\}$ ($i = 0, 1, \ldots, n$), $\sum_{i=0}^{n} p_i = 1$, and $P_i \equiv \Pr\{K \le i\} = \sum_{j=0}^{i} p_j$ ($i = 0, 1, 2, \ldots, n$), where $P_0 \equiv 0$ and $P_n \equiv 1$. Note that P_i represents the probability that when i units are operable, the system is operable, and increases with i from 0 to 1. Then, reliability of the system at time t is [37, p. 158, 45,46]

$$R_{n,P}(t) = \sum_{j=0}^{\infty} p_j \sum_{i=j}^{\infty} \binom{n}{i} \overline{F}(t)^i F(t)^{n-i} = \sum_{i=0}^{n} P_i \binom{n}{i} \overline{F}(t)^i F(t)^{n-i},$$

its failure distribution is

$$F_{n,P}(t) \equiv 1 - R_{n,P}(t) = \sum_{i=0}^{n} \overline{P}_i \binom{n}{i} \overline{F}(t)^i F(t)^{n-i}, \tag{1.36}$$

its mean time (MTTF) is when $F(t) = 1 - \mathrm{e}^{-\lambda t}$,

$$\mu_{n,P} = \frac{1}{\lambda} \sum_{i=1}^{n} \frac{P_i}{i}, \tag{1.37}$$

and its failure rate is

$$h_{n,P}(t) = \frac{-R'_{n,P}(t)}{R_{n,P}(t)} = \frac{nh(t)\sum_{i=1}^{n} p_i \binom{n-1}{i-1} \overline{F}(t)^i F(t)^{n-i}}{\sum_{i=1}^{n} P_i \binom{n}{i} \overline{F}(t)^i F(t)^{n-i}}. \tag{1.38}$$

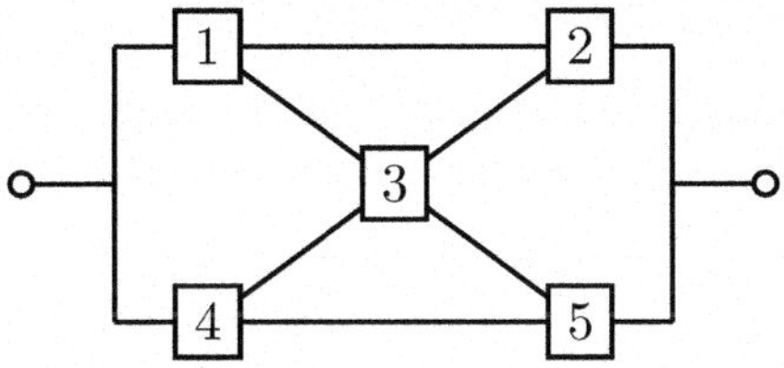

Figure 1.2 Bridge system with 5 units.

This shows that any redundant systems with identical units have a failure distribution $F_{n,P}(t)$ in (1.36), and when $F(t) = 1 - \mathrm{e}^{-\lambda t}$, MTTF is easily computed in (1.37). Furthermore, using the entropy model [35, p. 199], a system complexity is defined as

$$H_p = -\sum_{i=1}^{n} p_i \log_2 p_i, \tag{1.39}$$

where $p_i \log_2 p_i = 0$ when $p_i = 0$.

As an example of redundant systems, we give a bridge system with 5 units in Fig. 1.2. Then,

$$P_1 = 0, \quad P_2 = \frac{1}{5}, \quad P_3 = \frac{4}{5}, \quad P_4 = P_5 = 1,$$

and

$$p_1 = 0, \quad p_2 = \frac{1}{5}, \quad p_3 = \frac{3}{5}, \quad p_4 = \frac{1}{5}, \quad p_5 = 0.$$

Thus, reliability of the system is

$$\begin{aligned} R_5(t) &= \frac{1}{5}\binom{5}{2}\overline{F}(t)^2 F(t)^3 + \frac{4}{5}\binom{5}{3}\overline{F}(t)^3 F(t)^2 + \binom{5}{4}\overline{F}(t)^4 F(t) + \binom{5}{5}\overline{F}(t)^5 \\ &= \overline{F}(t)^2[1 + 2F(t) + F(t)^2 - 2F(t)^3], \end{aligned}$$

and when $F(t) = 1 - \mathrm{e}^{-\lambda t}$, MTTF is, from (1.37),

$$\mu_5 = \frac{1}{\lambda}\left(\frac{1}{5} \times \frac{1}{2} + \frac{4}{5} \times \frac{1}{3} + \frac{1}{4} + \frac{1}{5}\right) = \frac{49}{60\lambda}.$$

Furthermore, the complexity is

$$H_5 = -\frac{1}{5}\log_2\frac{1}{5} - \frac{3}{5}\log_2\frac{3}{5} - \frac{1}{5}\log_2\frac{1}{5} \fallingdotseq 1.371.$$

Optimal maintenance policies have been discussed theoretically under the assumption that an operating unit has a failure distribution $F(t)$. So that, replacing $F(t)$ with $F_{n,P}(t)$ in (1.36), we could discuss optimal policies for any redundant systems with identical units [47].

1.10 ASYMPTOTIC CALCULATIONS

Using an exponential distribution, we can calculate approximately some reliability measures of systems with a Weibull distribution [37, p.143, 48, 49, 50].

(1) Failure Times

When failures occur at a nonhomogeneous Poisson distribution with mean-value function $H(t)$, i.e., the unit undergoes minimal repair at failures in Section 1.5, the mean time to jth $(j = 1, 2, \dots)$ failure is

$$\mu_j = \sum_{i=0}^{j-1} \int_0^{\infty} \frac{H(t)^i}{i!} \mathrm{e}^{-H(t)} \mathrm{d}t. \tag{1.40}$$

In particular, when $F(t) = 1 - \mathrm{e}^{-t^{\alpha}}$ $(\alpha > 0)$, (1.40) is

$$\mu_j(\alpha) = \frac{\Gamma(j + 1/\alpha)}{\Gamma(j)}, \qquad \mu_j(1) = j. \tag{1.41}$$

Then, by Jensen's inequality which is widely used for convex and concave functions,

$$\mu_j(\alpha) \begin{cases} < j^{1/\alpha} & \text{for } \alpha > 1, \\ = j & \text{for } \alpha = 1, \\ > j^{1/\alpha} & \text{for } \alpha < 1. \end{cases} \tag{1.42}$$

(2) MTTF of Parallel System

When a parallel system with n units $(n = 1, 2, \dots)$ each of which has an identical failure distribution $F(t)$, MTTF is

$$\mu_n = \int_0^{\infty} [1 - F(t)^n] \mathrm{d}t.$$

When $F(t) = 1 - \mathrm{e}^{-t^{\alpha}}$ for $\alpha > 0$,

$$\mu_n(\alpha) = \int_0^{\infty} [1 - (1 - \mathrm{e}^{-t^{\alpha}})^n] \mathrm{d}t \begin{cases} < \left(\sum_{j=1}^{n} \frac{1}{j}\right)^{1/\alpha} & \text{for } \alpha > 1, \\ = \sum_{j=1}^{n} \frac{1}{j} & \text{for } \alpha = 1, \\ > \left(\sum_{j=1}^{n} \frac{1}{j}\right)^{1/\alpha} & \text{for } \alpha < 1. \end{cases} \tag{1.43}$$

MTTF of a general redundant system with a Weibull distribution $F(t) = 1 - \mathrm{e}^{-t^{\alpha}}$ is, from (1.37), approximately

$$\mu_{n,P}(\alpha) = \left(\sum_{j=1}^{n} \frac{P_j}{j} \right)^{1/\alpha}. \tag{1.44}$$

Table 1.2
Optimal $\mathbf{T}_k^*$ and approximate $\widetilde{\mathbf{T}}_k$ when $\mathbf{F}(t) = 1 - e^{-t^\alpha}$ and $c_1/c_2 = 0.02$.

k	$\alpha = 1.0$	$\alpha = 1.5$		$\alpha = 2.0$		$\alpha = 2.5$	
	T_k^*	T_k^*	$\widetilde{T}_k$	T_k^*	$\widetilde{T}_k$	T_k^*	$\widetilde{T}_k$
1	0.194	0.310	0.335	0.411	0.440	0.495	0.518
2	0.387	0.516	0.531	0.615	0.622	0.691	0.684
3	0.581	0.699	0.696	0.785	0.762	0.846	0.805
4	0.774	0.869	0.843	0.935	0.880	0.980	0.903
5	0.968	1.031	0.978	1.073	0.984	1.100	0.987
6	1.161	1.185	1.105	1.202	1.078	1.210	1.062
7	1.355	1.335	1.224	1.323	1.164	1.312	1.129
8	1.548	1.479	1.338	1.439	1.244	1.408	1.191
9	1.742	1.620	1.448	1.549	1.320	1.498	1.249
10	1.935	1.758	1.553	1.656	1.391	1.585	1.302

This shows that when the failure time has a Weibull distribution, if P_j is obtained in Section 1.9, MTTFs of any redundant systems are obtained approximately by the above simple function.

(3) Inspection Policy
When the unit with a failure distribution $F(t)$ is checked at successive times T_k $(k = 1, 2, \dots)$ to detect failure, the total expected cost until failure detection is

$$C(\mathbf{T}_k) = \sum_{k=0}^{\infty} [c_1 + c_2(T_{k+1} - T_k)]\overline{F}(T_k) - c_2\mu, \tag{1.45}$$

where $\mu \equiv \int_0^\infty \overline{F}(t)\mathrm{d}t$, c_1 = cost of check and c_2 = downtime cost per unit of time elapsed between failure and its detection at the next checking time, and optimal T_k to minimize $C(\mathbf{T}_k)$ satisfies

$$T_{k+1} - T_k = \frac{F(T_k) - F(T_{k-1})}{f(T_k)} - \frac{c_1}{c_2} \qquad (k = 1, 2, \dots). \tag{1.46}$$

In particular, $F(t) = 1 - e^{-\lambda t}$, $T_k^* = kT^*$ and T^* satisfies

$$e^{\lambda T} - 1 - \lambda T = \frac{c_1}{c_2/\lambda}. \tag{1.47}$$

When $F(t) = 1 - e^{-t^\alpha}$, optimal T_k^* are approximately $\widetilde{T}_k \equiv (kT^*)^{1/\alpha}$ where T^* is given in (1.47). Table 1.2 presents optimal T_k^* and approximate $\widetilde{T}_k$, which give good approximations of optimal T_k^*.

This asymptotic method could be applied to the estimation of any redundant systems with Weibull failure times. As further studies, we try to estimate replacement time for an exponential distribution from a Weibull distribution, because some optimal finite replacement time does not exist in case of an exponential failure time [51].

REFERENCES

1. Osaki, S., & Nakagawa, T. (1971). On a two-unit standby redundant system with standby failure. *Operations Research*, 19, 510-523.

2. Nakagawa, T. (2014). Studies on reliability and maintenance. In Nakamura, S., Qian, C. H., & Chen, M., Eds., *Reliability Modeling with Applications*, World Scientific, Singapore, 349-364.

3. Epstein, B., & Hosford, J. (1960). Reliability of some two unit redundant system. *Proceeding* 6th *National Symposium in Reliability and Quality Control*, 469-476, 1960.

4. Dohi, T., & Nakagawa, T. Eds. (2013). *Stochastic Reliability and Maintenance Modeling*, Springer, London.

5. Nakagawa, T. (2002). Two-unit redundant models. In Osaki, S., Eds., *Stochastic Models in Reliability and Maintenance*, Springer, Berlin, 165-191.

6. Nakagawa, T., & Osaki, S. (1974). *Stochastic behaviour of a two-unit standby redundant system. INFOR: Information Systems and Operational Research*, 12, 66-70.

7. Nakagawa, T. (2011). *Stochastic Processes with Applications to Reliability Theory*, Springer, London.

8. Nakagawa, T. (2005). *Maintenance Theory of Reliability*, Springer, London.

9. Kimura, M., Zhao, X., & Nakagawa, T. (2016). Reliability analysis of a cloud computing system with replication. In Fiondella, L., & Puliafito, A., Eds., *Principles of Performance and Reliability Modeling and Evaluation*, Springer, London, 401-423.

10. Nakagawa, T. (1977). Optimum preventive maintenance policies for repairable systems. *IEEE Transactions on Reliability*, R-26, 168-173.

11. Nakagawa, T. (1979). Optimal policies when preventive maintenance is imperfect. *IEEE Transactions on Reliability*, R-28, 331-332.

12. Brown, M., & Proschan, F. (1983). Imperfect repair. *Journal of Applied Probability*, 20, 851-859.

13. Kijima, M. (1988). Some results for repairable systems with general repair. *Journal of Applied Probability*, 26, 89-102.

14. Nakagawa, T. (1988). Sequential imperfect preventive maintenance policies. *IEEE Transactions on Reliability*, 37, 295-298.

15. Wang, H., & Pham, H. (2003). Optimal imperfect maintenance models. In Pham, H., Eds., *Handbook of Reliability Engineering*, Springer, London.

16. Nakagawa, T., & Osaki, S. (1975). The discrete Weibull distribution. *IEEE Transactions on Reliability*, R-24, 300-301.

17. Murthy, D. N. P., Xie, M., & Jiang, R. (2004). *Weibull Models*, Wiley, Hoboken, N.J..

18. Padgett, W. J., & Spurrier, J. D. (1985). Discrete failure models. *IEEE Transactions on Reliability*, 34, 253-256.

19. Krishna, H., & Pundir, P. S. (2009). Discrete Burr and discrete Pareto distributions. *Statistical Methodology*, 6, 177-18.

20. Barlow, R. E., & Proschan, F. (1965). *Mathematical Theory of Reliability*, Wiley, New York.

21. Nakagawa, T., & Kowada, M. (1983). Analysis of a system with minimal repair and its application to replacement policy. *European Journal of Operational Research*, 12, 176-182.

22. Sheu, S. H., Griffith, W. S., & Nakagawa, T. (1996) Extended optimal replacement model with random minimal repair costs. *European Journal of Operational Research*, 85, 636-649.

23. Tadj, L., Ouali, M.S., Yacout, S., & Ait-Kadi, D., Eds. (2011). *Replacement Models with Minimal Repair*, Springer, London.

24. Cox, D. R. (1962). *Renewal Theory*, Methuen, London.

25. Esary, J. D., Marshall, A. W., & Proschan, F. (1973). Shock models and wear processes. *Annals of Probability*, 1, 627-649.

26. Nakagawa, T., & Osaki, S. (1974). Some aspects of damage model. *Microelectronics Reliability*, 13, 253-257.

27. Taylor, H. M. (1975). Optimal replacement under additive damage and other failure models. *Naval Research Logistics Quarterly*, 22, 1-18.

28. Feldman, R. M. (1976). Optimal replacement with semi-Markov shock models. *Journal of Applied Probability*, 13, 108-117.

29. Zuckerman, D. (1977). Replacement models under additive damage. *Naval Research Logistics Quarterly*, 24, 549-558.

30. Nakagawa, T. (1976). On a replacement problem of a cumulative damage model. *Journal of the Operational Research Society*, 27, 895-900.

31. Nakagawa, T. (2007). *Shock and Damage Models in Reliability Theory*, Springer, London.

32. Zhao, X., Qian, C. H., & Sheu, S. H. (2014). Cumulative damage models with random working times. In Nakamura, S., Qian, C. H., & Chen, M., Eds., *Reliability Modeling with Applications*, World Scientific, Singapore, 79-98.

33. Zhao, X., & Nakagawa, T. (2018). *Advanced Maintenance Policies for Shock and Damage Models*, Springer, London.

34. Nakamura, S., Zhao, X., & Nakagawa, T. (2017). Constant and random full backup models with incremental and differential backup schemes. *International Journal of Reliability, Quality and Safety Engineering*, 24(3), 1750015.

35. Nakagawa, T. (2008). *Advanced Reliability Models and Maintenance Policies*, Springer, London.

36. Nakagawa, T., & Mizutani, S. (2009). A summary of maintenance policies for a finite interval. *Reliability Engineering and System Safety*, 94, 89-96.

37. Nakagawa, T. (2014). *Random Maintenance Policies*, Springer, London.

38. Mizutani, S., Zhao, X., & Nakagawa, T. (2022). Optimal inspection policies to minimize expected cost rates. To appear in *International Journal of Reliability, Quality and Safety Engineering*, 29.

39. Naruse, K., & Nakagawa, T. (2020). Optimal checkpoint intervals, schemes and structures for computing models. In Pham, H., Eds., *Reliability and Statistical Computing*, Springer, London, 265-287.

40. Zhao, X., & Nakagawa, T. (2012). Optimization problems of replacement first or last in reliability theory. *European Journal of Operational Research*, 223(1), 141-149.

41. Zhao, X., Al-Khalifa, K. N., Hamouda, A. M. S., & Nakagawa, T. (2015). First and last triggering event approaches for replacement with minimal repairs. *IEEE Transactions on Reliability*, 65, 197-207.

42. Nakagawa, T., & Zhao, X. (2015). *Maintenance Overtime Policies in Reliability Theory*, Springer, London.

43. Chen, M., Zhao, X., & Nakagawa, T. (2019). Replacement policies with general models. *Annals of Operations Research*, 277(1), 47-61.

44. Mizutani, S., Zhao, X., & Nakagawa, T. (2020). Which replacement is better at working cycles or number of failures. *IEICE TRANSACTIONS on Fundamentals of Electronics, Communications and Computer Sciences*, 103, 523-532.

45. Ito, K., Zhao, X., & Nakagawa, T. (2017). Random number of units for k-out-of-n. *Applied Mathematics and Computation*, 45, 563-572.

46. Ito, K., & Nakagawa, T. (2019). Reliability properties of K-out-of-N:G systems. In Ram, M., & Dohi, T., Eds., *System Engineering, Reliability Analysis Using k-out-of-n Structures*, CRC Press, Boca Raton, FL, 25-40.

47. Zhou, L., Nakamura, T., Nakagawa, T., Xiao, X., & Yamamoto, H. (2019) A summary of maintenance policies for K-out-of-N models and their applications to consecutive systems. In Ram, M., & Dohi, T., Eds., *System Engineering, Reliability Analysis Using k-out-of-n Structures*, CRC Press, Boca Raton, FL, 41-65.

48. Nakagawa, T., & Yun, W. Y. (2011). Note on MTTF of a parallel system. *International Journal of Reliability, Quality and Safety Engineering*, 18, 43-50.

49. Zhao, X., Al-Khalifa, K. N., & Nakagawa, T. (2015). Approximate methods for optimal replacement, and inspection policies. *Reliability Engineering and System Safety*, 144, 68-73.

50. Nakagawa, T., Mizutani, S., & Zhao, X. (2021). Approximate calculations of maintenance policies for Weibull distribution using exponential distribution. In Karanki, D. R., Eds., *Advanced in Performability Engineering*, Springer, London.

51. Zhao, X., Li, B., Mizutani, S., & Nakagawa, T. (2022). A revisit of age-based replacement models with exponential failure distributions. To appear in *IEEE Transactions on Reliability*, 71.

2 Replacement First and Last Policies with Random Times for Redundant Systems

Satoshi Mizutani
Aichi Institute of Technology, Japan

Mingchih Chen
Fu Jen Catholic University, Taiwan

CONTENTS

2.1 INTRODUCTION

Failures of complex systems such as networks and database systems sometimes cause great damage in society. Therefore, suitable maintenance schedules are very important to assure the safety and reliability of systems to avoid their failures. There exist several factors to plan maintenance policies. For example, some maintenance might be done after work, or after occurrence of serious damage due to a natural disaster, which would be difficult to predict previously. Therefore, replacement times are assumed to be random variables. We discuss several maintenance policies which minimize the expected cost rates and compare them.

We consider optimal maintenance policies for some complex systems such as industry equipment, network systems, database systems and computer processes. There have been many studies of maintenance policies using stochastic models [1, 2]. If the failure rate increases with age and usage frequency then it would be wise to make its maintenance at periodic times or at a certain number of uses [3]. Some random maintenance policies where the unit is replaced at a random time with a probability distribution have been proposed [3, 4, 5, 6, 7, 8, 9]. *Replacement first* and *replacement last* in which the unit is replaced at times such as planned time T and working cycle N, whichever occurs first and last, respectively, were considered by Nakagawa and Zhao [10]. *Replacement overtime* was proposed as applications of random maintenance policies in which the unit is replaced at the first completion of working cycles over a planned time T [11]. Some replacement overtime models in which the unit is replaced after the working or mission time were considered [11, 12, 13]. As a typical redundant system, K-out-of-n systems have been studied [14, 15, 16, 17], and their replacement policies were proposed [17, 18, 19]. Recently, *random K-out-of-n systems* which can present several redundant systems such as series, parallel, series-parallel and parallel-series system were proposed [20, 21].

We propose generalized replacement first and last policies with random times for redundant systems: (a) Replacement First, (b) Modified Replacement First, (c) Replacement Last and (d) Modified Replacement Last. We obtain the expected cost rates of each policy, make analytical discussions and compare them numerically, and decide which policy is better among the above replacement policies.

Section 2.2 shows general random age replacement and replacement first and last. Section 2.3 takes up some redundant systems such as series system, parallel system, and random K-out-of-n system, and Section 2.4 considers four replacement policies for each system. The expected cost rates are obtained and optimal policies which minimize them are derived. Section 2.5 gives numerical examples for the replacement policies in Section 2.4.

2.2 RANDOM AGE REPLACEMENT

2.2.1 RANDOM REPLACEMENT DISTRIBUTION

Consider a system with n $(n = 1, 2, \dots)$ units which is replaced preventively at random replacement times Y_j $(j = 1, 2, \dots, N)$ or at a planned time T $(0 < T \leq \infty)$. It is assumed that Y_j are independent and have a distribution $G_j(t) \equiv \Pr\{Y_j \leq t\}$ with finite $1/\theta_j \equiv \int_0^\infty \overline{G}_j(t)\mathrm{d}t$, where $\overline{\Phi}(t) \equiv 1 - \Phi(t)$ for any function $\Phi(t)$.

Denoting $Y_m \equiv \min\{Y_1, Y_2, \dots, Y_N\}$ and $Y_M \equiv \max\{Y_1, Y_2, \dots, Y_N\}$, respectively,

$$\Pr\{Y_{\mathrm{m}} \leq t\} = 1 - \prod_{j=1}^{N} \overline{G}_j(t), \qquad \Pr\{Y_{\mathrm{M}} \leq t\} = \prod_{j=1}^{N} G_j(t). \tag{2.1}$$

When the system is replaced immediately at failure, we consider the following four replacement policies in Table 2.1:

Table 2.1
Replacement Times for Random Age Replacement.

	Replacement First	Replacement Last
Normal	(a) $Y_F = \min\{T, Y_m\}$	(c) $Y_L = \max\{T, Y_M\}$
Modified	(b) $\widetilde{Y}_F = \min\{T, Y_M\}$	(d) $\widetilde{Y}_L = \max\{T, Y_m\}$

(a) REPLACEMENT FIRST
The system is replaced at time T or time Y_{m}, whichever occurs first, i.e., it is replaced at time $Y_F \equiv \min\{T, Y_{\mathrm{m}}\}$, and Y_F has a distribution

$$G_F(t) \equiv \Pr\{Y_F \le t\} = \begin{cases} 1 - \prod_{j=1}^{N} \overline{G}_j(t) & t < T, \\ 1 & t \ge T. \end{cases} \tag{2.2}$$

(b) MODIFIED REPLACEMENT FIRST
The system is replaced at time T or time Y_{M}, whichever occurs first, i.e., it is replaced at time $\widetilde{Y}_F \equiv \min\{T, Y_{\mathrm{M}}\}$, and $\widetilde{Y}_F$ has a distribution

$$\widetilde{G}_F(t) \equiv \Pr\{\widetilde{Y}_F \le t\} = \begin{cases} \prod_{j=1}^{N} G_j(t) & t < T, \\ 1 & t \ge T. \end{cases} \tag{2.3}$$

(c) REPLACEMENT LAST
The system is replaced at time T or time Y_{M}, whichever occurs last, i.e., it is replaced at time $Y_L \equiv \max\{T, Y_{\mathrm{M}}\}$, and Y_L has a distribution

$$G_L(t) \equiv \Pr\{Y_L \le t\} = \begin{cases} 0 & t < T, \\ \prod_{j=1}^{N} G_j(t) & t \ge T. \end{cases} \tag{2.4}$$

(d) MODIFIED REPLACEMENT LAST
The system is replaced at time T or time Y_{m} whichever occurs last, i.e., it is replaced at time $\widetilde{Y}_L \equiv \max\{T, Y_{\mathrm{m}}\}$, and $\widetilde{Y}_L$ has a distribution

$$\widetilde{G}_L(t) \equiv \Pr\{\widetilde{Y}_L \le t\} = \begin{cases} 0 & t < T, \\ 1 - \prod_{j=1}^{N} \overline{G}_j(t) & t \ge T. \end{cases} \tag{2.5}$$

Note that from the definition of Y_i and $\widetilde{Y}_i$ $(i = F, L)$, $Y_F \le \widetilde{Y}_F \le \widetilde{Y}_L \le Y_L$ and $G_F(t) \ge \widetilde{G}_F(t) \ge \widetilde{G}_L(t) \ge G_L(t)$ for $t \ge 0$.

2.2.2 REPLACEMENT POLICIES FOR SINGLE UNIT SYSTEM

When $n = 1$ and $N = 1$, i.e., the system consists of one unit with a failure distribution $F(t)$ and failure rate $h(t) \equiv f(t)/\overline{F}(t)$, where $f(t) \equiv \mathrm{d}F(t)/\mathrm{d}t$ is a density function of $F(t)$, it is replaced preventively at time Y with a distribution $G(t) \equiv \Pr\{Y \leq t\}$. Then, the expected rate is [4]

$$C_A(G) = \frac{c_R + (c_F - c_R)\int_0^\infty \overline{G}(t)\mathrm{d}F(t)}{\int_0^\infty \overline{F}(t)\overline{G}(t)\mathrm{d}t}, \tag{2.6}$$

where c_R =replacement cost at time Y and c_F =replacement cost at failure with $c_F > c_R$.

We consider the above four replacement policies for a single unit system and discuss their optimal policies when failure rate $h(t)$ increases strictly with t from $h(0) = 0$ to $h(\infty) = \infty$.

(a) REPLACEMENT FIRST

Putting $G(t) = G_F(t)$ in (2.2), from (2.6), the expected cost rate is

$$C_F(T) = \frac{c_R + (c_F - c_R)\int_0^T \prod_{j=1}^N \overline{G}_j(t)\mathrm{d}F(t)}{\int_0^T \prod_{j=1}^N \overline{G}_j(t)\overline{F}(t)\mathrm{d}t}. \tag{2.7}$$

Differentiating $C_F(T)$ with respect to T and setting it equal to zero,

$$\int_0^T \prod_{j=1}^N \overline{G}_j(t)\overline{F}(t)[h(T) - h(t)]\mathrm{d}t = \frac{c_R}{c_F - c_R}, \tag{2.8}$$

whose left-hand side increases strictly with T from 0 to ∞. Thus, there exists a finite and unique T_F^* $(0 < T_F^* < \infty)$ which satisfies (2.8), and the resulting cost rate is

$$C_F(T_F^*) = (c_F - c_R)h(T_F^*). \tag{2.9}$$

(b) MODIFIED REPLACEMENT FIRST

Putting $G(t) = \widetilde{G}_F(t)$ in (2.3), from (2.6), the expected cost rate is

$$\widetilde{C}_F(T) = \frac{c_R + (c_F - c_R)\int_0^T [1 - \prod_{j=1}^N G_j(t)]\mathrm{d}F(t)}{\int_0^T [1 - \prod_{j=1}^N G_j(t)]\overline{F}(t)\mathrm{d}t}. \tag{2.10}$$

Differentiating $\widetilde{C}_F(T)$ with respect to T and setting it equal to zero,

$$\int_0^T \left[1 - \prod_{j=1}^N G_j(t)\right] \overline{F}(t)[h(T) - h(t)]\mathrm{d}t = \frac{c_R}{c_F - c_R}, \tag{2.11}$$

whose left-hand side increases strictly with T from 0 to ∞. Thus, there exists a finite and unique $\widetilde{T}_F$ $(0 < \widetilde{T}_F < \infty)$ which satisfies (2.11), and the resulting cost rate is

$$\widetilde{C}_F(\widetilde{T}_F) = (c_F - c_R)h(\widetilde{T}_F). \tag{2.12}$$

Comparing (2.8) and (2.11), $T_F^* \geq \widetilde{T}_F$. Thus, from (2.9) and (2.12), modified replacement first is better than replacement first.

(c) REPLACEMENT LAST

Putting $G(t) = G_L(t)$ in (2.4), from (2.6), the expected cost rate is

$$C_L(T) = \frac{c_F - (c_F - c_R)\int_T^\infty \prod_{j=1}^N G_j(t)\mathrm{d}F(t)}{\int_0^T \overline{F}(t)\mathrm{d}t + \int_T^\infty [1 - \prod_{j=1}^N G_j(t)]\overline{F}(t)\mathrm{d}t}. \tag{2.13}$$

Differentiating $C_L(T)$ with respect to T and setting it equal to 0,

$$\int_0^T \overline{F}(t)[h(T) - h(t)]\mathrm{d}t - \int_T^\infty \left[1 - \prod_{j=1}^N G_j(t)\right] \overline{F}(t)[h(t) - h(T)]\mathrm{d}t = \frac{c_R}{c_F - c_R}, \tag{2.14}$$

whose left-hand side increases strictly with T from $-\int_0^\infty [1 - \prod_{j=1}^N G_j(t)]\mathrm{d}F(t)$ to ∞. Thus, there exists a finite and unique T_L^* $(0 < T_L^* < \infty)$ which satisfies (2.14), and the resulting cost rate is

$$C_L(T_L^*) = (c_F - c_R)h(T_L^*). \tag{2.15}$$

(d) MODIFIED REPLACEMENT LAST

Putting $G(t) = \widetilde{G}_L(t)$ in (2.5), from (2.6), the expected cost rate is

$$\widetilde{C}_L(T) = \frac{c_F - (c_F - c_R)\int_T^\infty [1 - \prod_{j=1}^N \overline{G}_j(t)]\mathrm{d}F(t)}{\int_0^T \overline{F}(t)\mathrm{d}t + \int_T^\infty \prod_{j=1}^N \overline{G}_j(t)\overline{F}(t)\mathrm{d}t}. \tag{2.16}$$

Differentiating $\widetilde{C}_L(T)$ with respect to T and setting it equal to 0,

$$\int_0^T \overline{F}(t)[h(T) - h(t)]\mathrm{d}t - \int_T^\infty \prod_{j=1}^N \overline{G}_j(t)\overline{F}(t)[h(t) - h(T)]\mathrm{d}t = \frac{c_R}{c_F - c_R}, \tag{2.17}$$

whose left-hand side increases strictly with T from $-\int_T^\infty \prod_{j=1}^N \overline{G}_j(t)\mathrm{d}F(t)$ to ∞. Thus, there exists a finite unique $\widetilde{T}_L$ $(0 < \widetilde{T}_L < \infty)$, which satisfies (2.17), and the resulting cost rate is

$$\widetilde{C}_L(\widetilde{T}_L) = (c_F - c_R)h(\widetilde{T}_L). \tag{2.18}$$

Comparing (2.14) and (2.17), $T_L^* \geq \widetilde{T}_L$. Thus, from (2.15) and (2.18), modified replacement last is better than replacement last.

2.3 REPLACEMENT POLICIES FOR REDUNDANT SYSTEM

Suppose some systems consist of n $(n = 1, 2, \dots)$ units, and unit i $(i = 1, 2, \dots, n)$ has an independent failure time X_i with distribution $F_i(t)$, its failure rate $h_i(t)$ and the cumulative hazard function is $H_i(t) \equiv \int_0^t h_i(u)\mathrm{d}u$, i.e., $F_i(t) = 1 - \mathrm{e}^{-H_i(t)} = 1 - \exp[-\int_0^t h_i(u)\mathrm{d}u]$. It is assumed that $h_i(t)$ increases strictly with t from 0, and at least one of $h_i(t)$ increases to ∞.

(1) SERIES SYSTEM

Suppose that the system consists of a series system. Denoting $X_{\mathrm{m}} \equiv \min\{X_1, X_2, \dots, X_n\}$, the system has a failure distribution

$$F_{\mathrm{m}}(t) \equiv \Pr\{X_{\mathrm{m}} \leq t\} = 1 - \prod_{i=1}^{n} \overline{F}_i(t). \tag{2.19}$$

(2) PARALLEL SYSTEM

Suppose that the system consists of a parallel system. Denoting $X_{\mathrm{M}} \equiv \max\{X_1, X_2, \dots, X_n\}$, the system has a failure distribution

$$F_{\mathrm{M}}(t) \equiv \Pr\{X_{\mathrm{M}} \leq t\} = \prod_{i=1}^{n} F_i(t). \tag{2.20}$$

(3) RANDOM K-OUT-OF-n SYSTEM

Consider a K-out-of-n system: Suppose that K is a random variable with probability function $p_k \equiv \Pr\{K = k\}$ $(k = 1, 2, \dots, n)$ for a specified n, where $p_0 \equiv 0$. The system consists of unit i $(i = 1, 2, \dots, n)$ and fails when at least K units has failed, which is called *Random K-out-of-n system* [20].

It is assumed that each unit has an identical failure distribution $F(t)$. The probability distribution of K is $P_k \equiv \Pr\{K \leq k\} = \sum_{j=1}^{k} p_j$, where $P_n = 1$ and P_k increases with k from p_1 to 1. Then, the reliability of the system at time t is [21]

$$\begin{aligned}\sum_{k=0}^{n} p_k \sum_{j=k}^{n} \binom{n}{j} [\overline{F}(t)]^j [F(t)]^{n-j} &= \sum_{j=0}^{n} \sum_{k=0}^{j} p_k \binom{n}{j} [\overline{F}(t)]^j [F(t)]^{n-j} \\ &= \sum_{k=1}^{n} P_k \binom{n}{k} [\overline{F}(t)]^k [F(t)]^{n-k},\end{aligned} \tag{2.21}$$

and the failure distribution is

$$F_{\mathrm{P}}(t) \equiv 1 - \sum_{k=1}^{n} P_k \binom{n}{k} [\overline{F}(t)]^k [F(t)]^{n-k} = \sum_{k=1}^{n} \overline{P}_k \binom{n}{k} [\overline{F}(t)]^k [F(t)]^{n-k}. \tag{2.22}$$

Random K-out-of-n system can represent some redundant systems such as a series system and parallel system. For example, when $p_n = 1$, $p_k = 0$ $(k = 1, 2, \ldots, n-1)$, i.e., when $P_n = 1$, $P_k = 0$ $(k = 1, 2, \ldots, n-1)$, the system is a series system with the failure distribution

$$F_P(t) = 1 - [\overline{F}(t)]^n,$$

and when $p_1 = 1$, $p_k = 0$ $(k = 2, 3, \ldots, n)$, i.e., when $P_k = 1$ $(k = 1, 2, \ldots, n)$, the system is a parallel system with the failure distribution

$$F_P(t) = F(t)^n.$$

When $F(t) = F_i(t)$, $F_P(t)$ correspond to (2.19) and (2.20), respectively. Especially, when

$$P_k = \begin{cases} 0 & k \leq K-1, \\ 1 & k \geq K, \end{cases}$$

the system corresponds to a K-out-of-n:G system.

We discuss optimal replacement policies for each system in the following sections.

2.4 SERIES SYSTEM

Consider a series system with a failure distribution $F(t) = 1 - \prod_{i=1}^{n}\overline{F}_i(t)$.

(a) REPLACEMENT FIRST

Putting $G(t) = G_F(t)$ in (2.2), from (2.19), the expected cost rate is

$$C_{SF}(T) = \frac{c_R + (c_F - c_R)\int_0^T \prod_{j=1}^{N}\overline{G}_j(t)\mathrm{d}[1 - \prod_{i=1}^{n}\overline{F}_i(t)]}{\int_0^T \prod_{j=1}^{N}\overline{G}_j(t)\prod_{i=1}^{n}\overline{F}_i(t)\mathrm{d}t}. \tag{2.23}$$

Differentiating $C_{SF}(T)$ with respect to T and setting it equal to zero,

$$\int_0^T \prod_{j=1}^{N}\overline{G}_j(t)\prod_{i=1}^{n}\overline{F}_i(t)\left\{\sum_{k=1}^{n}[h_k(T) - h_k(t)]\right\}\mathrm{d}t = \frac{c_R}{c_F - c_R}, \tag{2.24}$$

whose left-hand side increases strictly with T from 0 to ∞. Thus, there exists a finite and unique T_{SF}^* $(0 < T_{SF}^* < \infty)$ which satisfies (2.24), and the resulting cost rate is

$$C_{SF}(T_{SF}^*) = (c_F - c_R)\sum_{i=1}^{n} h_i(T_{SF}^*). \tag{2.25}$$

(b) MODIFIED REPLACEMENT FIRST

Putting $G(t) = \widetilde{G}_F(t)$ in (2.3), from (2.19), the expected cost rate is

$$\widetilde{C}_{SF}(T) = \frac{c_R + (c_F - c_R)\int_0^T [1 - \prod_{j=1}^N G_j(t)]\mathrm{d}[1 - \prod_{i=1}^n \overline{F}_i(t)]}{\int_0^T [1 - \prod_{j=1}^N G_j(t)] \prod_{i=1}^n \overline{F}_i(t)\mathrm{d}t}. \tag{2.26}$$

Differentiating $\widetilde{C}_{SF}(T)$ with respect to T and setting it equal to zero,

$$\int_0^T \left[1 - \prod_{j=1}^N G_j(t)\right] \prod_{i=1}^n \overline{F}_i(t) \left\{\sum_{k=1}^n [h_k(T) - h_k(t)]\right\} \mathrm{d}t = \frac{c_R}{c_F - c_R}, \tag{2.27}$$

whose left-hand side increases strictly with T from 0 to ∞. Thus, there exists a finite and unique $\widetilde{T}_{SF}$ $(0 < \widetilde{T}_{SF} < \infty)$, which satisfies (2.27), and the resulting cost rate is

$$\widetilde{C}_{SF}(\widetilde{T}_{SF}) = (c_F - c_R) \sum_{i=1}^n h_i(\widetilde{T}_{SF}). \tag{2.28}$$

Comparing (2.24) and (2.27), $T_{SF}^* \geq \widetilde{T}_{SF}$.

(c) REPLACEMENT LAST

Putting $G(t) = G_L(t)$ in (2.4), from (2.19), the expected cost rate is

$$C_{SL}(T) = \frac{c_F - (c_F - c_R)\int_T^\infty \prod_{j=1}^N G_j(t)\mathrm{d}[1 - \prod_{i=1}^n \overline{F}_i(t)]}{\int_0^T \prod_{i=1}^n \overline{F}_i(t)\mathrm{d}t + \int_T^\infty [1 - \prod_{j=1}^N G_j(t)] \prod_{i=1}^n \overline{F}_i(t)\mathrm{d}t}. \tag{2.29}$$

Differentiating $C_{SL}(T)$ with respect to T and setting it equal to zero,

$$\int_0^T \prod_{i=1}^n \overline{F}_i(t) \left\{\sum_{k=1}^n [h_k(T) - h_k(t)]\right\} \mathrm{d}t$$
$$- \int_T^\infty \left[1 - \prod_{j=1}^N G_j(t)\right] \prod_{i=1}^n \overline{F}_i(t) \left\{\sum_{k=1}^n [h_k(t) - h_k(T)]\right\} \mathrm{d}t = \frac{c_R}{c_F - c_R}, \tag{2.30}$$

whose left-hand side increases strictly with T to ∞. Thus, there exists a finite and unique T_{SL}^* $(0 < T_{SL}^* < \infty)$ which satisfies (2.30), and the resulting cost rate is

$$C_{SL}(T_{SL}^*) = (c_F - c_R) \sum_{k=1}^n h_k(T_{SL}^*). \tag{2.31}$$

(d) MODIFIED REPLACEMENT LAST

Putting $G(t) = \widetilde{G}_L(t)$ in (2.5), from (2.19), the expected cost rate is

$$\widetilde{C}_{SL}(T) = \frac{c_F - (c_F - c_R)\int_T^\infty [1 - \prod_{j=1}^N \overline{G}_j(t)]\mathrm{d}[1 - \prod_{i=1}^n \overline{F}_i(t)]}{\int_0^T \prod_{i=1}^n \overline{F}_i(t)\mathrm{d}t + \int_T^\infty \prod_{j=1}^N \overline{G}_j(t)\prod_{i=1}^n \overline{F}_i(t)\mathrm{d}t}. \tag{2.32}$$

Differentiating $\widetilde{C}_{SL}(T)$ with respect to T and setting it equal to zero,

$$\int_0^T \prod_{i=1}^n \overline{F}_i(t)\left\{\sum_{k=1}^n [h_k(T) - h_k(t)]\right\}\mathrm{d}t$$
$$-\int_T^\infty \prod_{j=1}^N \overline{G}_j(t)\prod_{i=1}^n \overline{F}_i(t)\left\{\sum_{k=1}^n [h_k(t) - h_k(T)]\right\}\mathrm{d}t = \frac{c_R}{c_F - c_R}, \tag{2.33}$$

whose left-hand side increases strictly with T to ∞. Thus, there exists a finite and unique $\widetilde{T}_{SL}$ $(0 < \widetilde{T}_{SL} < \infty)$ which satisfies (2.33), and the resulting cost rate is

$$\widetilde{C}_{SL}(\widetilde{T}_{SL}) = (c_F - c_R)\sum_{k=1}^n h_k(\widetilde{T}_{SL}). \tag{2.34}$$

Comparing (2.30) and (2.33), $T_{SL}^* \geq \widetilde{T}_{SL}$.

2.5 PARALLEL SYSTEM

Consider a parallel system with a failure distribution $F(t) = \prod_{i=1}^n F_i(t)$, and the failure rate is

$$h_P(t) \equiv \frac{F_M'(t)}{1 - F_M(t)} = \frac{\sum_{i=1}^n h_i(t)\mathrm{e}^{-H_i(t)}\prod_{j=1,\neq i}^n [1 - \mathrm{e}^{-H_j(t)}]}{1 - \prod_{i=1}^n [1 - \mathrm{e}^{-H_i(t)}]}, \tag{2.35}$$

which increases strictly to ∞ when at least one of $h_i(t)$ increases to ∞.

(a) REPLACEMENT FIRST

Putting $G(t) = G_F(t)$ in (2.2), from (2.20), the expected cost rate is

$$C_{PF}(T) = \frac{c_R + (c_F - c_R)\int_0^T \prod_{j=1}^N \overline{G}_j(t)\mathrm{d}\prod_{i=1}^n F_i(t)}{\int_0^T \prod_{i=1}^N \overline{G}_j(t)[1 - \prod_{i=1}^n F_i(t)]\mathrm{d}t}. \tag{2.36}$$

Differentiating $C_{PF}(T)$ with respect to T and setting it equal to zero,

$$\int_0^T \prod_{j=1}^N \overline{G}_j(t)\left[1 - \prod_{i=1}^n F_i(t)\right][h_P(T) - h_P(t)]\mathrm{d}t = \frac{c_R}{c_F - c_R}, \tag{2.37}$$

whose left-hand side increases strictly with T from 0 to ∞. Thus, there exists a finite and unique T^*_{PF} $(0 < T^*_{PF} < \infty)$ which satisfies (2.37), and the resulting cost rate is

$$C_{PF}(T^*_{PF}) = (c_F - c_R)h_P(T^*_{PF}). \tag{2.38}$$

(b) MODIFIED REPLACEMENT FIRST

Putting $G(t) = \widetilde{G}_F(t)$ in (2.3), from (2.20), the expected cost rate is

$$\widetilde{C}_{PF}(T) = \frac{c_R + (c_F - c_R)\int_0^T [1 - \prod_{j=1}^N G_j(t)]\mathrm{d}\prod_{i=1}^n F_i(t)}{\int_0^T [1 - \prod_{j=1}^N G_j(t)][1 - \prod_{i=1}^n F_i(t)]\mathrm{d}t}. \tag{2.39}$$

Differentiating $\widetilde{C}_{PF}(T)$ with respect to T and setting it equal to zero,

$$\int_0^T \left[1 - \prod_{j=1}^n G_j(t)\right]\left[1 - \prod_{i=1}^n F_i(t)\right][h_P(T) - h_P(t)]\mathrm{d}t = \frac{c_R}{c_F - c_R}, \tag{2.40}$$

whose left-hand side increases strictly with T from 0 to ∞. Thus, there exists a finite and unique $\widetilde{T}_{PF}$ $(0 < \widetilde{T}_{PF} < \infty)$ which satisfies (2.40), and the resulting cost rate is

$$\widetilde{C}_{PF}(\widetilde{T}_{PF}) = (c_F - c_R)h_P(\widetilde{T}_{PF}). \tag{2.41}$$

Comparing (2.37) and (2.40), $T^*_{PF} \geq \widetilde{T}_{PF}$.

(c) REPLACEMENT LAST

Putting $G(t) = G_{PL}(t)$ in (2.4), from (2.20), the expected cost rate is

$$C_{PL}(T) = \frac{c_F - (c_F - c_R)\int_T^\infty \prod_{j=1}^N G_j(t)\mathrm{d}\prod_{i=1}^n F_i(t)}{\int_0^T [1 - \prod_{i=1}^n F_i(t)]\mathrm{d}t + \int_T^\infty [1 - \prod_{j=1}^N G_j(t)][1 - \prod_{i=1}^n F_i(t)]\mathrm{d}t}. \tag{2.42}$$

Differentiating $C_{PL}(T)$ with respect to T and setting it equal to zero,

$$\int_0^T \left[1 - \prod_{i=1}^n F_i(t)\right][h_P(T) - h_P(t)]\mathrm{d}t$$
$$- \int_T^\infty \left[1 - \prod_{j=1}^N G_j(t)\right]\left[1 - \prod_{i=1}^n F_i(t)\right][h_P(t) - h_P(T)]\mathrm{d}t = \frac{c_R}{c_F - c_R}, \tag{2.43}$$

whose left-hand side increases strictly with T to ∞. Thus, there exists a finite and unique T^*_{PL} $(0 < T^*_{PL} < \infty)$ which satisfies (2.43), and the resulting cost rate is

$$C_{PL}(T^*_{PL}) = (c_F - c_R)h_P(T^*_{PL}). \tag{2.44}$$

(d) MODIFIED REPLACEMENT LAST

Putting $G(t) = \widetilde{G}_L(t)$ in (2.5), from (2.20), the expected cost rate is

$$\widetilde{C}_{PL}(T) = \frac{c_F - (c_F - c_R)\int_T^{\infty}[1 - \prod_{j=1}^{N}\overline{G}_j(t)]\mathrm{d}\prod_{i=1}^{n}F_i(t)}{\int_0^T[1 - \prod_{i=1}^{n}F_i(t)]\mathrm{d}t + \int_T^{\infty}\prod_{j=1}^{N}\overline{G}_j(t)[1 - \prod_{i=1}^{n}F_i(t)]\mathrm{d}t}. \tag{2.45}$$

Differentiating $\widetilde{C}_{PL}(T)$ with respect to T and setting it equal to zero,

$$\int_0^T \left[1 - \prod_{i=1}^{n} F_i(t)\right][h_P(T) - h_P(t)]\mathrm{d}t - \int_T^{\infty}\prod_{j=1}^{N}\overline{G}_j(t)\left[1 - \prod_{i=1}^{n} F_i(t)\right][h_P(t) - h_P(T)]\mathrm{d}t = \frac{c_R}{c_F - c_R}, \tag{2.46}$$

whose left-hand side increases strictly with T to ∞. Thus, there exists a finite and unique $\widetilde{T}_{PL}$ $(0 < \widetilde{T}_{PL} < \infty)$ which satisfies (2.46), and the resulting cost rate is

$$\widetilde{C}_{PL}(\widetilde{T}_{PL}) = (c_F - c_R)h_P(\widetilde{T}_{PL}). \tag{2.47}$$

Comparing (2.43) and (2.46), $T_{PL}^* \geq \widetilde{T}_{PL}$.

2.6 RANDOM K-OUT-OF-n SYSTEM

As examples of random K-out-of-n systems, we consider the following two systems with 4 units:

(1) PARALLEL-SERIES SYSTEM

When the system consists of a parallel-series system with 4 units in Fig. 2.1, $P_1 = 0$, $P_2 = 2/3$, $P_3 = 1$, $P_4 = 1$. Thus, the failure distribution of the system is, from (2.23),

$$\begin{aligned} F_1(t) &= 1 - \frac{2}{3}\binom{4}{2}\overline{F}(t)^2F(t)^2 - \binom{4}{3}\overline{F}(t)^3F(t) - \binom{4}{4}\overline{F}(t)^4 \\ &= 1 - 4\overline{F}(t)^2F(t)^2 - 4\overline{F}(t)^3F(t) - \overline{F}(t)^4 = 1 - \overline{F}(t)^2[1 + F(t)]^2, \end{aligned}$$

and the failure rate is

$$h_1(t) \equiv \frac{4h(t)F(t)}{1 + F(t)},$$

which increases with t from 0 to ∞.

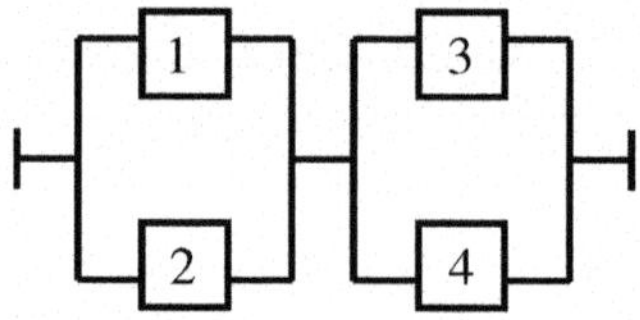

Figure 2.1 Parallel-Series System with 4 Units.

(a) REPLACEMENT FIRST
The expected cost rate is, from (2.7),

$$C_{1F}(T) = \frac{c_R + (c_F - c_R)\int_0^T \prod_{j=1}^N \overline{G}_j(t)\mathrm{d}F_1(t)}{\int_0^T \prod_{j=1}^N \overline{G}_j(t)\overline{F}_1(t)\mathrm{d}t}. \tag{2.48}$$

Differentiating $C_{1F}(T)$ with respect to T and setting it equal to zero,

$$4\int_0^T \prod_{j=1}^N \overline{G}_j(t)\overline{F}(t)^2[1+F(t)^2]\left[\frac{h(T)F(T)}{1+F(T)} - \frac{h(t)F(t)}{1+F(t)}\right]\mathrm{d}t = \frac{c_R}{c_F - c_R}, \tag{2.49}$$

whose left-hand side increases strictly with T from 0 to ∞. Thus, there exists a finite and unique T_{1F}^* $(0 < T_{1F}^* < \infty)$ which satisfies (2.49). Especially, when $F(t) = 1 - \mathrm{e}^{-\lambda t}$, (2.49) is

$$4\lambda\int_0^T \prod_{j=1}^N \overline{G}_j(t)\mathrm{e}^{-2\lambda t}[1+(1-\mathrm{e}^{-\lambda t})^2]\left(\frac{1}{2-\mathrm{e}^{-\lambda t}} - \frac{1}{2-\mathrm{e}^{-\lambda T}}\right)\mathrm{d}t = \frac{c_R}{c_F - c_R}, \tag{2.50}$$

and the resulting cost rate is

$$C_{1F}(T_{1F}^*) = (c_F - c_R)h_1(T_{1F}^*). \tag{2.51}$$

(b) MODIFIED REPLACEMENT FIRST
The expected cost rate is, from (2.10),

$$\widetilde{C}_{1F}(T) = \frac{c_R + (c_F - c_R)\int_0^T [1 - \prod_{j=1}^N G_j(t)]\mathrm{d}F_1(t)}{\int_0^T [1 - \prod_{j=1}^N G_j(t)]\overline{F}_1(t)\mathrm{d}t}. \tag{2.52}$$

Differentiating $\widetilde{C}_{1F}(T)$ with respect to T and setting it equal to zero,

$$4\int_0^T \left[1 - \prod_{j=1}^N G_j(t)\right]\overline{F}(t)^2[1+F(t)^2]\left[\frac{h(T)F(T)}{1+F(T)} - \frac{h(t)F(t)}{1+F(t)}\right]\mathrm{d}t = \frac{c_R}{c_F - c_R}, \tag{2.53}$$

whose left-hand side increases strictly with T from 0 to ∞. Thus, there exists a finite and unique $\widetilde{T}_{1F}$ $(0 < \widetilde{T}_{1F} < \infty)$ which satisfies (2.53). Especially, when $F(t) = 1 - \mathrm{e}^{-\lambda t}$, (2.53) is

$$4\lambda \int_0^T \left[1 - \prod_{j=1}^N G_j(t)\right] \mathrm{e}^{-2\lambda t}[1 + (1 - \mathrm{e}^{-\lambda t})^2] \left(\frac{1}{2 - \mathrm{e}^{-\lambda t}} - \frac{1}{2 - \mathrm{e}^{-\lambda T}}\right) \mathrm{d}t$$
$$= \frac{c_R}{c_F - c_R}, \tag{2.54}$$

and the resulting cost rate is

$$\widetilde{C}_{1F}(\widetilde{T}_{1F}) = (c_F - c_R) h_1(\widetilde{T}_{1F}). \tag{2.55}$$

Comparing (2.49) and (2.53), $T^*_{1F} \geq \widetilde{T}_{1F}$.

(c) REPLACEMENT LAST

The expected cost rate is from (2.13),

$$C_{1L}(T) = \frac{c_F - (c_F - c_R) \int_T^\infty \prod_{j=1}^N G_j(t) \mathrm{d}F_1(t)}{\int_0^T \overline{F}_1(t) \mathrm{d}t + \int_T^\infty [1 - \prod_{j=1}^N G_j(t)] \overline{F}_1(t) \mathrm{d}t}. \tag{2.56}$$

Differentiating $C_{1L}(T)$ with respect to T and setting it equal to zero,

$$4 \int_0^T \overline{F}(t)^2 [1 + F(t)^2] \left[\frac{h(T)F(T)}{1 + F(T)} - \frac{h(t)F(t)}{1 + F(t)}\right] \mathrm{d}t$$
$$- 4 \int_T^\infty \left[1 - \prod_{j=1}^N G_j(t)\right] \overline{F}(t)^2 [1 + F(t)^2] \left[\frac{h(t)F(t)}{1 + F(t)} - \frac{h(T)F(T)}{1 + F(T)}\right] \mathrm{d}t$$
$$= \frac{c_R}{c_F - c_R}, \tag{2.57}$$

whose left-hand side increases strictly with T to ∞. Thus, there exists a finite and unique T^*_{1L} $(0 < T^*_{1L} < \infty)$ which satisfies (2.57). Especially, when $F(t) = 1 - \mathrm{e}^{-\lambda t}$, (2.57) is

$$4\lambda \int_0^T \mathrm{e}^{-2\lambda t} [1 + (1 - \mathrm{e}^{-\lambda t})^2] \left(\frac{1}{2 - \mathrm{e}^{-\lambda t}} - \frac{1}{2 - \mathrm{e}^{-\lambda T}}\right) \mathrm{d}t$$
$$- 4\lambda \int_T^\infty \left[1 - \prod_{j=1}^N G_j(t)\right] \mathrm{e}^{-2\lambda t} [1 + (1 - \mathrm{e}^{-\lambda t})^2] \left(\frac{1}{2 - \mathrm{e}^{-\lambda T}} - \frac{1}{2 - \mathrm{e}^{-\lambda t}}\right) \mathrm{d}t$$
$$= \frac{c_R}{c_F - c_R}, \tag{2.58}$$

and the resulting cost rate is

$$C_{1L}(T_{1L}^*) = (c_F - c_R)h_1(T_{1L}^*). \tag{2.59}$$

(d) MODIFIED REPLACEMENT LAST

The expected cost rate is, from (2.16),

$$\widetilde{C}_{1L}(T) = \frac{c_F - (c_F - c_R)\int_T^\infty \left[1 - \prod_{j=1}^N \overline{G}_j(t)\right] \mathrm{d}F_1(t)}{\int_0^T \overline{F}_1(t)\mathrm{d}t + \int_T^\infty \prod_{j=1}^N \overline{G}_j(t)\overline{F}_1(t)\mathrm{d}t}. \tag{2.60}$$

Differentiating $\widetilde{C}_{1L}(T)$ with respect to T and setting it equal to zero,

$$\begin{aligned}&4\int_0^T \overline{F}(t)^2[1+F(t)^2]\left[\frac{h(T)F(T)}{1+F(T)} - \frac{h(t)F(t)}{1+F(t)}\right] \mathrm{d}t \\ &- 4\int_T^\infty \prod_{j=1}^N \overline{G}_j(t)\overline{F}(t)^2[1+F(t)^2]\left[\frac{h(t)F(t)}{1+F(t)} - \frac{h(T)F(T)}{1+F(T)}\right] \mathrm{d}t = \frac{c_R}{c_F - c_R},\end{aligned} \tag{2.61}$$

whose left-hand side increases strictly with T to ∞. Thus, there exists a finite and unique $\widetilde{T}_{1L}$ $(0 < \widetilde{T}_{1L} < \infty)$ which satisfies (2.61). Especially, when $F(t) = 1 - \mathrm{e}^{-\lambda t}$, (2.61) is

$$\begin{aligned}&4\lambda\int_0^T \mathrm{e}^{-2\lambda t}[1+(1-\mathrm{e}^{-\lambda t})^2]\left(\frac{1}{2-\mathrm{e}^{-\lambda t}} - \frac{1}{2-\mathrm{e}^{-\lambda T}}\right)\mathrm{d}t \\ &- 4\lambda\int_T^\infty \prod_{j=1}^N \overline{G}_j(t)\mathrm{e}^{-2\lambda t}[1+(1-\mathrm{e}^{-\lambda t})^2]\left(\frac{1}{2-\mathrm{e}^{-\lambda T}} - \frac{1}{2-\mathrm{e}^{-\lambda t}}\right)\mathrm{d}t \\ &= \frac{c_R}{c_F - c_R},\end{aligned} \tag{2.62}$$

and the resulting cost rate is

$$\widetilde{C}_{1L}(\widetilde{T}_{1L}) = (c_F - c_R)h_1(\widetilde{T}_{1L}). \tag{2.63}$$

Comparing (2.57) and (2.61), $T_{1L}^* \geq \widetilde{T}_{1L}$.

(2) SERIES-PARALLEL SYSTEM

When the system consists of a series-parallel system with 4 units in Fig. 2.2, $P_1 = 0$, $P_2 = 1/3$, $P_3 = 1$, $P_4 = 1$. Thus, the failure distribution of the system is, from (2.22),

$$\begin{aligned}F_2(t) &= 1 - \frac{1}{3}\binom{4}{2}\overline{F}(t)^2F(t)^2 - \binom{4}{3}\overline{F}(t)^4F(t) - \binom{4}{4}\overline{F}(t)^4 \\ &= 1 - \overline{F}(t)^2[1 + F(t) + F(t)\overline{F}(t)],\end{aligned} \tag{2.64}$$

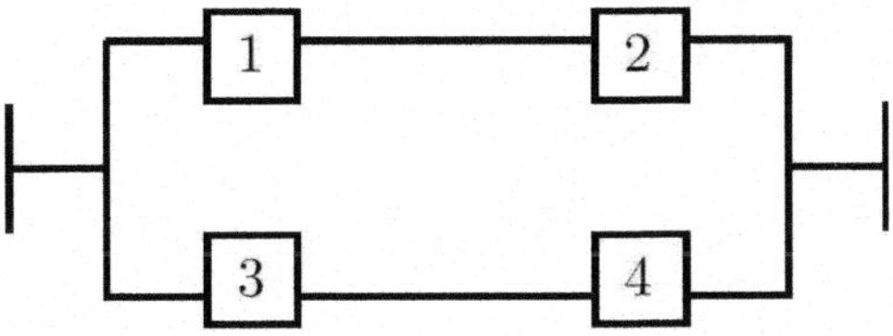

Figure 2.2 Series-parallel system with 4 units.

and the failure rate is

$$h_2(t) = \frac{4h(t)F(t)[1+\overline{F}(t)]}{1+F(t)+F(t)\overline{F}(t)}, \tag{2.65}$$

which increases strictly with t from 0 to ∞.

(a) REPLACEMENT FIRST

The expected cost rate is, from (2.7),

$$C_{2F}(T) = \frac{c_R + (c_F - c_R)\int_0^T \prod_{j=1}^N \overline{G}_j(t)\mathrm{d}F_2(t)}{\int_0^T \prod_{j=1}^N \overline{G}_j(t)\overline{F}_2(t)\mathrm{d}t}. \tag{2.66}$$

Differentiating $C_{2F}(T)$ with respect to T and setting it equal to zero,

$$4\int_0^T \prod_{j=1}^N \overline{G}_j(t)\overline{F}(t)^2[1+F(t)+F(t)\overline{F}(t)]$$
$$\times\left[\frac{h(T)F(T)[1+\overline{F}(T)]}{1+F(T)+F(T)\overline{F}(T)} - \frac{h(t)F(t)[1+\overline{F}(t)]}{1+F(t)+F(t)\overline{F}(t)}\right]\mathrm{d}t = \frac{c_R}{c_F - c_R}, \tag{2.67}$$

whose left-hand side increases strictly with T from 0 to ∞. Thus, there exists a finite and unique T_{2F}^* $(0 < T_{2F}^* < \infty)$ which satisfies (2.67), and the resulting cost rate is

$$C_{2F}(T_{2F}^*) = (c_F - c_R)h_2(T_{2F}^*). \tag{2.68}$$

(b) MODIFIED REPLACEMENT FIRST

The expected cost rate is, from (2.10),

$$\widetilde{C}_{2F}(T) = \frac{c_R + (c_F - c_R)\int_0^T [1-\prod_{j=1}^N G_j(t)]\mathrm{d}F_2(t)}{\int_0^T [1-\prod_{j=1}^N \overline{G}_j(t)]\overline{F}_2(t)\mathrm{d}t}. \tag{2.69}$$

Differentiating $\widetilde{C}_{2F}(T)$ with respect to T and setting it equal to zero,

$$4\int_0^T \left[1-\prod_{j=1}^N G_j(t)\right]\overline{F}(t)^2[1+F(t)+F(t)\overline{F}(t)]$$
$$\times\left[\frac{h(T)F(T)[1+\overline{F}(T)]}{1+F(T)+F(T)\overline{F}(T)}-\frac{h(t)F(t)[1+\overline{F}(t)]}{1+F(t)+F(t)\overline{F}(t)}\right]\mathrm{d}t=\frac{c_R}{c_F-c_R}, \tag{2.70}$$

whose left-hand side increases strictly with T to ∞. Thus, there exists a finite and unique $\widetilde{T}_{2F}$ $(0<\widetilde{T}_{2F}<\infty)$ which satisfies (2.70), and the resulting cost rate is

$$\widetilde{C}_{2F}(\widetilde{T}_{2F})=(c_F-c_R)h_2(\widetilde{T}_{2F}). \tag{2.71}$$

Comparing (2.67) and (2.70), $T_{2F}^*\geq\widetilde{T}_{2F}$.

(c) REPLACEMENT LAST

From (2.13), the expected cost rate is

$$C_{2L}(T)=\frac{c_F-(c_F-c_R)\int_T^\infty\prod_{j=1}^N G_j(t)\mathrm{d}F_2(t)}{\int_0^T\overline{F}_2(t)\mathrm{d}t+\int_T^\infty[1-\prod_{j=1}^N G_j(t)]\overline{F}_2(t)\mathrm{d}t}. \tag{2.72}$$

Differentiating $C_{2L}(T)$ with respect to T and setting it equal to zero,

$$4\int_0^T\overline{F}(t)^2[1+F(t)+F(t)\overline{F}(t)]\left\{\frac{h(T)F(T)[1+\overline{F}(T)]}{1+F(T)+F(T)\overline{F}(T)}-\frac{h(t)F(t)[1+\overline{F}(t)]}{1+F(t)+F(t)\overline{F}(t)}\right\}\mathrm{d}t$$
$$-4\int_T^\infty\left[1-\prod_{j=1}^N G_j(t)\right]\overline{F}(t)^2[1+F(t)+F(t)\overline{F}(t)]$$
$$\times\left\{\frac{h(t)F(t)[1+\overline{F}(t)]}{1+F(t)+F(t)\overline{F}(t)}-\frac{h(T)F(T)[1+\overline{F}(T)]}{1+F(T)+F(T)\overline{F}(T)}\right\}\mathrm{d}t=\frac{c_R}{c_F-c_R}, \tag{2.73}$$

whose left-hand side increases strictly with T to ∞. Thus, there exists a finite and unique T_{2L}^* $(0<T_{2L}^*<\infty)$ which satisfies (2.73), and the resulting cost rate is

$$C_{2L}(T_{2L}^*)=(c_F-c_R)h_2(T_{2L}^*). \tag{2.74}$$

(d) MODIFIED REPLACEMENT LAST

The expected cost rate is, from (2.16),

$$\widetilde{C}_{2L}(T)=\frac{c_F-(c_F-c_R)\int_T^\infty[1-\prod_{j=1}^N\overline{G}_j(t)]\mathrm{d}F_2(t)}{\int_0^T\overline{F}_2(t)\mathrm{d}t+\int_T^\infty\prod_{j=1}^N\overline{G}_j(t)\overline{F}_2(t)\mathrm{d}t}. \tag{2.75}$$

Table 2.2
Optimal Replacement Time and Cost for Single Unit System.

c_F/c_R	T_F^*	$C(T_F^*)/c_R$	$\widetilde{T}_F$	$C(\widetilde{T}_F)/c_R$	T_L^*	$C(T_L^*)/c_R$	$\widetilde{T}_L$	$C(\widetilde{T}_L)/c_R$
5	442.38	8.09	7.55	5.39	5.13	5.18	5.13	5.18
6	128.20	8.94	4.43	6.38	3.41	6.22	3.41	6.22
7	53.26	9.82	3.06	7.38	2.57	7.25	2.57	7.25
8	27.27	10.72	2.32	8.38	2.09	8.29	2.09	8.29
9	16.05	11.62	1.87	9.37	1.78	9.32	1.79	9.33
10	10.42	12.51	1.57	10.36	1.57	10.35	1.58	10.36
11	7.28	13.42	1.36	11.34	1.41	11.38	1.43	11.40
12	5.37	14.32	1.20	12.32	1.29	12.41	1.32	12.44
13	4.14	15.22	1.08	13.30	1.20	13.44	1.23	13.47
14	3.31	16.12	0.98	14.27	1.12	14.46	1.16	14.51
15	2.72	17.02	0.90	15.23	1.06	15.49	1.10	15.55

Differentiating $\widetilde{C}_{2L}(T)$ with respect to T and setting it equal to zero,

$$4\int_0^T \overline{F}(t)^2[1+F(t)+\overline{F}(t)\overline{F}(t)]\left\{\frac{h(T)F(T)[1+\overline{F}(T)]}{1+F(T)+F(T)\overline{F}(T)}-\frac{h(t)F(t)[1+\overline{F}(t)]}{1+F(t)+F(t)\overline{F}(t)}\right\}\mathrm{d}t$$

$$-4\int_T^\infty \prod_{j=1}^N \overline{G}_j(t)\overline{F}(t)^2[1+F(t)+F(t)\overline{F}(t)]$$

$$\times\left\{\frac{h(t)F(t)[1+\overline{F}(t)]}{1+F(t)+F(t)\overline{F}(t)}-\frac{h(T)F(T)[1+\overline{F}(T)]}{1+F(T)+F(T)\overline{F}(T)}\right\}\mathrm{d}t=\frac{c_R}{c_F-c_R}, \tag{2.76}$$

whose left-hand side increases strictly with T to ∞. Thus, there exists a finite and unique $\widetilde{T}_{2L}$ $(0<\widetilde{T}_{2L}<\infty)$ which satisfies (2.76), and the resulting cost rate is

$$\widetilde{C}_{2L}(\widetilde{T}_{2L})=(c_F-c_R)h_2(\widetilde{T}_{2L}). \tag{2.77}$$

Comparing (2.73) and (2.76), $T_{2L}^*\geq\widetilde{T}_{2L}$.

2.7 NUMERICAL EXAMPLES OF FOUR REDUNDANT SYSTEMS WITH 4 UNITS

Suppose that the number of random replacement times is $N=3$, and random replacement times Y_j $(j=1,2,3)$ have exponential distribution $G_j(t)=1-\mathrm{e}^{-\theta_i t}$ with $\theta_1=1.0$, $\theta_2=1.2$ and $\theta_3=1.3$, respectively.

At first, we present optimal T_F^*, $\widetilde{T}_F$, T_L^*, $\widetilde{T}_L$ for a single system in Table 2.2 when the unit has an identical Weibull distribution $F(t)=1-\mathrm{e}^{-t^{1.1}}$. It can be seen that T_F^*, $\widetilde{T}_F$, T_L^*, $\widetilde{T}_L$ decrease with c_F/c_R, and their cost rates increase with c_F/c_R.

Table 2.3
Optimal Replacement Time and Cost for Series System.

c_F/c_R	T^*_{SF}	$C(T^*_{SF})/c_R$	$\widetilde{T}_{SF}$	$C(\widetilde{T}_{SF})/c_R$	T^*_{SL}	$C(T^*_{SL})/c_R$	$\widetilde{T}_{SL}$	$C(\widetilde{T}_{SL})/c_R$
5	5.05	20.69	1.50	18.33	1.46	18.27	1.46	18.27
6	2.52	24.13	0.98	21.96	0.97	21.92	0.96	21.92
7	1.54	27.57	0.73	25.58	0.73	25.57	0.72	25.55
8	1.07	31.00	0.58	29.17	0.59	29.21	0.58	29.16
9	0.80	34.44	0.48	32.74	0.50	32.85	0.48	32.73
10	0.64	37.87	0.42	36.28	0.44	36.49	0.42	36.29
11	0.53	41.30	0.37	39.79	0.40	40.12	0.37	39.83
12	0.45	44.72	0.33	43.27	0.36	43.76	0.33	43.35
13	0.40	48.12	0.30	46.73	0.34	47.39	0.30	46.86
14	0.35	51.52	0.27	50.17	0.32	51.03	0.28	50.35
15	0.32	54.90	0.25	53.59	0.30	54.66	0.26	53.84

Table 2.4
Optimal Replacement Time and Cost for Parallel System.

c_F/c_R	T^*_{PF}	$C(T^*_{PF})/c_R$	$\widetilde{T}_{PF}$	$C(\widetilde{T}_{PF})/c_R$	T^*_{PL}	$C(T^*_{PL})/c_R$	$\widetilde{T}_{PL}$	$C(\widetilde{T}_{PL})/c_R$
5	3.71	3.85	1.00	1.77	1.08	1.93	0.97	1.71
6	1.96	3.92	0.89	1.91	1.00	2.22	0.87	1.87
7	1.52	3.99	0.81	2.04	0.96	2.51	0.80	2.00
8	1.29	4.06	0.76	2.14	0.92	2.80	0.75	2.12
9	1.15	4.13	0.72	2.24	0.90	3.09	0.71	2.23
10	1.04	4.19	0.68	2.33	0.88	3.38	0.68	2.33
11	0.97	4.25	0.65	2.41	0.86	3.67	0.65	2.42
12	0.91	4.31	0.63	2.48	0.85	3.96	0.63	2.51
13	0.86	4.36	0.60	2.55	0.84	4.25	0.61	2.60
14	0.82	4.42	0.59	2.62	0.83	4.53	0.59	2.68
15	0.78	4.47	0.57	2.68	0.82	4.82	0.58	2.77

Next, we propose the number of units is $n = 4$ and each unit has an identical Weibull distribution $F(t) = 1 - \mathrm{e}^{-t^{1.1}}$. Note that all of four systems have increasing failure rates. Table 2.3–2.6 present optimal T^*_{iF}, $\widetilde{T}_{iF}$, T^*_{iL}, $\widetilde{T}_{iL}$ and their expected cost rates $C(T^*_{iF})$, $C(\widetilde{T}_{iF})$, $C(T^*_{iL})$, $C(\widetilde{T}_{iL})$ $(i = S, P, 1, 2)$. All of optimal T^*_{iF}, $\widetilde{T}_{iF}$, T^*_{iL}, $\widetilde{T}_{iL}$ decrease with c_F/c_R, and their cost rates increase with c_F/c_R as in Table 2.2. In addition, $T^*_{iF} \geq \widetilde{T}_{iF}$, $T^*_{iL} \geq \widetilde{T}_{iL}$ $(i = S, P, 1, 2)$, as shown previously.

Furthermore, comparing T^*_{iF} with $\widetilde{T}_{iF}$ and T^*_{iL} with $\widetilde{T}_{iL}$ $(i = S, F, 1, 2)$, $T^*_{iF} \geq \widetilde{T}_{iF}$ and $T^*_{iL} \geq \widetilde{T}_{iL}$, as shown previously, which means that modified replacement first and last are better than the respective replacement first and last. Comparing $\widetilde{T}_{iF}$ and $\widetilde{T}_{iL}$, when c_F/c_R is large, modified replacement first is better than modified replacement last. This means that if replacement cost

Table 2.5
Optimal Replacement Time and Cost for Parallel-Series System.

c_F/c_R	T_{1F}^*	$C(T_{1F}^*)/c_R$	$\widetilde{T}_{1F}$	$C(\widetilde{T}_{1F})/c_R$	T_{1L}^*	$C(T_{1L}^*)/c_R$	$\widetilde{T}_{1L}$	$C(\widetilde{T}_{1L})/c_R$
5	1.02	6.88	0.59	4.99	0.65	5.35	0.59	4.98
6	0.75	7.31	0.49	5.53	0.58	6.20	0.50	5.56
7	0.62	7.73	0.43	6.01	0.54	7.06	0.44	6.10
8	0.53	8.12	0.39	6.46	0.51	7.91	0.40	6.61
9	0.47	8.49	0.36	6.87	0.49	8.76	0.37	7.10
10	0.42	8.84	0.33	7.25	0.47	9.61	0.35	7.58
11	0.39	9.18	0.31	7.62	0.46	10.46	0.33	8.05
12	0.36	9.50	0.29	7.96	0.45	11.32	0.32	8.51
13	0.34	9.81	0.28	8.29	0.44	12.17	0.30	8.97
14	0.32	10.11	0.27	8.60	0.43	13.02	0.29	9.42
15	0.30	10.40	0.25	8.90	0.43	13.87	0.28	9.87

Table 2.6
Optimal Replacement Time and Cost for Series-Parallel System.

c_F/c_R	T_{2F}^*	$C(T_{2F}^*)/c_R$	$\widetilde{T}_{2F}$	$C(\widetilde{T}_{2F})/c_R$	T_{2L}^*	$C(T_{2L}^*)/c_R$	$\widetilde{T}_{2L}$	$C(\widetilde{T}_{2L})/c_R$
5	0.70	7.22	0.44	5.77	0.53	6.39	0.44	5.82
6	0.52	7.90	0.37	6.51	0.48	7.60	0.38	6.65
7	0.43	8.53	0.32	7.17	0.45	8.82	0.34	7.45
8	0.37	9.12	0.29	7.78	0.43	10.03	0.31	8.22
9	0.33	9.67	0.27	8.35	0.42	11.25	0.29	8.98
10	0.30	10.18	0.25	8.87	0.41	12.46	0.28	9.72
11	0.27	10.67	0.23	9.37	0.40	13.68	0.27	10.46
12	0.25	11.14	0.22	9.84	0.39	14.89	0.26	11.20
13	0.24	11.58	0.21	10.29	0.39	16.11	0.25	11.92
14	0.23	12.00	0.20	10.72	0.38	17.32	0.24	12.65
15	0.21	12.41	0.19	11.13	0.38	18.53	0.24	13.37

c_F is larger, we should replace the unit earlier.

We concludes from these results that we should determine "which replacement is better" [22] by planning appropriate maintenance policies for an objective redundant system.

2.8 CONCLUSIONS

We have proposed four replacement first and last policies with random times for redundant systems. Especially, we have considered replacement policies for the system with 4 units such as series system, parallel system, parallel-series and series-parallel systems, and have discussed their optimal replacement policies to minimize the expected cost rates. We have obtained the expected cost rates by putting the failure distributions of each system and random replacement distributions into those of a single unit system. We have numerically compared the optimal replacement times and the resulting cost rates. From the above discussions, we have gotten that modified replacement first and last are better than the respective replacement first and last. This means that when replacement times are redundant random ones, we should preventively replace the system later for replacement first and earlier for replacement last. As a future work, we should apply these results to real systems such as network and database systems.

ACKNOWLEDGMENT

This work was supported by JSPS KAKENHI Grant Number 18K01713 and 20K04992.

REFERENCES

1. Nakagawa, T. (2005). *Maintenance Theory of Reliability*, Springer, Londo.
2. Nakagawa, T. (2007). *Advanced Reliability Models and Maintenance Policies*, Springer, London.
3. Nakagawa, T. (2014). *Random Maintenance Policies*, Springer, London.
4. Barlow, R. E., & Proschan, F. (1965). *Mathematical Theory of Reliability*, Wiley, New York.
5. Yun, W., & Choi, C. (2000). Optimum replacement intervals with random time horizon. *Journal of Quality in Maintenance Engineering*, 6, 269-274.
6. Yun, W., & Nakagawa, T. (2000). Replacement and inspection policies for products with random life cycle. *Reliability Engineering & System Safety*, 95, 161-165.
7. Chen, M., Mizutani, S., & Nakagawa, T. (2010). Random and age replacement policies. *International Journal of Reliability, Quality and Safety Engineering*, 17, 1, 27-39.
8. Chen, M., Nakamura, S., & Nakagawa, T. (2010). Replacement and preventive maintenance models with random working times. *IEICE TRNASACTIONS on Fundamentals of Electronics, Communications and Computer Sciences*, E93-A, 500-507.
9. Nakagawa, T., Zhao, X., & Yun, W. (2011). Optimal age replacement and inspection policies with random failure and replacement times. *International Journal of Reliability, Quality and Safety Engineering*, 18, 405-416.
10. Zhao, X., & Nakagawa, T. (2012). Optimization problems of replacement first or last in reliability theory. *European Journal of Operational Research*, 223, 141-149.

11. Nakagawa, T. & Zhao, X. (2015). *Maintenance Overtime Policies in Reliability Theory*, Springer, London.

12. Zhao, X., & Nakagawa, T. (2013). Optimal periodic and random inspection with first, last, and overtime policies. *International Journal of Systems Science*, DOI: 10.1080/00207721.2013.827263.

13. Zhao, X., Qian, C., & Nakagawa, T. (2014). Optimal age and periodic replacement with overtime policies. *International Journal of Reliability, Quality and Safety Engineering*, 21, 1450016.

14. Bollinger, R. C., & Salvia, A. A. (1985). Consecutive-k-out-of-n:F systems with sequential failures. *IEEE Transactions on Reliability*, R-34, 43-45.

15. Zuo, M. J., & Kuo, W. (1990). Design and performance analysis of consecutive-k-out-of-n structure. *Naval Research Logistics*, 37, 203-230.

16. Pham, H., Eds. (2003). Handbook of Reliability Engineering. Springer, London.

17. Ram, M., Dohi, T., Eds. (2019). *System Engineering, Reliability Analysis Using k-out-of-n Structures*, CRC Press, Boca Raton, FL.

18. Bentolhoda, J., & Lance, F. (2019). Impact of correlated failure on the maintenance of multi-state consecutive 2-out-of-n: Failed systems. In Ram, M., & Dohi, T., Eds., *System Engineering, Reliability Analysis Using k-out-of-n Structures*, CRC Press, Boca Raton, FL, 105-123.

19. Zhou, L., Yamamoto, H., Nakamura, T., & Xiao, X. (2020). Optimization Problems for Consecutive-k-out-of-n:G Systems. *IEICE TRNASACTIONS on Fundamentals of Electronics, Communications and Computer Sciences*, E103-A, 741-748.

20. Ito, K., Zhao, X., & Nakagawa, T. (2017). Random number of unit for k-out-of-n. *Applied Mathematics and Computation*, 45, 563-572.

21. Ito, K., & Nakagawa, T. (2019). Reliability Properties of K-out-of-N:G Systems. In Ram, M., & Dohi, T., Eds., *System Engineering, Reliability Analysis Using k-out-of-n Structures*, CRC Press, Boca Raton, FL, 25-40.

22. Mizutani, S., Zhao, X., & Nakagawa, T. (2020). Which replacement is better at working cycles or numbers of failures. *IEICE TRNASACTIONS on Fundamentals of Electronics, Communications and Computer Sciences*, E103-A(2), 523-532.

3 Backup Policies with Random Data Updates

Xufeng Zhao
Nanjing University of Aeronautics and Astronautics, China

Jiajia Cai
Nanjing University of Aeronautics and Astronautics, China

Cunhua Qian
Nanjing Tech University, China

Syouji Nakamura
Kinjo Gakuin University, Japan

CONTENTS

3.1 INTRODUCTION

Database, a set of data and the way they are organized [1], has now become the lifeblood for some organizations in this modern society. For instance, the transactional systems of commercial banks grind to a halt within few minutes if something goes wrong with their databases. For the super critical databases in commercial airports and nuclear plants, it is expected to back up the data several times a day, or even to use database replication techniques [2] for real-time backups to achieve high data security.

Hardware crashes and human errors aren't the only reasons to corrupt the digital media and data documents, there are also some traditional disasters such as fires, floods, and earthquakes that can destroy the whole database system [3]. Normally database management system (DBMS) can be set to implement a hierarchy of daily, weekly and monthly backups, including mix and match modes between full backup and differential/incremental backup, and transaction log backup [4], whose backup frequency depends on the factors, as we know, such as rate of data update, database availability, criticality of data, etc. The transaction logs can be backed up continually, e.g., online backup [5, 6], which uploads local changed part of the files to its servers in a continuous backup setup. A full daily backup may suffice for the low sensitive data and the database with a few dozens of transactions a day. Those backup tapes are transferred to a geographically separate location that is monitored 24/7 for safe keeping.

All backup processes themselves will introduce some locks on the database and consume resources, so that the following limitations of the situations should be considered while scheduling backups: (i) The period of time to run backups on the system, which is called backup window [7] and will have the least amount of interference with normal operations of database. (ii) The hard drive is busy reading files and exporting updated data for the purpose of backup, and its full bandwidth is no longer available for other data transactions. (iii) Data storage cost that depends on the design of different backup schemes, and data transfer cost for a distributed backup when the network bandwidth is limited. (iv) Constant costs for labor requirement, backup software, etc. In this paper, we take up a database system that must be up and running 24 hours a day, 7 days a week, and try to consider the above situations into models, using the random maintenance approach in reliability theory [8], that is, backups are scheduled in random ways, putting their backup windows in non-busy states with user's convenience and balancing the total costs of data backup, and data restoration.

We use the following three widely accepted backup modes for our modelings and discussions, i.e., full backup, incremental backup and differential backup [9, 10, 11, 12]:

> *Full backup*, a lazy but simple mode that exports all the data files updated since the last full backup. When a full recovery after database breakdown is needed, the data restoration needs only the last full

backup. Obviously, this mode means that many periodic full backups will need to be implemented, requiring long periods of database downtime on a regular basis.

Incremental backup, the mode that exports only the data files updated since the last backup (a full or incremental backup). Incremental backups are much smaller and quicker than full backups, and the data restoration after failure needs the last full backup plus all the incremental backups until the point-in-time of breakdown.

Differential backup, the mode that exports all data files updated since the last full backup. The advantage to this mode is quicker recovery time, requiring only a full backup and the last differential backup to restore the entire updated data.

In this chapter, we suppose for a 24/7 database system that the above full, incremental and differential backups are only implemented after a large volume of data files have been updated, in other words, we keep the database running when it is busy with data transactions in large volumes and back up the updated data files when the large update is over or when the database goes into a relative non-busy state. This is different from the traditional discussions of periodic backup schemes [13, 14] as it would be unreasonable to implement any backup when lots of data transactions are waiting for processing even the scheduled backup time has arrived. However, we don't know the exact times when data updates in large volumes, i.e., the large updates occur randomly, and we also cannot know the exact volume of data for this update, but it can be assumed as a random variable. We give the above assumptions as the costs of backup and recovery depend on the total exported/imported data. Random updates with random volumes of updated data forms a compound stochastic process, which has similar formulation of shock and damage process in reliability [15], so that the technique of cumulative damage models can be used for modelings. In addition, we apply the overtime technique in random maintenance [16, 17] into models, i.e., delaying backup at the completion of the forthcoming update over a scheduled backup time to make sure that backup window could begin right after update.

Advantages of adopting non-periodic backups over periodic ones for mission success probability have been illustrated recently [18], however, to the best of our knowledge, there is no research works on the merit of the proposed random incremental and differential backup models. In Section 3.2, we firstly obtain the total costs and the expected cost rates of data backup and data restoration for respective incremental and differential backups when full backups are considered as renewal points [19]. In Section 3.3, optimum full backup times are found for the models when the full backup is scheduled at a number N of updates. When full backup is scheduled at the completion of the forthcoming update over a planned time T, the overtime models are considered in Sections 3.4 and 3.5. Comparisons of the backup models with respective decision variables N and T are discussed in Section 3.6. Finally, concluding remarks are given in Section 3.7.

3.2 EXPECTED COST RATES

Suppose in a database system that the data stored in files updates randomly in large volumes at a renewal process according to an identical distribution $F(t)$ with a density function $f(t) \equiv \mathrm{d}F(t)/\mathrm{d}t$ and finite mean $1/\lambda \equiv \int_0^\infty \overline{F}(t)\mathrm{d}t$, where $\overline{\Phi}(t) \equiv 1 - \Phi(t)$ for any function $\Phi(t)$. A volume W_j ($j = 1, 2, \cdots$) of updated data due to the jth update has an identical distribution $G(x) \equiv \Pr\{W_j \leq x\}$ with finite mean $1/\omega \equiv \int_0^\infty \overline{G}(x)\mathrm{d}x$.

It is assumed that the database failure occurs randomly with a general distribution $D(t)$ with a density function $d(t) \equiv \mathrm{d}D(t)/\mathrm{d}t$ and finite mean $1/\mu \equiv \int_0^\infty \overline{D}(t)\mathrm{d}t$, and the failure rate $r(t) \equiv d(t)/\mathrm{d}\overline{D}(t)$ [19] is supposed to be increasing with t strictly to $r(\infty)$ that might be infinity.

A full backup should be made immediately after database breakdown as a renewal point for incremental and differential backups. In order to protect the security of data and prevent the enormous restoration cost due to failure, a full backup is scheduled preventively at a number N ($N = 1, 2, \cdots$) of updates, i.e., at a number N of incremental/differential backups. Then, the probability that a full backup is implemented at failure is

$$\int_0^\infty [1 - F^{(N)}(t)]\mathrm{d}D(t) = \int_0^\infty D(t)\mathrm{d}F^{(N)}(t), \tag{3.1}$$

and the probability that it is implemented at update N is

$$\int_0^\infty \overline{D}(t)\mathrm{d}F^{(N)}(t) = \int_0^\infty F^{(N)}(t)\mathrm{d}D(t), \tag{3.2}$$

where $\phi^{(j)}(t)$ ($j = 1, 2, \cdots$) is the j-fold convolution of $\phi(t)$ and $\phi^{(0)}(t) \equiv 1$ for any $t \geq 0$, and note that (3.1) + (3.2) $\equiv 1$.

We introduce the following costs for data backup and data restoration schemes: c_F is a constant cost of full backup, $c_B + c_0 x$ is the cost for incremental/differential backup when a total volume x of data has been updated, $c_R + c_0 x + jc_N$ is the data restoration cost after failure when a number j of incremental backups have been implemented, and $c_R + c_0 x$ is the data restoration cost after failure when the differential backup is scheduled. We denote that, for $j = 1, 2, \cdots, N$,

$$M_j \equiv \int_0^\infty (c_B + c_0 x)\mathrm{d}G^{(j)}(x) = c_B + \frac{jc_0}{\omega},$$

$$N_j \equiv \int_0^\infty (c_R + c_0 x)\mathrm{d}G^{(j)}(x) = c_R + \frac{jc_0}{\omega}.$$

Then, jM_1 = the expected cost of j incremental backups, $\sum_{i=1}^{j} M_i$ = the expected cost of j differential backups, $N_j + jc_N$ = the data restoration cost when j incremental backups are implemented to import saved data files, and N_j = the data restoration cost when the jth differential backup is implemented. Thus, the expected cost until full backup when incremental backup is scheduled is

$$
\begin{aligned}
\widetilde{C}_I(N) =&(c_F + NM_1)\int_0^\infty F^{(N)}(t)\mathrm{d}D(t) \\
&+ \sum_{j=0}^{N-1}(c_F + jM_1 + N_j + jc_N)\int_0^\infty [F^{(j)}(t) - F^{(j+1)}(t)]\mathrm{d}D(t) \\
=&c_F + \left(c_B + \frac{c_O}{\omega}\right)\sum_{j=1}^{N}\int_0^\infty F^{(j)}(t)\mathrm{d}D(t) \\
&+ \sum_{j=0}^{N-1}\left[c_R + j\left(c_N + \frac{c_O}{\omega}\right)\right]\int_0^\infty [F^{(j)}(t) - F^{(j+1)}(t)]\mathrm{d}D(t), \quad (3.3)
\end{aligned}
$$

and the expected cost until full backup when differential backup is scheduled is

$$
\begin{aligned}
\widetilde{C}_D(N) =& \left(c_F + \sum_{i=1}^{N} M_i\right)\int_0^\infty F^{(N)}(t)\mathrm{d}D(t) \\
&+ \sum_{j=0}^{N-1}\left(c_F + \sum_{i=1}^{j} M_i + N_j\right)\int_0^\infty [F^{(j)}(t) - F^{(j+1)}(t)]\mathrm{d}D(t) \\
=&c_F + \sum_{j=1}^{N}\left(c_B + \frac{jc_O}{\omega}\right)\int_0^\infty F^{(j)}(t)\mathrm{d}D(t) \\
&+ \sum_{j=0}^{N-1}\left(c_R + \frac{jc_O}{\omega}\right)\int_0^\infty [F^{(j)}(t) - F^{(j+1)}(t)]\mathrm{d}D(t). \quad (3.4)
\end{aligned}
$$

The difference of $\widetilde{C}_I(N)$ and $\widetilde{C}_D(N)$ is

$$
\begin{aligned}
\widetilde{C}_I(N) - \widetilde{C}_D(N) =& \frac{c_O}{\omega}\sum_{j=0}^{N-1} j\int_0^\infty F^{(j+1)}(t)\mathrm{d}D(t) \\
&- c_N\sum_{j=0}^{N-1} j\int_0^\infty [F^{(j)}(t) - F^{(j+1)}(t)]\mathrm{d}D(t), \quad (3.5)
\end{aligned}
$$

that is, if

$$
\left(c_N + \frac{c_O}{\omega}\right)\sum_{j=0}^{N-1} j\int_0^\infty F^{(j+1)}(t)\mathrm{d}D(t) > c_N\sum_{j=0}^{N-1} j\int_0^\infty F^{(j)}(t)\mathrm{d}D(t)
$$

holds, then incremental backup would save more cost than differential backup.

In particular, when $D(t) = 1 - \mathrm{e}^{-\mu t}$, if

$$\frac{c_0}{\omega}\left[\frac{F^*(\mu)}{1-F^*(\mu)}\right] > c_N,$$

then incremental backup saves more cost, where $F^*(\mu)$ is a Laplace-Stieltjes transform of $F(t)$, i.e., $F^*(\mu) \equiv \int_0^\infty \mathrm{e}^{-\mu t}\mathrm{d}F(t)$.

The mean time to full backup is, from (3.1) and (3.2),

$$\int_0^\infty t\overline{D}(t)\mathrm{d}F^{(N)}(t) + \int_0^\infty t[1-F^{(N)}(t)]\mathrm{d}D(t) = \int_0^\infty \overline{D}(t)[1-F^{(N)}(t)]\mathrm{d}t. \tag{3.6}$$

Using the renewal theory [19], the expected cost rate for incremental backup is

$$C_I(N) = \frac{\begin{array}{c} c_F + (c_B + c_0/\omega)\sum_{j=1}^N \int_0^\infty F^{(j)}(t)\mathrm{d}D(t) \\ +\sum_{j=0}^{N-1}[c_R + j(c_N + c_0/\omega)]\int_0^\infty [F^{(j)}(t) - F^{(j+1)}(t)]\mathrm{d}D(t)\end{array}}{\int_0^\infty \overline{D}(t)[1-F^{(N)}(t)]\mathrm{d}t}, \tag{3.7}$$

and the expected cost rate for differential backup is

$$C_D(N) = \frac{\begin{array}{c} c_F + \sum_{j=1}^N (c_B + jc_0/\omega)\int_0^\infty F^{(j)}(t)\mathrm{d}D(t) \\ +\sum_{j=0}^{N-1}(c_R + jc_0/\omega)\int_0^\infty [F^{(j)}(t) - F^{(j+1)}(t)]\mathrm{d}D(t)\end{array}}{\int_0^\infty \overline{D}(t)[1-F^{(N)}(t)]\mathrm{d}t}. \tag{3.8}$$

3.3 OPTIMUM BACKUP TIMES

We discuss optimum full backup times N_I^* and N_D^* to minimize $C_I(N)$ and $C_D(N)$, respectively.

3.3.1 INCREMENTAL BACKUP

We find optimum N_I^* to minimize $C_I(N)$ in (3.7). Forming the inequality $C_I(N+1) - C_I(N) \geq 0$,

$$Q_I(N)\int_0^\infty \overline{D}(t)[1-F^{(N)}(t)]\mathrm{d}t - \left(c_B + \frac{c_0}{\omega}\right)\sum_{j=1}^N \int_0^\infty F^{(j)}(t)\mathrm{d}D(t)$$
$$-\sum_{j=0}^{N-1}\left[c_R + j\left(c_N + \frac{c_0}{\omega}\right)\right]\int_0^\infty [F^{(j)}(t) - F^{(j+1)}(t)]\mathrm{d}D(t) \geq c_F, \tag{3.9}$$

where

$$Q_I(N) = \frac{\begin{array}{c}(c_B + c_0/\omega)\int_0^\infty F^{(N+1)}(t)\mathrm{d}D(t) \\ +[c_R + N(c_N + c_0/\omega)]\int_0^\infty [F^{(N)}(t) - F^{(N+1)}(t)]\mathrm{d}D(t)\end{array}}{\int_0^\infty \overline{D}(t)[F^{(N)}(t) - F^{(N+1)}(t)]\mathrm{d}t}.$$

Let $L_I(N)$ denote the left-hand side of (3.9) and $M_F(t) \equiv \sum_{j=1}^{\infty} F^{(j)}(t)$. If $Q_I(N)$ increases strictly with N to $Q_I(\infty)$, and

$$L_I(\infty) \equiv \frac{Q_I(\infty)}{\mu} - \left(c_B + c_N + \frac{2c_0}{\omega}\right)\int_0^{\infty} M_F(t)\mathrm{d}D(t) - c_R > c_F,$$

then there exists a finite and unique minimum N_I^* $(1 \le N_I^* < \infty)$ which satisfies (3.9).

3.3.1.1 Case I

When $D(t) = 1 - \mathrm{e}^{-\mu t}$,

$$\frac{Q_I(N)}{\mu} \equiv \frac{(c_B + c_0/\omega)F^*(\mu)}{1 + F^*(\mu)} + N\left(c_N + \frac{c_0}{\omega}\right) + c_R,$$

which increases strictly with N to ∞. Thus, optimum N_I^* $(1 \le N_I^* < \infty)$ exists and satisfies

$$\sum_{j=1}^{N}\{1 - [F^*(\mu)]^j\} \ge \frac{c_F}{c_N + c_0/\omega}. \tag{3.10}$$

3.3.1.2 Case II

When $F(t) = 1 - \mathrm{e}^{-\lambda t}$, i.e., $F^{(j)}(t) - \sum_{i=j}^{\infty}[(\lambda t)^i/i!]\mathrm{e}^{-\lambda t}$ $(j - 0, 1, 2, \cdots)$, we obtain

$$\sum_{j=1}^{N}\int_0^{\infty} F^{(j)}(t)\mathrm{d}D(t) = \sum_{j=1}^{N}\int_0^{\infty}\overline{D}(t)\mathrm{d}F^{(j)}(t) = \lambda\sum_{j=0}^{N-1}\int_0^{\infty}\overline{D}(t)\frac{(\lambda t)^j}{j!}\mathrm{e}^{-\lambda t}\mathrm{d}t.$$

Thus, the expected cost rate in (3.7) is

$$C_I(N) = \frac{c_F + \sum_{j=0}^{N-1}[c_R + j(c_N + c_0/\omega)]\int_0^{\infty}[(\lambda t)^j/j!]\mathrm{e}^{-\lambda t}\mathrm{d}D(t)}{\sum_{j=0}^{N-1}\int_0^{\infty}[(\lambda t)^j/j!]\mathrm{e}^{-\lambda t}\overline{D}(t)\mathrm{d}t} + \lambda\left(c_B + \frac{c_0}{\omega}\right). \tag{3.11}$$

Forming the inequality $C_I(N+1) - C_I(N) \ge 0$,

$$\begin{aligned} & c_R\left[Q(N)\sum_{j=0}^{N-1}\int_0^{\infty}\frac{(\lambda t)^j}{j!}\mathrm{e}^{-\lambda t}\overline{D}(t)\mathrm{d}t - \sum_{j=0}^{N-1}\int_0^{\infty}\frac{(\lambda t)^j}{j!}\mathrm{e}^{-\lambda t}\mathrm{d}D(t)\right] \\ & + \left(c_N + \frac{c_0}{\omega}\right)\left\{NQ(N)\sum_{j=0}^{N-1}\int_0^{\infty}\frac{(\lambda t)^j}{j!}\mathrm{e}^{-\lambda t}\overline{D}(t)\mathrm{d}t - \sum_{j=0}^{N-1} j\int_0^{\infty}\frac{(\lambda t)^j}{j!}\mathrm{e}^{-\lambda t}\mathrm{d}D(t)\right\} \\ & \ge c_F. \end{aligned} \tag{3.12}$$

where $Q(N) = \lim_{T\to\infty} Q(N,T)$ and

$$Q(N,T) \equiv \frac{\int_0^T (\lambda t)^N \mathrm{e}^{-\lambda t}\mathrm{d}D(t)}{\int_0^T (\lambda t)^N \mathrm{e}^{-\lambda t}\overline{D}(t)\mathrm{d}t}.$$

We next prove that when $r(t)$ increases strictly with t, $Q(N,T)$ increases strictly with N to $r(T)$ [19]. Forming $Q(N+1,T) - Q(N,T) > 0$ and letting

$$\begin{aligned} L_I(T) \equiv & \int_0^T (\lambda t)^{N+1}\mathrm{e}^{-\lambda t}\mathrm{d}D(t) \int_0^T ((\lambda t)^N \mathrm{e}^{-\lambda t}\overline{D}(t)\mathrm{d}t) \\ & - \int_0^T (\lambda t)^N \mathrm{e}^{-\lambda t}\mathrm{d}D(t) \int_0^T (\lambda t)^{N+1}\mathrm{e}^{-\lambda t}\overline{D}(t)\mathrm{d}t, \end{aligned}$$

we have $L_I(0) = 0$ and

$$L_I'(T) = (\lambda T)^N \mathrm{e}^{-\lambda T}\overline{D}(T) \int_0^T (\lambda t)^N \mathrm{e}^{-\lambda t}\overline{D}(t)(\lambda T - \lambda t)[r(T) - r(t)]\mathrm{d}t > 0,$$

which follows that $Q(N,T)$ increases strictly with N to $r(T)$ for any $T > 0$. Thus, the left-hand side of (3.12) increases strictly with N to ∞.

In particular, when $D(t) = 1 - \mathrm{e}^{-\mu t}$, (3.12) becomes

$$\sum_{j=1}^{N}\left[1 - \left(\frac{\lambda}{\lambda+\mu}\right)^j\right] \geq \frac{c_F}{c_N + c_O/\omega},$$

which agrees with (3.10) when $F^*(\mu) = \lambda/(\lambda+\mu)$.

3.3.2 DIFFERENTIAL BACKUP

We find optimum N_D^* to minimize $C_D(N)$ in (3.8). Forming the inequality $C_D(N+1) - C_D(N) \geq 0$,

$$\begin{aligned} & Q_D(N) \int_0^\infty \overline{D}(t)[1 - F^{(N)}(t)]\mathrm{d}t - \sum_{j=1}^{N}\left(c_B + \frac{jc_O}{\omega}\right)\int_0^\infty F^{(j)}(t)\mathrm{d}D(t) \\ & - \sum_{j=0}^{N-1}\left(c_R + \frac{jc_O}{\omega}\right)\int_0^\infty [F^{(j)}(t) - F^{(j+1)}(t)]\mathrm{d}D(t) \geq c_F, \end{aligned} \tag{3.13}$$

where

$$Q_D(N) \equiv \frac{\begin{array}{c}[c_B + (N+1)c_O/\omega]\int_0^\infty F^{(N+1)}(t)\mathrm{d}D(t) \\ +[c_R + Nc_O/\omega]\int_0^\infty [F^{(N)}(t) - F^{(N+1)}(t)]\mathrm{d}D(t)\end{array}}{\int_0^\infty \overline{D}(t)[F^{(N)}(t) - F^{(N+1)}(t)]\mathrm{d}t}.$$

Let $L_D(T)$ denote the left-hand side of (3.13). If $Q_D(N)$ increases strictly with N to $Q_D(\infty)$, and

$$L_D(\infty) = \frac{Q_D(\infty)}{\mu} - \sum_{j=1}^{\infty}\left(c_B + \frac{jc_O}{\omega}\right)\int_0^{\infty} F^{(j)}(t)\mathrm{d}D(t)$$
$$-\frac{c_O}{\omega}\int_0^{\infty} M_F(t)\mathrm{d}D(t) - c_R > c_F,$$

then there exists a finite and unique minimum N_D^* $(1 \le N_D^* < \infty)$ which satisfies (3.13).

3.3.2.1 Case I

When $D(t) = 1 - \mathrm{e}^{-\mu t}$,

$$\frac{Q_D(N)}{\mu} = \frac{[c_B + (N+1)c_O/\omega]F^*(\mu)}{1 - F^*(\mu)} + \frac{Nc_O}{\omega} + c_R,$$

which increases strictly with N to ∞. Thus, optimum N_D^* $(1 \le N_D^* < \infty)$ exists and satisfies

$$\frac{1}{1 - F^*(\mu)}\sum_{j=1}^{N}\{1 - [F^*(\mu)]^j\} \ge \frac{c_F}{c_O/\omega}, \tag{3.14}$$

whose left-hand side increases with N from 1 to ∞.

3.3.2.2 Case II

When $F(t) = 1 - \mathrm{e}^{-\lambda t}$, optimum N_D^* satisfies, from (3.13),

$$\left(c_R + \frac{Nc_O}{\omega}\right)\left\{Q(N)\int_0^{\infty}\overline{D}(t)[1 - F^{(N)}(t)]\mathrm{d}t - \int_0^{\infty}[1 - F^{(N)}(t)]\mathrm{d}D(t)\right\}$$
$$+\frac{c_O}{\omega}\sum_{j=0}^{N-1}(N-j)\int_0^{\infty} F^{(j)}(t)\mathrm{d}D(t) \ge c_F, \tag{3.15}$$

where $Q(N)$ is given in (3.12), and the left-hand side of (3.15) increases strictly with N to ∞.

In particular, when $D(t) = 1 - \mathrm{e}^{-\mu t}$, (3.15) becomes

$$\sum_{j=0}^{N-1}(N-j)\left(\frac{\lambda}{\lambda+\mu}\right)^j \ge \frac{c_F}{c_O/\omega},$$

which agrees with (3.14) when $F^*(\mu) = \lambda/(\lambda+\mu)$.

3.3.3 NUMERICAL EXAMPLE

When $D(t) = 1 - e^{-\mu t}$, $F(t) = 1 - e^{-t}$, $c_B = 0.1, c_R = 0.2, c_N = 0.5$ and $c_F = 5.0$, Table 3.1 presents optimum N_I^*, N_D^* and their cost rates $C_I(N_I^*)/\mu$ $C_D(N_D^*)/\mu$. We conclude from Table 3.1 that,

(a) The cost c_O/ω denotes the variable backup or restoration cost of the expected volume of data for each update. When c_O/ω increases, the expected cost rates for all backup schemes should be increasing, which agrees with those in Table 3.1. In order to prevent the total backup and restoration costs, the optimum number N_I^* for incremental backups and N_D^* for differential backups will decrease.

(b) When the failure rate μ increases to the data update rate $\lambda = 1.0$, N_I^* decreases while N_D^* increases. This indicates that the incremental backup scheme costs more for data restoration after failure so that we should decrease N_I^* to save the number of incremental backups for restoration. For the differential backup model, the expected number of differential backups will be limited by the increasing μ so that we may increase N_D^* as far as possible due to the less restoration cost.

(c) It also can be found from Table 3.1 that the incremental backup is more economical when the failure rate is lower, while the differential backup is more economical when the failure rate is higher.

Table 3.1
Optimum N_I^*, N_D^*, and their cost rates $C_I(N_I^*)/\mu$ and $C_D(N_D^*)/\mu$.

μ	$c_O/\omega = 0.5$				$c_O/\omega = 1.0$			
	N_I^*	$C_I(N_I^*)/\mu$	N_D^*	$C_D(N_D^*)/\mu$	N_I^*	$C_I(N_I^*)/\mu$	N_D^*	$C_D(N_D^*)/\mu$
0.01	35	94.334	6	313.843	28	151.724	4	480.286
0.05	18	29.342	6	66.684	14	42.838	4	101.340
0.1	14	19.615	6	35.990	11	27.199	4	54.142
0.5	11	11.619	8	12.927	8	13.927	5	17.330
1.0	12	11.892	12	11.892	9	13.308	7	14.199
5.0	13	13.173	21	12.767	9	13.780	11	13.476

3.4 OVERTIME BACKUP MODELS

Suppose that a full backup is scheduled at the completion of the forthcoming update over a planned time T $(0 \leq T < \infty)$ or at database failure, whichever occurs first. Then, the probability that the full backup is made over time T is

$$\sum_{j=0}^{\infty}\int_0^T\left[\int_{T-t}^{\infty}\overline{D}(t+u)\mathrm{d}F(u)\right]\mathrm{d}F^{(j)}(t), \tag{3.16}$$

and the probability that it is made at failure is

$$\begin{aligned}D(T)&+\sum_{j=0}^{\infty}\int_0^T\left[\int_{T-t}^{\infty}\overline{F}(u)\mathrm{d}D(t+u)\right]\mathrm{d}F^{(j)}(t)\\&=\sum_{j=0}^{\infty}\int_0^T\left[\int_{T-t}^{\infty}D(t+u)\mathrm{d}F(u)\right]\mathrm{d}F^{(j)}(t).\end{aligned} \tag{3.17}$$

Noting that

$$\begin{aligned}&\sum_{j=0}^{\infty}\int_0^T\left[\int_0^{T-t}\overline{D}(t+u)\overline{F}(u)\mathrm{d}u\right]\mathrm{d}F^{(j)}(t)\\&=\sum_{j=0}^{\infty}\int_0^T\overline{D}(t)[F^{(j)}(t)-F^{(j+1)}(t)]\mathrm{d}t=\int_0^T\overline{D}(t)\mathrm{d}t,\end{aligned}$$

the mean time to full backup is

$$\begin{aligned}&\sum_{j=0}^{\infty}\int_0^T\left[\int_{T-t}^{\infty}(t+u)\overline{D}(t+u)\mathrm{d}F(u)\right]\mathrm{d}F^{(j)}(t)+\int_0^T t\mathrm{d}D(t)\\&+\sum_{j=0}^{\infty}\int_0^T\left[\int_{T-t}^{\infty}(t+u)\overline{F}(u)\mathrm{d}(t+u)\right]\mathrm{d}F^{(j)}(t)\\&=\sum_{j=0}^{\infty}\int_0^T\left[\int_0^{\infty}\overline{D}(t+u)\overline{F}(u)\mathrm{d}u\right]\mathrm{d}F^{(j)}(t).\end{aligned} \tag{3.18}$$

Noting that

$$\int_0^{\infty}\overline{F}(u)\mathrm{d}D(t+u)=\int_0^{\infty}[D(t+u)-D(t)]\mathrm{d}F(u),$$

the expected cost until full backup when incremental backup is scheduled is

$$\begin{aligned}\widetilde{C}_{OI}(T)=&\sum_{j=0}^{\infty}[c_F+(j+1)M_1]\int_0^T\left[\int_{T-t}^{\infty}\overline{D}(t+u)\mathrm{d}F(u)\right]\mathrm{d}F^{(j)}(t)\\&+\sum_{j=0}^{\infty}(c_F+jM_1+N_j+jc_N)\left\{\int_0^T[F^{(j)}(t)-F^{(j+1)}(t)]\mathrm{d}D(t)\right.\\&\left.+\int_0^T\left[\int_{T-t}^{\infty}\overline{F}(u)\mathrm{d}D(t+u)\right]\mathrm{d}F^{(j)}(t)\right\}\end{aligned}$$

$$= c_F + \left(c_B + \frac{c_O}{\omega}\right)\sum_{j=0}^{\infty}\int_0^T\left[\int_0^{\infty}\overline{D}(t+u)\mathrm{d}F(u)\right]\mathrm{d}F^{(j)}(t)$$
$$+\sum_{j=0}^{\infty}\left[c_R+j\left(c_N+\frac{c_O}{\omega}\right)\right]\int_0^T\left\{\int_0^{\infty}[D(t+u)-D(t)]\mathrm{d}F(u)\right\}\mathrm{d}F^{(j)}(t), \tag{3.19}$$

and the expected cost until full backup when differential backup is scheduled is

$$\widetilde{C}_{OD}(T) = \sum_{j=0}^{\infty}\left(c_F + \sum_{i=1}^{j+1}M_i\right)\int_0^T\left[\int_{T-t}^{\infty}\overline{D}(t+u)\mathrm{d}F(u)\right]\mathrm{d}F^{(j)}(t)$$
$$+\sum_{j=0}^{\infty}\left(c_F + \sum_{i=1}^{j}M_i + N_j\right)\int_0^T\left[\int_0^{\infty}\overline{F}(u)\mathrm{d}D(t+u)\right]\mathrm{d}F^{(j)}(t)$$
$$= c_F + \sum_{j=0}^{\infty}\left[c_B + \frac{(j+1)c_O}{\omega}\right]\int_0^T\left[\int_0^{\infty}\overline{D}(t+u)\mathrm{d}F(u)\right]\mathrm{d}F^{(j)}(t)$$
$$+\sum_{j=0}^{\infty}\left(c_R + \frac{jc_O}{\omega}\right)\int_0^T\left\{\int_0^{\infty}[D(t+u)-D(t)]\mathrm{d}F(u)\right\}\mathrm{d}F^{(j)}(t). \tag{3.20}$$

The difference of $\widetilde{C}_{OD}(T)$ and $\widetilde{C}_{OI}(T)$ is

$$\widetilde{C}_{OD}(T) - \widetilde{C}_{OI}(T) = \frac{c_O}{\omega}\sum_{j=0}^{\infty}j\int_0^T\left[\int_0^{\infty}\overline{D}(t+u)\mathrm{d}F(u)\right]\mathrm{d}F^{(j)}(t)$$
$$- c_N\sum_{j=0}^{\infty}j\int_0^T\left\{\int_0^{\infty}[D(t+u)-D(t)]\mathrm{d}F(u)\right\}\mathrm{d}F^{(j)}(t), \tag{3.21}$$

that is, if

$$\left(c_N + \frac{c_O}{\omega}\right)\sum_{j=0}^{\infty}j\int_0^T\left[\int_0^{\infty}\overline{D}(t+u)\mathrm{d}F(u)\right]\mathrm{d}F^{(j)}(t)$$
$$> c_N\sum_{j=0}^{\infty}j\int_0^T\overline{D}(t)\mathrm{d}F^{(j)}(t)$$

holds, then incremental backup would save more cost than differential backup does.

Therefore, the expected cost rate for incremental backup is

$$C_{OI}(T) = \frac{\begin{array}{c} c_F + (c_B + c_O/\omega)\sum_{j=0}^{\infty}\int_0^T[\int_0^{\infty}\overline{D}(t+u)\mathrm{d}F(u)]\mathrm{d}F^{(j)}(t) \\ +\sum_{j=0}^{\infty}[c_R + j(c_N + c_O/\omega)]\int_0^T\{\int_0^{\infty}[D(t+u)-D(t)]\mathrm{d}F(u)\}\mathrm{d}F^{(j)}(t)\end{array}}{\sum_{j=0}^{\infty}\int_0^T[\int_0^{\infty}\overline{D}(t+u)\overline{F}(u)\mathrm{d}u]\mathrm{d}F^{(j)}(t)}, \tag{3.22}$$

and the expected cost rate for differential backup is

$$C_{OD}(T) = \frac{\begin{array}{c} c_F + \sum_{j=0}^{\infty}[c_B + (j+1)c_O/\omega)]\int_0^T[\int_0^{\infty}\overline{D}(t+u)\mathrm{d}F(u)]\mathrm{d}F^{(j)}(t) \\ +\sum_{j=0}^{\infty}[c_R + jc_O/\omega]\int_0^T\{\int_0^{\infty}[D(t+u)-D(t)]\mathrm{d}F(u)\}\mathrm{d}F^{(j)}(t)\end{array}}{\sum_{j=0}^{\infty}\int_0^T[\int_0^{\infty}\overline{D}(t+u)\overline{F}(u)\mathrm{d}u]\mathrm{d}F^{(j)}(t)}. \tag{3.23}$$

3.5 OPTIMUM BACKUP TIMES

We discuss optimum full backup times T_{OI}^* and T_{OD}^* to minimize $C_{OI}(T)$ and $C_{OD}(T)$, respectively.

3.5.1 INCREMENTAL BACKUP

3.5.1.1 Case I

When $D(t) = 1 - \mathrm{e}^{-\mu t}$, the expected cost rate in (3.22) is

$$\frac{C_{OI}(T)}{\mu} = \frac{c_F + (c_N + c_O/\omega)[1 - F^*(\mu)]\sum_{j=0}^{\infty} j\int_0^T \mathrm{e}^{-\mu t}\mathrm{d}F^{(j)}(t)}{[1 - F^*(\mu)]\sum_{j=0}^{\infty}\int_0^T \mathrm{e}^{-\mu t}\mathrm{d}F^{(j)}(t)} + \left(c_B + \frac{c_O}{\omega}\right)\frac{F^*(\mu)}{1 - F^*(\mu)} + c_R. \tag{3.24}$$

Differentiating $C_{OI}(T)$ with respect to T and setting it equal to zero,

$$Q(T)\sum_{j=0}^{\infty}\int_0^T \mathrm{e}^{-\mu t}\mathrm{d}F^{(j)}(t) - \sum_{j=0}^{\infty} j\int_0^T \mathrm{e}^{-\mu t}\mathrm{d}F^{(j)}(t) = \frac{c_F}{(c_N + c_O/\omega)[1 - F^*(\mu)]}, \tag{3.25}$$

where

$$Q(T) \equiv \frac{\sum_{j=1}^{\infty} jf^{(j)}(T)}{\sum_{j=1}^{\infty} f^{(j)}(T)}.$$

If $Q(T)$ increases strictly with T and

$$Q(\infty) - \frac{F^*(\mu)}{1 - F^*(\mu)} > \frac{c_F}{c_N + c_O/\omega},$$

then there exists a finite and unique T_{OI}^* $(0 < T_{OI}^* < \infty)$ which satisfies (3.25).

In particular, when $F(t) = 1 - \mathrm{e}^{-\lambda t}$, $Q(T) = 1 + \lambda T$, and (3.25) becomes

$$\frac{\mu}{\lambda+\mu} + \lambda T - \frac{\lambda^2}{(\lambda+\mu)\mu}(1 - \mathrm{e}^{-\mu T}) = \frac{c_F}{c_N + c_O/\omega}, \tag{3.26}$$

whose left-hand increases strictly with T from $\mu/(\lambda+\mu)$ to ∞. Thus, if

$$\frac{\mu}{\lambda+\mu}\left(c_N + \frac{c_O}{\omega}\right) \geq c_F,$$

then $T^*_{OI} = 0$, that is, the full backup should be made at the first update.

3.5.1.2 Case II

When $F(t) = 1 - \mathrm{e}^{-\lambda t}$, the expected cost rate in (3.22) is

$$C_{OI}(T) = \frac{\begin{array}{c} c_F + c_R[D(T) + \int_T^\infty \mathrm{e}^{-\lambda(t-T)}\mathrm{d}D(t)] \\ +(c_N + c_O/\omega)[\int_0^T \lambda t \mathrm{d}D(t) + \lambda T \int_T^\infty \mathrm{e}^{-\lambda(t-T)}\mathrm{d}D(t)] \end{array}}{\int_0^T \overline{D}(t)\mathrm{d}t + \int_T^\infty \mathrm{e}^{-\lambda(t-T)}\overline{D}(t)\mathrm{d}t} + \lambda\left(c_B + \frac{c_O}{\omega}\right). \tag{3.27}$$

Differentiating $C_{OI}(T)$ with respect to T and setting it equal to zero,

$$\begin{aligned} &\left[c_R + \left(c_N + \frac{c_O}{\omega}\right)(1+\lambda T)\right] \frac{\int_T^\infty \mathrm{e}^{-\lambda t}\mathrm{d}D(t)}{\int_T^\infty \mathrm{e}^{-\lambda t}\overline{D}(t)\mathrm{d}t} \int_0^T \overline{D}(t)\mathrm{d}t \\ &- c_R D(T) - \left(c_N + \frac{c_O}{\omega}\right)\left[\int_0^T \lambda t \mathrm{d}D(t) - \int_T^\infty \mathrm{e}^{-\lambda(t-T)}\mathrm{d}D(t)\right] = c_F, \end{aligned} \tag{3.28}$$

whose left-hand side increases strictly with T to ∞. Therefore, there exists a finite and unique T^*_{OI} $(0 < T^*_{OI} < \infty)$ which satisfies (3.28).

In particular, when $D(t) = 1 - \mathrm{e}^{-\mu t}$, (3.28) agrees with (3.26).

3.5.2 DIFFERENTIAL BACKUP

3.5.2.1 Case I

When $D(t) = 1 - \mathrm{e}^{-\mu t}$, the expected cost rate in (3.23) is

$$\frac{C_{OD}(T)}{\mu} = \frac{c_F + (c_O/\omega)\sum_{j=0}^\infty j \int_0^T \mathrm{e}^{-\mu t}\mathrm{d}F^{(j)}(t)}{[1 - F^*(\mu)]\sum_{j=0}^\infty \int_0^T \mathrm{e}^{-\mu t}\mathrm{d}F^{(j)}(t)} + \left(c_B + \frac{c_O}{\omega}\right)\frac{F^*(\mu)}{1 - F^*(\mu)} + c_R. \tag{3.29}$$

Differentiating $C_{OD}(T)$ with respect to T and setting it equal to zero,

$$Q(T)\sum_{j=0}^\infty \int_0^T \mathrm{e}^{-\mu t}\mathrm{d}F^{(j)}(t) - \sum_{j=0}^\infty j \int_0^T \mathrm{e}^{-\mu t}\mathrm{d}F^{(j)}(t) = \frac{c_F}{c_O/\omega}, \tag{3.30}$$

whose left-hand side agrees with that of (3.25).

If $Q(T)$ increases strictly with T and

$$Q(\infty) - \frac{F^*(\mu)}{1-F^*(\mu)} > \frac{c_F}{(c_0/\omega)[1-F^*(\mu)]},$$

then there exists a finite and unique T^*_{OD} $(0 < T^*_{OD} < \infty)$ which satisfies (3.30).

In particular, when $F(t) = 1 - \mathrm{e}^{-\lambda t}$, $Q(T) = 1 + \lambda T$, and (3.30) becomes

$$1 + \lambda T + \frac{\lambda^2 T}{\mu} - \left(\frac{\lambda}{\mu}\right)^2 (1 - \mathrm{e}^{-\mu T}) = \frac{c_F}{c_0/\omega}, \tag{3.31}$$

whose left-hand increases strictly with T from 1 to ∞. Thus, if $c_0/\omega \geq c_F$, then $T^*_{OD} = 0$,

3.5.2.2 Case II

When $F(t) = 1 - \mathrm{e}^{-\lambda t}$, the expected cost rate in (3.23) is

$$C_{OD}(T) = \frac{\begin{array}{c} c_F + c_R[D(T) + \int_T^\infty \mathrm{e}^{-\lambda(t-T)}\mathrm{d}D(t)] \\ +(c_0/\omega)\lambda \int_0^T (1+\lambda t)\overline{D}(t)\mathrm{d}t \end{array}}{\int_0^T \overline{D}(t)\mathrm{d}t + \int_T^\infty \overline{D}(t)\mathrm{e}^{-\lambda(t-T)}\mathrm{d}t} + \lambda\left(c_K + \frac{c_0}{\omega}\right). \tag{3.32}$$

Differentiating $C_{OD}(T)$ with respect to T and setting it equal to zero,

$$\frac{c_R \int_T^\infty \mathrm{e}^{-\lambda t}\mathrm{d}D(t) + (c_0/\omega)(1+\lambda T)\mathrm{e}^{-\lambda T}\overline{D}(T)}{\int_T^\infty \overline{D}(t)\mathrm{e}^{-\lambda t}\mathrm{d}t} \int_0^T \overline{D}(t)\mathrm{d}t - c_R D(T)$$
$$+ \frac{c_0}{\omega}\left[(1+\lambda T)\overline{D}(T) - \lambda \int_0^T (1+\lambda t)\overline{D}(t)\mathrm{d}t\right] = c_F, \tag{3.33}$$

whose left-hand side increases strictly with T to ∞. Therefore, there exists a finite and unique T^*_{OD} $(0 < T^*_{OD} < \infty)$ which satisfies (3.33).

In particular, when $D(t) = 1 - \mathrm{e}^{-\mu t}$, (3.33) agrees with (3.31).

When $D(t) = 1 - \mathrm{e}^{-\mu t}$, $F(t) = 1 - \mathrm{e}^{-t}$, $c_B = 0.1, c_R = 0.2, c_N = 0.5$ and $c_F = 5.0$, Table 3.2 presents optimum T^*_{OI}, T^*_{OD}, and their cost rates $C_{OI}(T^*_{OI})/\mu$ and $C_{OD}(T^*_{OD})/\mu$. Comparing Table 3.1 and Table 3.2, optimum T^*_{OI} and T^*_{OD} have the same properties with N^*_I and N^*_D for c_0/ω and μ. We may also conclude that the backup model with variable N is more economical than that is based on variable T, as $C_I(N^*_I)/\mu < C_{OI}(T^*_{OI})/\mu$ and $C_D(N^*_D)/\mu < C_{OD}(T^*_{OD})/\mu$, which will be approved analytically in the following section.

3.6 COMPARISONS OF UPDATE N AND OVERTIME T

In order to compare the backup models with respective decision variables N and T discussed above analytically, we proposed the model in which the full

Table 3.2

Optimum T^*_{OI}, T^*_{OD}, and their cost rates $C_{OI}(T^*_{OI})/\mu$ and $C_{OD}(T^*_{OD})/\mu$.

μ	$c_0/\omega = 0.5$				$c_0/\omega = 1.0$			
	T^*_{OI}	$C_{OI}(T^*_{OI})/\mu$	T^*_{OD}	$C_{OD}(T^*_{OD})/\mu$	T^*_{OI}	$C_{OI}(T^*_{OI})/\mu$	T^*_{OD}	$C_{OD}(T^*_{OD})/\mu$
0.01	33.556	94.757	4.505	338.239	27.122	152.381	3.179	532.282
0.05	16.474	29.674	4.646	71.474	13.123	43.385	3.248	111.405
0.1	12.688	19.888	4.832	38.275	9.980	27.671	3.338	58.919
0.5	9.481	11.880	6.937	13.305	6.937	14.305	4.263	18.187
1.0	10.989	12.788	10.989	12.788	7.662	14.293	5.996	15.291
5.0	26.434	27.753	38.735	24.161	18.818	30.147	22.771	28.946

backup is scheduled preventively at the completion of the forthcoming update over a planned time T $(0 \le T < \infty)$ or at a number N $(N = 1, 2, \cdots)$ of updates, whichever occurs first. Then, the probability that the full backup is made at update N is

$$\int_0^T \overline{D}(t)\mathrm{d}F^{(N)}(t), \tag{3.34}$$

and the probability that it is made over time T is

$$\sum_{j=0}^{N-1}\int_0^T \left[\int_{T-t}^{\infty} \overline{D}(t+u)\mathrm{d}F(u)\right] \mathrm{d}F^{(j)}(t), \tag{3.35}$$

and the probability that it is made at failure is

$$\int_0^T [1 - F^{(N)}(t)]\mathrm{d}D(t) + \sum_{j=0}^{N-1}\int_0^T \left[\int_{T-t}^{\infty} \overline{F}(u)\mathrm{d}D(t+u)\right] \mathrm{d}F^{(j)}(t), \tag{3.36}$$

The mean time to full backup is

$$\begin{aligned}
&\int_0^T t\overline{D}(t)\mathrm{d}F^{(N)}(t) + \sum_{j=0}^{N-1}\int_0^T \left[\int_{T-t}^{\infty} (t+u)\overline{D}(t+u)\mathrm{d}F(u)\right] \mathrm{d}F^{(j)}(t) \\
&+ \int_0^T t[1 - F^{(N)}(t)]\mathrm{d}D(t) + \sum_{j=0}^{N-1}\int_0^T \left[\int_{T-t}^{\infty} (t+u)\overline{F}(u)\mathrm{d}D(t+u)\right] \mathrm{d}F^{(j)}(t) \\
&= \sum_{j=0}^{N-1}\int_0^T \left[\int_0^{\infty} \overline{D}(t+u)\overline{F}(u)\mathrm{d}u\right] \mathrm{d}F^{(j)}(t),
\end{aligned} \tag{3.37}$$

which agrees with (3.6) when $T \to \infty$ and with (3.18) when $N \to \infty$.

Therefore, the expected cost until full backup when incremental backup is scheduled is

$$\begin{aligned}
\widetilde{C}_I(T,N) =&(c_F+NM_1)\int_0^T \overline{D}(t)\mathrm{d}F^{(N)}(t)\\
&+\sum_{j=0}^{N-1}[c_F+(j+1)M_1]\int_0^T\left[\int_{T-t}^{\infty}\overline{D}(t+u)\mathrm{d}F(u)\right]\mathrm{d}F^{(j)}(t)\\
&+\sum_{j=0}^{N-1}(c_F+jM_1+N_j+jc_N)\int_0^T\left[\int_0^{\infty}\overline{F}(u)\mathrm{d}D(t+u)\right]\mathrm{d}F^{(j)}(t)\\
=&c_F+\left(c_B+\frac{c_0}{\omega}\right)\sum_{j=0}^{N-1}\int_0^T\left[\int_0^{\infty}\overline{D}(t+u)\mathrm{d}F(u)\right]\mathrm{d}F^{(j)}(t)\\
&+\sum_{j=0}^{N-1}\left[c_R+j\left(c_N+\frac{c_0}{\omega}\right)\right]\int_0^T\left\{\int_0^{\infty}[D(t+u)-D(t)]\mathrm{d}F(u)\right\}\mathrm{d}F^{(j)}(t),
\end{aligned} \tag{3.38}$$

and the expected cost until full backup when differential backup is scheduled is

$$\begin{aligned}
\widetilde{C}_D(T,N) =&\left(c_F+\sum_{i=1}^{N}M_i\right)\int_0^T \overline{D}(t)\mathrm{d}F^{(N)}(t)\\
&+\sum_{j=0}^{N-1}\left(c_F+\sum_{i=1}^{j+1}M_i\right)\int_0^T\left[\int_{T-t}^{\infty}\overline{D}(t+u)\mathrm{d}F(u)\right]\mathrm{d}F^{(j)}(t)\\
&+\sum_{j=0}^{N-1}\left(c_F+\sum_{i=1}^{j}M_i+N_j\right)\left\{\int_0^T[F^{(j)}(t)-F^{(j+1)}(t)]\mathrm{d}t\right.\\
&\left.+\int_0^T\left[\int_{T-t}^{\infty}\overline{F}(u)\mathrm{d}D(t+u)\right]\mathrm{d}F^{(j)}(t)\right\}\\
=&c_F+\sum_{j=0}^{N-1}\left[c_B+\frac{(j+1)c_0}{\omega}\right]\int_0^T\left[\int_0^{\infty}\overline{D}(t+u)\mathrm{d}F(u)\right]\mathrm{d}F^{(j)}(t)\\
&+\sum_{j=0}^{N-1}\left(c_B+\frac{jc_0}{\omega}\right)\int_0^T\left\{\int_0^{\infty}[D(t+u)-D(t)]\mathrm{d}F(u)\right\}\mathrm{d}F^{(j)}(t).
\end{aligned} \tag{3.39}$$

The difference of $\widetilde{C}_I(T,N)$ and $\widetilde{C}_D(T,N)$ is

$$\widetilde{C}_D(T,N)-\widetilde{C}_I(T,N)=\frac{c_0}{\omega}\sum_{j=0}^{N-1}j\int_0^T\left[\int_0^\infty\overline{D}(t+u)\mathrm{d}F(u)\right]\mathrm{d}F^{(j)}(t)$$
$$-c_N\sum_{j=0}^{N-1}j\int_0^T\left\{\int_0^\infty[D(t+u)-D(t)]\mathrm{d}F(u)\right\}\mathrm{d}F^{(j)}(t),\tag{3.40}$$

that is, if

$$\left(c_N+\frac{c_0}{\omega}\right)\sum_{j=0}^{N-1}j\int_0^T\left[\int_0^\infty\overline{D}(t+u)\mathrm{d}F(u)\right]\mathrm{d}F^{(j)}(t)>c_N\sum_{j=0}^{N-1}j\int_0^T\overline{D}(t)\mathrm{d}F^{(j)}(t)$$

holds, then incremental backup would save more cost than differential backup.

The expected cost rate for incremental backup is

$$C_I(T,N)=$$
$$\frac{\begin{array}{c}c_F+(c_B+c_0/\omega)\sum_{j=0}^{N-1}\int_0^T[\int_0^\infty\overline{D}(t+u)\mathrm{d}F(u)]\mathrm{d}F^{(j)}(t)\\+\sum_{j=0}^{N-1}[c_R+j(c_N+c_0/\omega)]\int_0^T\{\int_0^\infty[D(t+u)-D(t)]\mathrm{d}F(u)\}\mathrm{d}F^{(j)}(t)\end{array}}{\sum_{j=0}^{N-1}\int_0^T[\int_0^\infty\overline{D}(t+u)\overline{F}(u)\mathrm{d}u]\mathrm{d}F^{(j)}(t)},\tag{3.41}$$

and the expected cost rate for differential backup is

$$C_D(T,N)=\frac{\begin{array}{c}c_F+\sum_{j=0}^{N-1}[c_B+(j+1)c_0/\omega]\int_0^T[\int_0^\infty\overline{D}(t+u)\mathrm{d}F(u)]\mathrm{d}F^{(j)}(t)\\+\sum_{j=0}^{N-1}(c_R+jc_0/\omega)\int_0^T\{\int_0^\infty[D(t+u)-D(t)]\mathrm{d}F(u)\}\mathrm{d}F^{(j)}(t)\end{array}}{\sum_{j=0}^{N-1}\int_0^T[\int_0^\infty\overline{D}(t+u)\overline{F}(u)\mathrm{d}u]\mathrm{d}F^{(j)}(t)}.\tag{3.42}$$

Clearly, $\lim_{T\to\infty}C_I(T,N)=C_I(N)$ in (3.7), $\lim_{T\to\infty}C_D(T,N)=C_D(N)$ in (3.8), $\lim_{N\to\infty}C_I(T,N)=C_{OI}(T)$ in (3.22), and $\lim_{N\to\infty}C_D(T,N)=C_{OD}(T)$ in (3.23).

3.6.1 INCREMENTAL BACKUP

We find optimum T_{IN}^* and N_{IT}^* to minimize $C_I(T,N)$ in (3.41) when $D(t)=1-\mathrm{e}^{-\mu t}$. Then, the expected cost rate in (3.41) is

$$\frac{C_I(T,N)}{\mu}=\frac{c_F+(c_N+c_0/\omega)[1-F^*(\mu)]\sum_{j=0}^{N-1}j\int_0^T\mathrm{e}^{-\mu t}\mathrm{d}F^{(j)}(t)}{[1-F^*(\mu)]\sum_{j=0}^{N-1}\int_0^T\mathrm{e}^{-\mu t}\mathrm{d}F^{(j)}(t)}$$
$$+\left(c_0+\frac{c_0}{\omega}\right)\frac{F^*(\mu)}{1-F^*(\mu)}+c_R.\tag{3.43}$$

In particular, when $N=1$,

$$\frac{C_I(T,1)}{\mu}=\frac{c_F+c_R[1-F^*(\mu)]+(c_B+c_0/\omega)F^*(\mu)}{1-F^*(\mu)}, \tag{3.44}$$

and $T^*_{IN}=\infty$.

Forming the inequality $C_I(T,N-1)-C_I(T,N)<0$ and $C_I(T,N+1)-C_I(T,N)\geq 0$,

$$\sum_{j=0}^{N-1}(N-1-j)\int_0^T \mathrm{e}^{-\mu t}\mathrm{d}F^{(j)}(t)<\frac{c_F}{(c_N+c_N/\omega)[1-F^*(\mu)]}$$
$$\leq \sum_{j=0}^{N}(N-j)\int_0^T \mathrm{e}^{-\mu t}\mathrm{d}F^{(j)}(t), \tag{3.45}$$

whose right-hand side increases strictly with N from 0 to ∞. Thus, there exists a finite and unique N^*_{IT} $(1<N^*_{IT}<\infty)$ which satisfies (3.45) for any $T>0$. Furthermore, the right-hand side increases strictly with T to the left-hand side of (3.10), i.e., optimum N^*_{IT} decreases with T to N^*_I given in (3.10).

Differentiating $C_I(T,N)$ with respect to T and setting it equal to zero,

$$\frac{\sum_{j=1}^{N-1} jf^{(j)}(t)}{\sum_{j=1}^{N-1} f^{(j)}(t)}\sum_{j=1}^{N-1}\int_0^T \mathrm{e}^{-\mu t}\mathrm{d}F^{(j)}(t)-\sum_{j=1}^{N-1} j\int_0^T \mathrm{e}^{-\mu t}\mathrm{d}F^{(j)}(t)$$
$$=\frac{c_F}{(c_N+c_0/\omega)[1-F^*(\mu)]}. \tag{3.46}$$

Substituting (3.45) for (3.46),

$$\frac{\sum_{j=1}^{N-1} jf^{(j)}(t)}{\sum_{j=1}^{N-1} f^{(j)}(t)}>N-1, \tag{3.47}$$

which does not hold for any T. Thus, there does not exist any finite T^*_{IN} which satisfies (3.46) for N^*_{IT}, and hence, $T^*_{IN}=\infty$. This concludes that optimum full backup time is $(T^*_{IN}=\infty, N^*_{IT}=N^*_I)$, where N^*_I is given in (3.10).

3.6.2 DIFFERENTIAL BACKUP

We find optimum T^*_{DN} and N^*_{DT} to minimize $C_D(T,N)$ in (3.42) when $D(t)=1-\mathrm{e}^{-\mu t}$. Then, the expected cost rate in (3.42) is

$$\frac{C_D(T,N)}{\mu}=\frac{c_F+(c_0/\omega)\sum_{j=0}^{N-1} j\int_0^T \mathrm{e}^{-\mu t}\mathrm{d}F^{(j)}(t)}{[1-F^*(\mu)]\sum_{j=0}^{N-1}\int_0^T \mathrm{e}^{-\mu t}\mathrm{d}F^{(j)}(t)}$$
$$+\left(c_B+\frac{c_0}{\omega}\right)\frac{F^*(\mu)}{1-F^*(\mu)}+c_R. \tag{3.48}$$

In particular, when $N=1$, $C_D(T,1)=C_I(T,1)$ in (3.44).

Forming the inequality $C_D(T,N-1)-C_D(T,N)<0$ and $C_D(T,N+1)-C_D(T,N)\geq 0$,

$$\sum_{j=0}^{N-1}(N-1-j)\int_0^T \mathrm{e}^{-\mu t}\mathrm{d}F^{(j)}(t) < \frac{c_F}{c_0/\omega} \leq \sum_{j=0}^{N}(N-j)\int_0^T \mathrm{e}^{-\mu t}\mathrm{d}F^{(j)}(t), \tag{3.49}$$

whose right-hand side increases strictly with N from 0 to ∞. Thus, there exists a finite and unique N_{DT}^* $(1\leq N_{DT}^*<\infty)$ which satisfies (3.49) for any $T>0$. Furthermore, the right-hand side increases strictly with T to the left-hand side of (3.14), i.e., optimum N_{DT}^* decreases with T to N_D^* given in (3.14).

Differentiating $C_D(T,N)$ with respect to T and setting it equal to zero,

$$\frac{\sum_{j=1}^{N-1} jf^{(j)}(t)}{\sum_{j=1}^{N-1} f^{(j)}(t)}\sum_{j=0}^{N-1}\int_0^T \mathrm{e}^{-\mu t}\mathrm{d}F^{(j)}(t)-\sum_{j=0}^{N-1} j\int_0^T \mathrm{e}^{-\mu t}\mathrm{d}F^{(j)}(t)=\frac{c_F}{c_0/\omega}. \tag{3.50}$$

Substituting (3.49) for (3.50),

$$\frac{\sum_{j=1}^{N-1} jf^{(j)}(t)}{\sum_{j=1}^{N-1} f^{(j)}(t)} > N-1,$$

which agrees with (3.47), i.e., the optimum full backup time is $(T_{DN}^*=\infty, N_{DT}^*=N_D^*)$, where N_D^* is given in (3.14).

It concludes from discussions of Sections 3.6.1 and 3.6.2 that when the same cost of full backup is supposed for models with N and T, both incremental and differential backups with update N and overtime T would degenerate into respective models with N in which only N_I^* and N_D^* are found. In other words, models with update N would be better than those with overtime T from point of cost savings.

3.6.3 NUMERICAL EXAMPLES

When $D(t)=1-\mathrm{e}^{-\mu t}$, $F(t)=1-\mathrm{e}^{-t}$, $c_0/\omega=0.5$, $c_B=0.1, c_R=0.2, c_N=0.5$ and $c_F=5.0$, optimum N_{IT}^* and N_{DT}^* in Table 3.3 are computed from (3.45) and (3.49) for given T, and optimum T_{IN}^* and T_{DN}^* are computed from (3.46) and (3.50) for given N. Comparing of Tables 3.1, 3.2, 3.3 and 3.4, we conclude that,

(a) Both N_{IT}^* and N_{DT}^* in Table 3.3 decrease with T to N_I^* and N_D^* in Table 3.1, and their cost rates $C_I(T,N_{IT}^*)/\mu$ and $C_D(T,N_{DT}^*)/\mu$ decrease with T to $C_I(N_I^*)/\mu$ and $C_D(N_D^*)/\mu$ in Table 3.1. It also means the models with update N are more economical than those with overtime T.

(b) In Table 3.4, when the failure rate μ becomes small, optimum $T_{IN}^* = \infty$ that means the full backup is made at the number $N = 10$ or $N = 20$ of incremental backups. In this case, $N = 10$ and $N = 20$ may be small enough to limit the data restoration costs, however, they are not the optimum ones.

(c) In Table 3.4, when the failure rate μ becomes high, optimum $T_{DN}^* = \infty$ that means the full backup is made at the number $N = 10$ differential backups. In this case, $N = 10$ may be small enough to limit the backup costs, however, they are not the optimum ones.

Table 3.3

Optimum N_{IT}^*, N_{DT}^*, and their cost rates $C_I(T, N_{IT}^*)/\mu$ and $C_D(T, N_{DT}^*)/\mu$ for given $T = 10.0, 20.0$.

μ	$T = 10.0$				$T = 20.0$			
	N_{IT}^*	$C_I(T, N_{IT}^*)/\mu$	N_{DT}^*	$C_D(T, N_{DT}^*)/\mu$	N_{IT}^*	$C_I(T, N_{IT}^*)/\mu$	N_{DT}^*	$C_D(T, N_{DT}^*)/\mu$
0.01	59	119.184	6	314.917	39	98.726	6	314.709
0.05	19	31.127	6	66.963	18	29.387	6	66.919
0.1	14	20.052	6	36.180	14	19.642	6	36.158
0.5	11	11.857	8	13.212	11	11.846	6	13.202
1.0	12	12.790	12	12.790	12	12.760	12	12.760
5.0	31	30.902	51	30.425	29	29.217	48	28.749

Table 3.4

Optimum T_{IN}^*, T_{DN}^*, and their cost rates $C_I(T_{IN}^*, N)/\mu$ and $C_D(T_{DN}^*, N)/\mu$ for given $N = 10, 20$.

μ	$N = 10$				$N = 20$			
	T_{IN}^*	$C_I(T_{IN}^*, N)/\mu$	T_{DN}^*	$C_D(T_{DN}^*, N)/\mu$	T_{IN}^*	$C_I(T_{IN}^*, N)/\mu$	T_{DN}^*	$C_D(T_{DN}^*, N)/\mu$
0.01	∞	124.024	4.749	336.459	∞	99.191	4.505	338.239
0.05	∞	31.558	4.939	71.069	23.576	29.442	4.646	71.465
0.1	∞	19.991	5.204	38.038	13.300	19.858	4.832	38.275
0.5	∞	10.975	11.707	13.236	9.525	11.880	6.939	13.305
1.0	∞	9.965	∞	9.965	11.172	12.788	11.172	12.788
5.0	150.912	9.265	∞	6.920	62.595	18.939	111.488	11.607

3.7 CONCLUSIONS

This chapter has proposed random backup models, where the incremental and differential backups are implemented randomly right after data updates in large volumes, and the full backups are scheduled at a number N of updates and at the completion of the forthcoming update over time T. It has been supposed that updates arrive at a renewal process with distribution $F(t)$, the volumes of updated data are random variables W_j, and the costs of backup and recovery depend on the accumulated volumes x of data of the total updates. We have obtained the expected costs of data backup and breakdown recovery for incremental and differential backup schemes, e.g., (3.3) and (3.4) in Section 3.2, and the expected cost rates, e.g., (3.7) and (3.8) in Section 3.2, using the renewal theory when full backups are considered as renewal points for the whole backup process. Optimum solutions of full backups, e.g., N_I^* in (3.10) and N_D^* in (3.14), have been discussed in analytical ways when the breakdown time and each volume of updated data are exponentially distributed.

We have compared the expected backup and recovery costs of incremental and differential backups in (3.5), (3.21) and (3.40). It has been shown that both incremental and differential backups have their advantages and their disadvantages in saving costs, i.e., differential backup simplifies data restoration but increases backup cost as compared to the incremental backup. We have also compared the backup models with respective decision variables N and T and shown that models with update N are better than those with overtime T. However, the implementation of overtime T backup may be simpler than that at update N as we don't care the number of updates, so that when the overtime backup models are better will become the follow-up studies.

ACKNOWLEDGEMENT

This work is supported by National Natural Science Foundation of China (NO. 71801126), Natural Science Foundation of Jiangsu Province (NO. BK20180412) and Fundamental Research Funds for the Central Universities (NO. NR2018003).

REFERENCES

1. Silberschatz, A., Korth, H.F., & Sudarshan, S. (2010). *Database System Concepts* (Sixth Edition). McGraw-Hill.

2. Kumar, A., & Segev, A. (1993). Cost and availability tradeoffs in replicated concurrency control. *ACM Transactions on Database Systems*, 18, 102-131.

3. Hawkins, S.M., Yen, D.C., & Chou, D.C. (2000). Disaster recovery planning: a strategy for data security. *Information Management & Computer Security*, 8, 222-230.

4. Mcdowall, R.D. Computer (In) security-2: computer system backup and recovery. *The Quality Assurance Journal*, 5, 149-155.

5. Clapperton, G. (2000). Understanding online backup. *PC Network Advisor*, 121, 15-18.

6. Zhang, X., Feng, W., & Qin, X. (2013). Performance evaluation of online backup cloud storage. *International Journal of Cloud Applications and Computing*, 3, 20-33.

7. Ricart, G., Epstein, M., & Laube, S. (2005). Data backup. US Patent 6,892,221.

8. Nakagawa, T. (2014) *Random Maintenance Policies*. Springer.

9. Fong, Y., & Manley, S. (2007). Efficient true image recovery of data from full, differential, and incremental backups. US Patent 7,251,749.

10. Microsoft Support. (2012). Description of full, incremental, and differential backups, Retrieved 21, http://support.microsoft.com/kb/136621.

11. Symantec Enterprise Technical Support. (2012). What are the differences between Differential and Incremental backups? Article: TECH7665. Created: 2000-01-27, Updated: 2012-05-12, Retrieved 21.

12. NovaStor. (2014). Differential and Incremental backups: Why should you care?, Retrieved 31.

13. Qian, C., Nakamura, S., & Nakagawa, T. (2002). Optimal backup policies for a database system with incremental backup. *Electronics and Communications in Japan*, Part 3, 85, 1-9.

14. Nakamura, S., Qian, C., Fukumoto, S., & Nakagawa, T. (2003). Optimal backup policy for a database system with incremental and full backup. *Mathematical and Computer Modelling*, 11, 1373-1379.

15. Nakagawa, T. (2007). *Shock and Damage Models in Reliability Theory*, Springer.

16. Zhao, X., Mizutani, S., & Nakagawa, T. (2015). Which is better for replacement policies with continuous or discrete scheduled times? *European Journal of Operational Research*, 242, 477-486.

17. Nakagawa, T., & Zhao, X. (2015). *Maintenance Overtime Policies in Reliability Theory*, Springer.

18. Levitin, G., Xing, L., & Dai, Y. (2017). Optimal distribution of nonperiodic full and incremental backups. *IEEE Transactions on Systems, Man, and Cybernetics: Systems*, 47, 3310-3320.

19. Nakagawa, T. (2005). *Maintenance Theory of Reliability*. Springer.

4 An Optimal Age Replacement Policy for a Reparable System Consisting of Main and Auxiliary Subsystems

Lirong Cui
Qingdao University, China

Jingyuan Shen
Nanjing University of Science and Technology, China

Fengming Kang
Beijing Institute of Technology, China

CONTENTS

4.1 INTRODUCTION

Any repairable systems can be maintained via corrective and preventive actions for failed and non-failed subsystems, respectively. Each maintenance action is executed according to some policies definitely. In literature there are many maintenance policies, for example, see Nakagawa (2005), Cui (2008) and Nakagawa (2014). The age replacement policy is one of the most common since it is convenient to be carried out in real world. The optimization on age replacement policy has much literature already, in which the basic target is to get the optimal age replacement time under some constraints. However, it has little discussion on repairable systems consisting major and minor/auxiliary components.

Early studies for systems with major and minor components could be dated back to Meeker & Escobar (1998), wherein component failures have two different influences on the system: some failures cause the system to stop operating while the others do not. In the literature, different types of minor components are studied, among them failures of major and minor components are almost assumed to be independent (e.g., Taghipour et al., 2010; Hajipour & Taghipour, 2016; Babishin & Taghipour, 2016). This is true when the failures of the minor components do not affect the reliability of the major components. However, if minor components are equipped to protect the main components, it is obvious that their protection capability would degrade when some of them fail and finally affect the reliability or performance of the main components. Thus it is reasonable to consider dependence between the main components and the protective minor components in the reliability model. Few researches could be found that have discussed systems with dependent major and minor components (e.g., Shen et al., 2020; Shen et al., 2021). In the present chapter, optimal age replacement policies will be considered for the repairable systems consisting dependent and independent main and minor subsystems, respectively. To our best knowledge, it is the first time to study the optimal age replacement policy for this kind of repairable systems. Especially, for the systems with independent major and minor subsystems, the derived results can reduce to a common situation which has been considered in many related textbooks and monographs if the repair costs for the minor/ auxiliary subsystems are negligible, i.e., the minor subsystem in the repairable systems is not considered.

The organization of the remainder of the chapter is as follows. In section 4.2, a reliability model is developed for systems with independent major and minor subsystems, based on which the expected costs per unit of time for infinite and finite time spans are derived, respectively. The optimal solution for age replacement time for the system is then given and discussed in section 4.3. Moreover, section 4.4 developed an optimal model for systems with dependent major and minor subsystems, where a simulation algorithm is proposed to numerically estimate the long-run average cost and further obtain an optimal age replacement policy. Finally, numerical examples are shown in section 4.5 to illustrate the results obtained in this chapter. A flowchart for the framework of this chapter is shown in Figure 4.1.

4.2 ASSUMPTIONS AND MODELLING

Before optimization modelling of the age replacement policy for main and auxiliary repairable systems, some assumptions are needed, which are presented in terms of practical and theoretical situations.

(1) A reparable system consists of two subsystems with any structure, one subsystem is a major part, and the other is a minor (auxiliary) part.

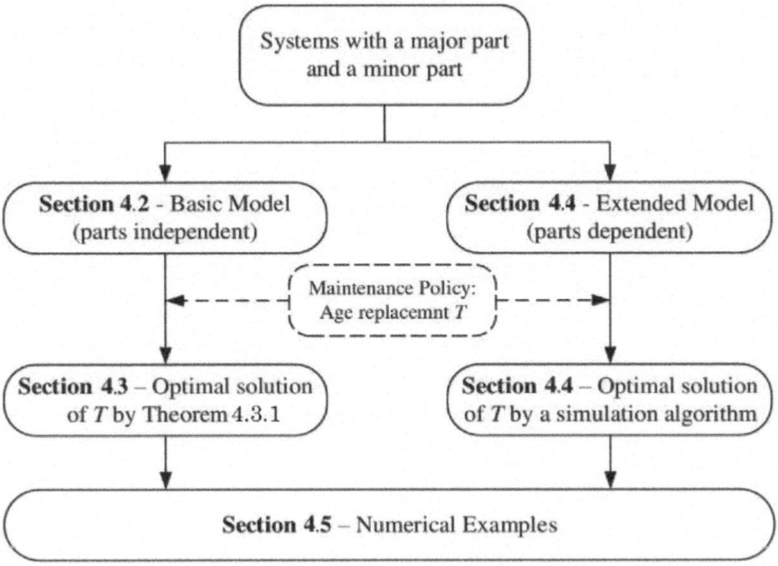

Figure 4.1 Flowchart for the framework of the chapter.

(2) The major and minor parts have lifetimes X_1 and X_2 with distribution functions $F_1(x)$ and $F_2(x)$, respectively.

(3) If a failure occurs for the minor subsystem, then a minimal repair action will be done, i.e., the repair action for the minor subsystem is "as good as old"; If a failure occurs for the major subsystem, then a replacement action will be done for two subsystems together, i.e., the replacement action for the system is "as good as new". All repair or replacement times are negligible.

(4) A planned system replacement will be done at a constant time T after its installation or at a failure of the major subsystem, whichever occurs first, where T ranges over $[0, \infty)$.

(5) Repair costs $c_{i,f}$ $(i = 1, 2)$ are incurred for each failed subsystems while $c_{i,p}$ $(i = 1, 2)$ are incurred for each non-failed subsystems because of preventively replacements. It is reasonable to assume that $c_{1,f} > c_{1,p} > c_{2,f} > c_{2,p} > 0$.

(6) Random variables X_1 and X_2 are independent of each other, and let $\lambda_i(x)$ be the failure rate function of X_i $(i = 1, 2)$. Both failure rate functions are increasing.

(7) It is assumed that the system is new at the beginning time $t = 0$.

The assumptions given above are reasonable, and the process of age replacement with planned time T can be seen in Figure 4.2.

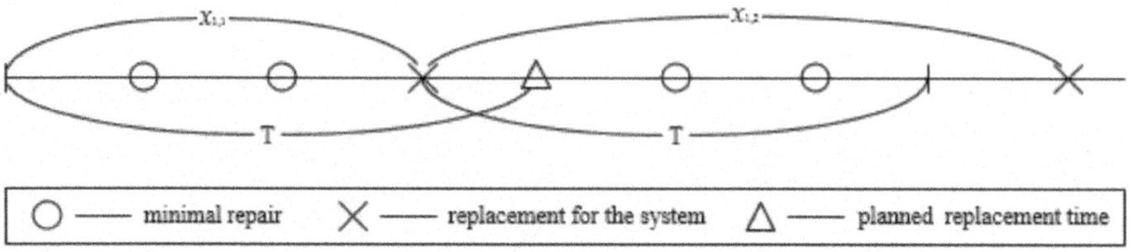

Figure 4.2 A possible path for the process of age replacement with planned time T

Note: $x_{1,1}$ and $x_{1,2}$ are two realizations of the lifetime of the major part X_1. In $x_{i,j}$, the second subscript represents the jth realization of X_i.

The reliability diagram of the system can be any structure for both subsystem, i.e., our model is structure-free for both subsystems. It is clear to know the evolution process of the system is a renewal process, and the renewal points are at the failures of the major subsystem or at each planned replacement time T after an installation, whichever occurs first. Thus based on the result of renewal process theory, for example, see Ross (1996), it has

$$C(T) := \lim_{t\to\infty} \frac{\text{expected cost during } [0,t]}{t} = \frac{E[\text{cost in a cycle}]}{E[\text{cycle}]},$$

where $C(T)$ is the expected cost per unit of time for an infinite time span.

The cost in a cycle is as follows.

$$\begin{aligned}\text{Cost}[T] := {} & [N_2(X_1)c_{2,f} + c_{1,f} + c_{2,p}]\, I\{T \geq X_1\} \\ & + [N_2(T)c_{2,f} + c_{1,p} + c_{2,p}]\, I\{T < X_1\},\end{aligned}$$

where $N_2(t)$ is the number of failures for the minor subsystem by time t, $I\{A\}$ is an indicator function of event A, i.e, $I\{A\} = 1$ if event A occurs, and $I\{A\} = 0$ otherwise.

There are $N_2(X_1)$ minimal repairs for the failed minor subsystem before a renewal point which results from a failure of the major subsystem, and in this situation, there is one replacement for failed major subsystem and non-failed minor subsystem, which leads to cost $c_{1,f} + c_{2,p}$. Similarly, before the renewal point which results from the planned time T, there are $N(T)$ minimal repairs for the failed minor subsystem before the renewal point, and there is replacement for the system, which costs $c_{1,p} + c_{2,p}$.

The cycle duration is as follows.

$$\text{cycle} := X_1 I\{T \geq X_1\} + T I\{T < X_1\}.$$

This is because the minor subsystem does not affect the renewal points which depend only on the major subsystem. Thus it has

$$
\begin{aligned}
&C(T)\\
&=\frac{E\{[N_2(X_1)c_{2,f}+c_{1,f}+c_{2,p}]I\{T\geq X_1\}+[N_2(T)c_{2,f}+c_{1,p}+c_{2,p}]I\{T<X_1\}\}}{E[X_1I\{T\geq X_1\}+TI\{T<X_1\}]}\\
&=\frac{\int_0^T[\Lambda_2(x)c_{2,f}+c_{1,f}+c_{2,p}]\mathrm{d}F_1(x)+[\Lambda_2(T)c_{2,f}+c_{1,p}+c_{2,p}]\overline{F}_1(T)}{\int_0^T x\mathrm{d}F_1(x)+\int_T^\infty T\mathrm{d}F_1(x)}\\
&=\frac{c_{1,f}F_1(T)+c_{1,p}\overline{F}_1(t)+c_{2,p}+c_{2,f}\Lambda_2(T)-c_{2,f}\int_0^T\lambda_2(x)F_1(x)\mathrm{d}x}{\int_0^T\overline{F}_1(x)\mathrm{d}x},
\end{aligned}
$$

where $\overline{F}_1(t)=1-F_1(t)$, $\lambda_2(t)=\mathrm{d}\Lambda_2(t)/\mathrm{d}t$ which is the failure rate function of X_2, and $\Lambda_2(t)=E[N_2(t)]$. The equality $\Lambda_2(t)=E[N_2(t)]$ holds due to the property of nonhomogeneous Poisson process or it can refer to Theorem 1 in Cui et al. (2020).

It can be seen when $c_{2,f}=c_{2,p}=0$, the model reduces to the well-known result for the age replacement of single-unit system, for example, see Nakagawa (2005, Chapter 3), thus the model and problem discussed in the present paper is an extension of age replacement for single-unit system.

In the end of this section, we derive the expected cost during interval time $[0,t]$. Let $C_T(t)$ be the cost for the planned age replacement constant time T during interval time $[0,t]$, and $\varphi(t,T):=E[C_T(t)]$.

$$
\begin{aligned}
\varphi(t,T)&=E[C_T(t)I\{X_1\leq T\}]+E[C_T(t)I\{X_1>T\}]\\
&=E\,[[N_2(t)c_{2,f}]I\{t<X_1\leq T\}\\
&\quad+[N_2(X_1)c_{2,f}+c_{1,f}+c_{2,p}+C_T(t-X_1)]I\{\min(t,T)\geq X_1\}]\\
&\quad+E\,[[N_2(t)c_{2,f}]I\{t<T<X_1\}\\
&\quad+[N_2(T)c_{2,f}+c_{1,p}+c_{2,p}+C_T(t-T)]I\{t\geq T,X_1>T\}]\\
&=c_{2,f}\Lambda_2(t)[F_1(T)-F_1(t)]I\{t<T\}\\
&\quad+\int_0^{t\wedge T}[c_{2,f}\Lambda_2(x)+c_{1,f}+c_{2,p}+\varphi(t-x,T)]\mathrm{d}F_1(x)I\{t\wedge T>0\}\\
&\quad+c_{2,f}\Lambda_2(t)\overline{F}_1(T)I\{t<t\}\\
&\quad+[\Lambda_2(T)c_{2,f}+c_{1,p}+c_{2,p}+\varphi(t-T,T)]\overline{F}_1(T)I\{t>T\},
\end{aligned}
$$

where $a\wedge b=\min(a,b)$, $\varphi(t,T)=0$ if $t\leq 0$ for any positive T, i.e., with boundary condition $\varphi(0,T)=0$. This is an integral functional equation which is difficult to be solved in analytic way, but a numerical way can be handled easily. We can use it to do a comparison between the optimal expected cost per unit of time for an infinite time span and average expected cost during interval time $[0, t]$, i.e.,

$$\varphi_{avg}(t, T) := \frac{\varphi(t, T)}{t}.$$

For a given time t, let $\Delta = t/n$, which makes Δ is small enough, i.e., taking n be large enough to ensure the accuracy of calculations. The calculations can be done via an iterative way as follows.

$$\begin{aligned}
\varphi(i\Delta, T) = {} & c_{2,f}\Lambda_2(i\Delta)[F_1(T) - F_1(i\Delta)]I\{i\Delta < T\} \\
& + \int_0^{i\Delta \wedge T} [c_{2,f}\Lambda_2(x) + c_{1,p} + c_{2,p}]\mathrm{d}F_1(x) \\
& + \sum_{j=1}^{i\wedge[T/\Delta]} \varphi((i-j)\Delta, T)[F_1(j\Delta) - F_1((j-1)\Delta)] \\
& + c_{2,f}\Lambda_2(i\Delta)\overline{F}_1(T)I\{i\Delta < T\} \\
& + [\Lambda_2(T)c_{2,f} + c_{1,p} + c_{2,p} + \varphi(i\Delta - T, T)\overline{F}_1(T)I\{i\Delta > T\},
\end{aligned}$$

with the initial value $\varphi(0, T) = 0$. Then the average expected cost during interval time $[0, t]$

$$\varphi_{avg}(t, T) \approx \frac{\varphi(n\Delta, T)}{t}.$$

4.3 OPTIMAL SOLUTION AND DISCUSSIONS

If it minimizes the expected cost per unit of time for an infinite time span, i.e.,

$$\begin{aligned}
& \min_{T\in[0,\infty)} C(T) \\
& = \min_{T\in[0,\infty)} \frac{c_{1,f}F_1(T) + c_{1,p}\overline{F}_1(T) + c_{2,p} + c_{2,f}\Lambda_2(T) - c_{2,f}\int_0^T \lambda_2(x)F_1(x)\mathrm{d}x}{\int_0^T \overline{F}_1(x)\mathrm{d}x},
\end{aligned}$$

then the optimal replacement planned time T is given, if it exists.

In fact, it has that $C(T) = C_1(T) + C_2(T)$, where

$$C_1(T) = \frac{c_{1,f}F_1(T) + c_{1,p}\overline{F}_1(T)}{\int_0^T \overline{F}_1(x)\mathrm{d}x},$$

$$\text{and } C_2(T) = \frac{c_{2,p} + c_{2,f}\Lambda_2(T) - c_{2,f}\int_0^T \lambda_2(x)F_1(x)\mathrm{d}x}{\int_0^T \overline{F}_1(x)\mathrm{d}x}.$$

For $C_1(T)$, we know that see Cao & Cheng (2006, Theorem 8.1.1) if $\lambda_1(\infty) > c_{1,f}\int_0^\infty \overline{F}_1(x)\mathrm{d}x/(c_{1,f} - c_{1,p})$, then there exists a unique solution to make $C_1(T) = min$, otherwise, it has not any limited optimal solution for $C_1(T)$. On the other hand, it has

$$\frac{\mathrm{d}C_2(T)}{\mathrm{d}T} \propto c_{2,f}\left[\lambda_2(T)\overline{F}_1(T)\int_0^T \overline{F}_1(x)\mathrm{d}x - \overline{F}_1(T)\int_0^T \lambda_2(x)\overline{F}_1(x)\mathrm{d}x\right] - c_{2,p}\overline{F}_1(T).$$

If $c_{2,f}[\lambda_2(T)\int_0^t \overline{F}_1(x)\mathrm{d}x - \int_0^T \lambda_2(x)\overline{F}_1(x)\mathrm{d}x] - c_{2,p} = 0$ has a unique finite solution, and it is clear that $\lambda_2(T)\int_0^T \overline{F}_1(x)\mathrm{d}x - \int_0^T \lambda_2(x)\overline{F}_1(x)\mathrm{d}x$ is an increasing function of T due to the increasing property of $\lambda_2(T)$. Thus a theorem is gotten as follows.

Theorem 4.3.1 *If* $\lambda_1(\infty) > c_{1,f}\int_0^\infty T\overline{F}_1(x)\mathrm{d}x/(c_{1,f} - c_{1,p})$ *and* $c_{2,f}[\lambda_2(T)\int_0^T \overline{F}_1(x)\mathrm{d}x - \int_0^T \lambda_2(x)\overline{F}_1(x)\mathrm{d}x] - c_{2,p} = 0$ *has a unique solution, then there exists finite optimal solution (s) for* $\min_{T\in[0,\infty)} C(T)$, *and the optimal solution(s) ranges over* $[0, \max(T_1^*, T_2^*)]$.

The understanding may be supported by Figure 4.3 at below.

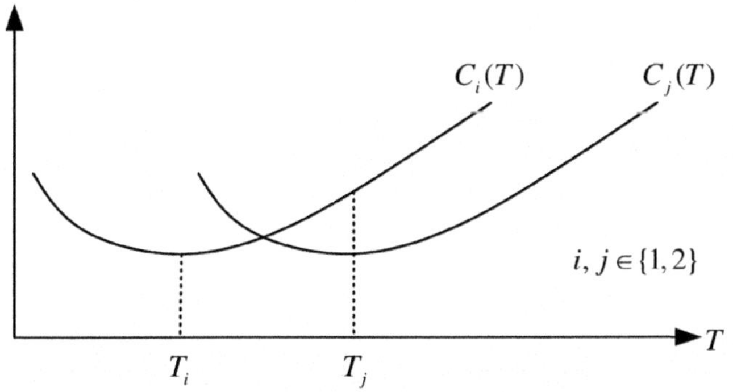

Figure 4.3 General curves for $C_1(T)$ and $C_2(T)$.

In general, it is assumed that both failure rate functions $\lambda_1(x)$ and $\lambda_2(x)$ are increasing functions, which makes a sense for age replacement policy for main and auxiliary component systems.

4.4 EXTENDED MODEL FOR SYSTEMS WITH DEPENDENT PARTS

The above sections discuss reliability modelling and maintenance cost optimization for systems with independent major and minor parts. In this section, a situation that the two parts being dependent in a system is considered. Related examples could be seen in many real systems where the minor parts are used to protect the major parts. Under this situation, the expected lifetimes of the major parts are obviously different when it equipped with or without minor parts. To model such reliability systems, some basic assumptions are listed as follows.

(1) A reparable system consists of two subsystems with any structure, one subsystem is a major part, and the other is a minor (auxiliary) part. The system fails if and only if the major part fails. The system is totally new at the beginning time $t = 0$.

(2) The minor part has lifetime X_2 with distribution $F_2(x)$. When the minor part is working, the major part has lifetime X_1 with distribution function $F_1(x)$. Otherwise, it has lifetime X_0 with distribution function $F_0(x)$. Variables X_0, X_1 and X_2 are independent each other.

(3) Failures of the minor part are hidden, while failures of the major part is self-announcing since the whole system halts under the situation. For a failed system, a corrective replacement action will be done for the two parts together. Besides, a planned system replacement will be done at a constant time T ($T \in [0, \infty)$) after its installation or at a failure of the major part, whichever occurs first. Each replacement makes the system as good as new.

(4) Time for each replacement is negligible. Replacement costs $c_{i,f}$ ($i = 1, 2$) are incurred for each failed part while $c_{i,p}$ ($i = 1, 2$) are incurred for each non-failed part because of planned replacements. It is reasonable to assume that $c_{1,f} > c_{1,p} > c_{2,f} > c_{2,p} > 0$.

A possible path of the systems is illustrated in Figure 4.4.

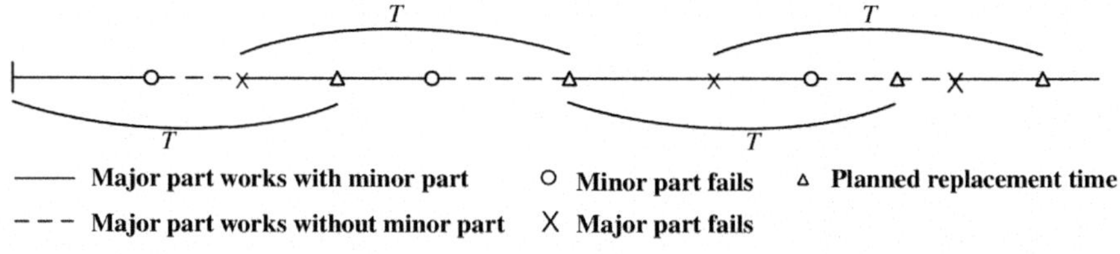

Figure 4.4 A possible path for system with dependent major and minor parts.

The main object of this section is to optimize the presented maintenance policy. For which we choose the planned replacement time T and the long-run average replacement cost respectively as the decision variable and objective function, and then formulate the optimization problem. Let $C_0(T)$ be the long-run average cost of the system with dependent parts under the

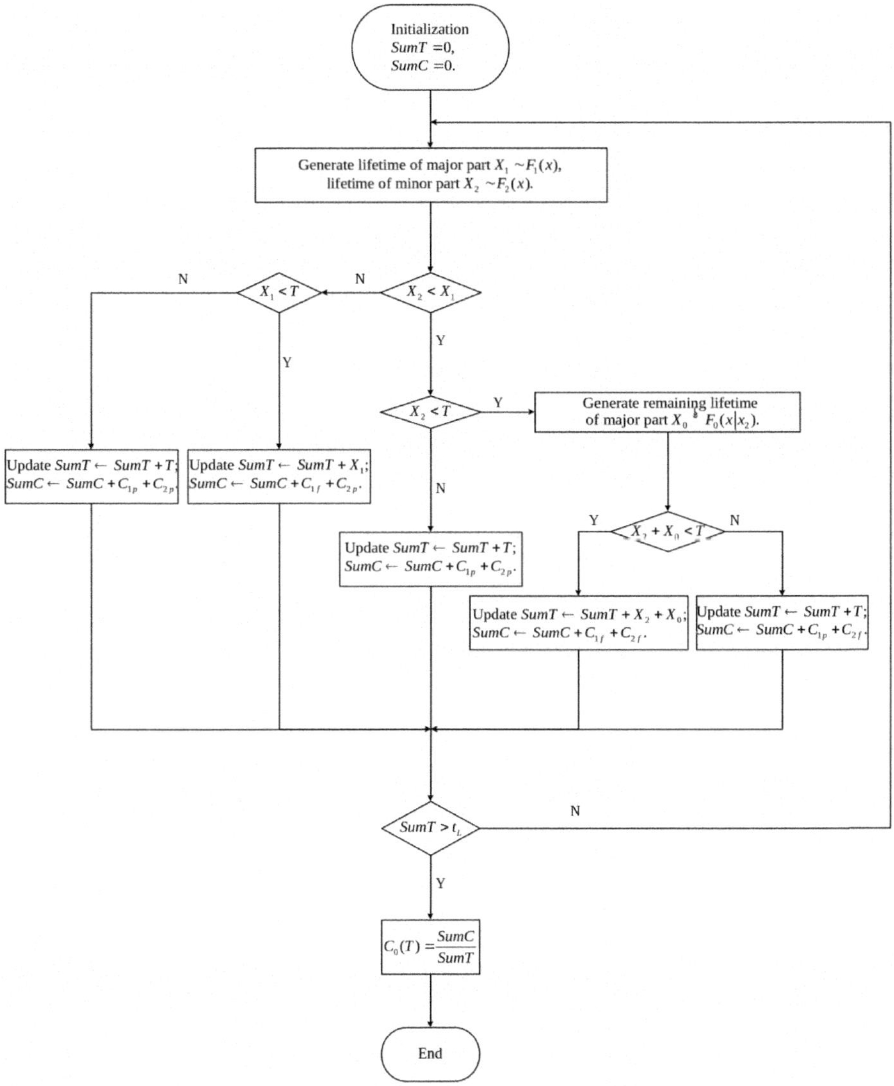

Figure 4.5 The flow chart of the algorithm proposed for estimating $C_0(T)$.

presented maintenance policy. An algorithm is proposed for numerically estimating $C_0(T)$. The flow chart of the proposed algorithm is shown in Figure 4.5.

By executing the algorithm, we could simulate the cost per unit of time for a long time span $[0, t_L]$ given that the planned replacement time is T. Repeat the simulation for $N = 10000$ times, then the average of the N results could be used to estimate the long-run average replacement cost of the system. The optimal maintenance policy could then be achieved by varying the planned replacement time T.

4.5 NUMERICAL EXAMPLES

In this section, first a numerical example is presented to illustrate the optimal results for systems with independent parts. Then a numerical example is for the system with dependent parts by using the algorithm proposed in Section 4.4. For the second model, sensitivity analysis is made to show how the parameters influence the optimization results.

4.5.1 SYSTEM WITH INDEPENDENT PARTS

Suppose the four parameters are: $c_{1,f} = 3$, $c_{1,p} = 1$, $c_{2,p} = 0.5$. The major subsystem has a lifetime with Gamma distribution, i.e., $F_1(x) = 1-(1+x)e^{-x}$. The minor subsystem has a lifetime with Weibull distribution, i.e., $F_2(x) = 1 - e^{-x^2}$.

The curves for $C_1(t)$, $C_2(T)$ and $C(T) = C_1(T)+C_2(T)$ are given in Figure 4.6 as follows, respectively.

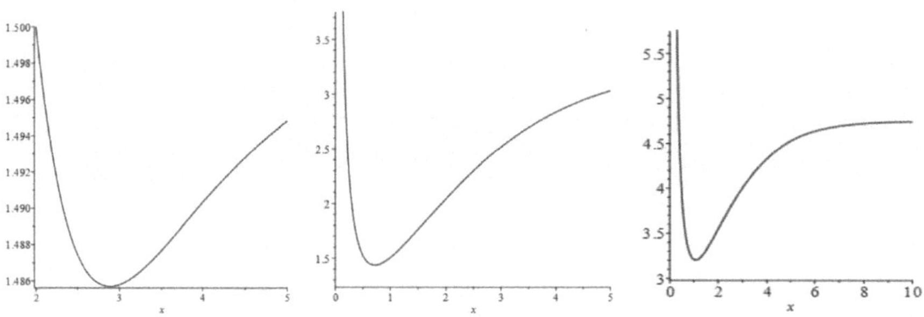

Figure 4.6 The curves for $C_1(T)$, $C_2(T)$ and $C(T)$ (from left to right, respectively).

In this example, the optimal age replacement time is $T^* = 1.083$ and the optimal expected cost per unit of time for an infinite time span is $C(T^*) = 3.204$. In order to compare the expected costs per unit of time between finite interval and infinite span, the comparisons between $\varphi_{avg}(t,T)$ and $C(T)$ are shown in Figure 4.7 for $\Delta = 0.01$ and $T = 0.5, 1.08, 1.5$ (from left to right in Figure 4.7), respectively. From Figure 4.7, we can know that the present expected cost per unit of time for an infinite time span is greater than average expected cost during interval time $[0,t]$, but the latter will approach to the former as time goes to infinity.

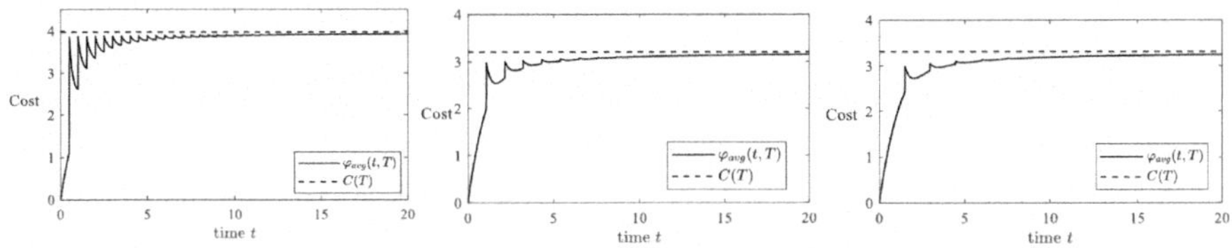

Figure 4.7 Curves for finite interval and infinite span for expected costs per unit of time.

4.5.2 SYSTEM WITH DEPENDENT PARTS

Suppose the cost parameters $c_{1,f}$, $c_{1,p}$, $c_{2,f}$, $c_{2,p}$ and the lifetime distribution functions $F_1(x)$ and $F_2(x)$ are the same as that in Section 4.5.1. We further assume that the lifetime of the major part without a minor part is exponentially distributed with distribution function $F_0(x) = 1 - e^{-3x}$. Based on the algorithm proposed in Section 4.4, the long-run average costs of the system with different planned replacement times are shown in Figure 4.8. It shows that the optimal planned replacement time could be derived numerically, which is $T^* = 0.9$ with minimal long-run average cost 3.5318.

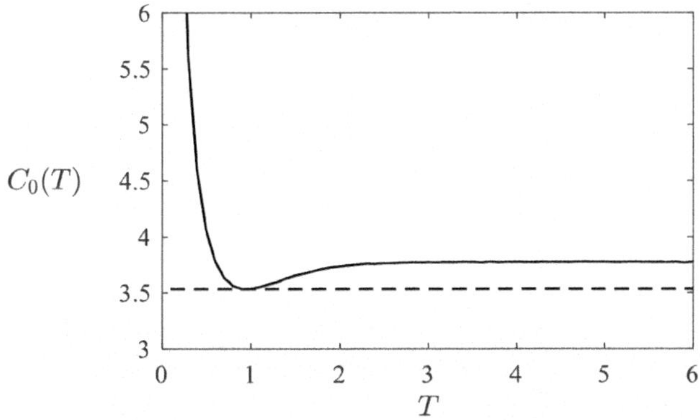

Figure 4.8 Long-run average costs of the system with different planned replacement times.

In the following, sensitivity analysis is made to show how the cost parameters and lifetime distributions influence the optimization results. First, consider the replacement cost of the failed major part C_{1f}. We fix other parameters and let C_{1f} vary from 3 to 11 with step 2. The long-run average costs of the system with different planned replacement times are shown in Figure 4.9.

Based on the results in Figure 4.9, the optimal policies $(T^*, C_0(T^*))$ for different C_{1f} are extracted and summarized in Table 4.1.

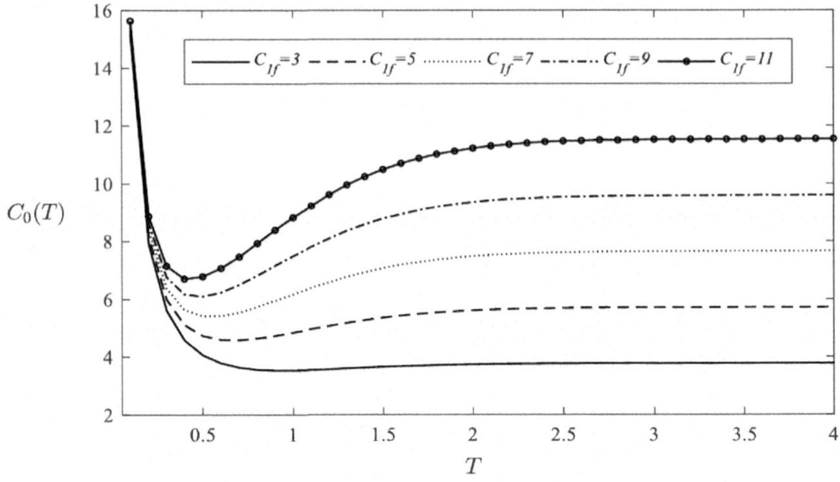

Figure 4.9 Results of the sensitivity analysis of C_{1f}.

Table 4.1
The optimal policies $(\mathbf{T^*}, \mathbf{C_0(T^*)})$ for different $\mathbf{C_{1f}}$.

C_{1f}	3	5	7	9	11
T^*	1	0.7	0.5	0.5	0.4
$C_0(T^*)$	3.5319	4.5893	5.4194	6.0976	6.7064

From the results above, it is obvious that the optimal long-run average costs increase with C_{1f}. In contrast, the optimal planned replacement time T^* decreases since more frequently preventive replacements could help to avoid unprepared system failures, especially when the corrective replacement becomes more and more costly.

Similarly, for the preventive replacement cost of the major part C_{1p}, we fix other parameters and let $C_{1f} = 11$ and C_{1p} vary from 3 to 11 with step 2. The long-run average costs of the system with different planned replacement times are shown in Figure 4.10. Under this situation, the optimal planned replacement time T^* increases with C_{1p}.

Finally, consider the rate parameter of the lifetime distribution of the major part without a minor part $F_0(x)$, denoted by λ_0. We fix other parameters and let $C_{1f} = 7$ and λ_0 vary from 1 to 9 with step 2. The long-run average costs of the system with different λ_0 are shown in Figure 4.11. It can be seen that the optimal planned replacement time T^* shows a decreasing trend with λ_0.

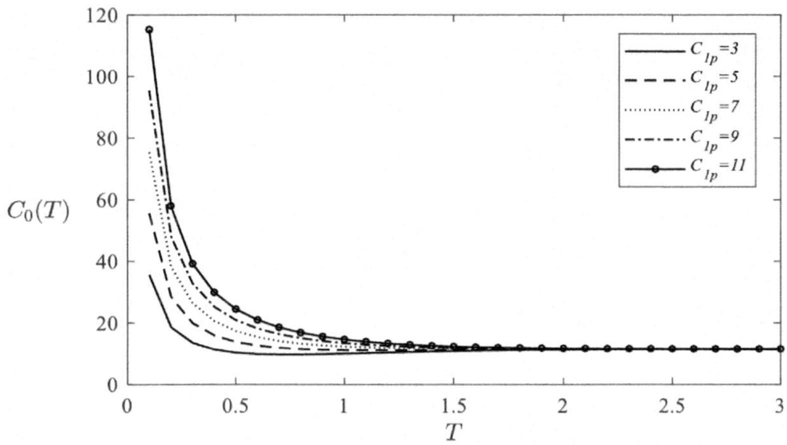

Figure 4.10 Results of the sensitivity analysis of C_{1p}.

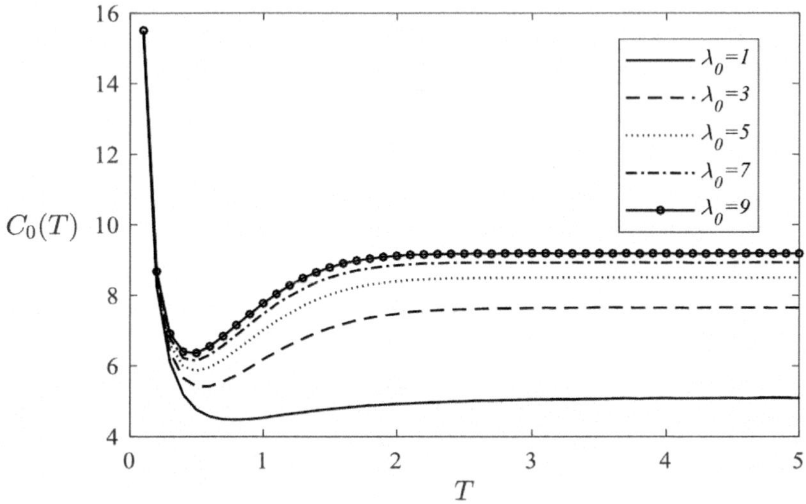

Figure 4.11 Results of the sensitivity analysis of λ_0.

4.6 CONCLUSION

The optimal policy of age replacement for main and auxiliary repairable systems is given, which is an extension work of the well-known result in the age replacement policy. The expected cost during interval time $[0, t]$ is presented as well, which is rarely studied in literature. Based on this result, the comparison work between finite interval and infinite span for expected costs per unit of time has been done in the paper, which may be useful to judge the error between finite interval and infinite span for expected costs per unit of time. Because most works appeared in literature considered the infinite span for expected cost per unit of time rather than expected cost per unit of time or average expected cost for finite interval. Our results show the difference between both, sometimes, is large, which must be paid some attention when the expected cost per unit of time for infinite span is used in both theory and practice.

ACKNOWLEDGEMENTS

This work is supported by the National Natural Science Foundation of China under grants 71871021 and 71801128.

REFERENCES

1. Babishin, V., & Taghipour, S. (2016). Optimal maintenance policy for multi-component systems with periodic and opportunistic inspections and preventive replacements, *Applied Mathematical Modelling*, 40(23-24), 10480-10505.

2. Cui, L.R. (2008). Maintenance models and optimization, *Handbook Performability Engineering* (Chapter 48), 789-805. Springer, London.

3. Cao, J.H., & Cheng, K. (2006). *Introduction to Reliability Mathematics* (In Chinese), Higher Education Press, Beijing.

4. Hajipour, Y., & Taghipour, S. (2016). Non-periodic inspection optimization of multicomponent and k-out-of-m systems, *Reliability Engineering & System Safety*, 156, 228-243.

5. Cui, L.R., Hawkes, A., & Yi, H. (2020). An elementary derivation of moments of Hawkes processes, *Advances in Applied Probability*, 52, 102-137.

6. Meeker, W.Q., & Escobar, L.A. (1998). *Statistical Methods for Reliability Data*, John Wiley & Sons, New York.

7. Nakagawa, T. (2005). *Maintenance Theory of Reliability* , Springer, London.

8. Nakagawa, T. (2014). *Random Maintenance Polices*, Springer, London.

9. Ross, S.M. (1996). *Stochastic Processes* (2nd Edition), John Wiley & Sons, New York.

10. Shen, J.Y., Hu, J.W., & Ma, Y.Z. (2020). Two preventive replacement strategies for systems with protective auxiliary parts subject to degradation and economic dependence, *Reliability Engineering & System Safety*, 204, 107144.

11. Shen, J.Y., Zhang, Y.J., Ma, Y.Z., & Lin, C. (2021). A novel opportunistic maintenance strategy for systems with dependent main and auxiliary components, *IMA Journal of Management Mathematics*, 32(1), 69-90.

12. Taghipour, S., Banjevic, D., & Jardine, A.K. (2010). Periodic inspection optimization model for a complex repairable system, *Reliability Engineering & System Safety*, 95(9), 944-952.

5 Extended Replacement Policy in Damage Models

Shey-Heui Sheu
Asia University, Taiwan
China Medical University Hospital, China Medical University, Taiwan

Tzu-Hsin Liu
Chaoyang University of Technology, Taiwan

Wei-Teng Sheu
National Taiwan University of Science and Technology, Taiwan

Zhe-George Zhang
Western Washington University, USA

Jau-Chuan Ke
National Taichung University of Science and Technology, Taiwan

CONTENTS

5.1 INTRODUCTION

Generally, the system will degenerate with age and repairable failures that appear randomly during operations. Hence, it is necessary to study the optimal replacement policy to lessen the occurrence of system failures based on the reliability theory. A considerably common and easy to perform is the age replacement policy that replaces a system at age T or at failure, whichever comes first. Barlow and Hunter (1960) discussed a maintenance policy with minimal repairs and replacements. Some extensions of these previous policies can be found in Zhao et al. (2018a, 2018b), Eryilmaz and Kan (2019), Sheu et al. (2019), Huang et al. (2019), Zarezadeh & Asadi (2019), Dong et al. (2021).

In most literature, the system either performs minimal repair or perfect repair at failures based on repair cost or types of failures. This is the concept of the imperfect repair model advanced by Brown and Proschan (1983). A clear circumscription of minimal repair is given by Nakagawa and Kowada (1983). Recently, many authors such as Kitagawa et al. (2017), Jafary et al. (2017), Tsai et al. (2017), Mizutani & Nakagawa (2018), Sheu et al. (2021a), Zhao et al. (in press), study the replacement policies on minimal repair models in the literature.

Consider a system that suffers damage due to shocks. The system will fail once the additive damage exceeds a pre-specified level. Then as described in Cox (1962), the system generates a cumulative process. The development of maintenance theory, advanced maintenance techniques with shock and damage models are surveyed by Zhao & Nakagawa (2017). In the literature, many researchers such as Tsai et al. (2017), Lai and Chen (2017), Kaio (2018), Chang & Chen (2019), Sheu et al. (2016, 2020, 2021b) have investigated replacement policies under damage models.

This paper proposes a bivariate replacement policy for a two-unit system under damage models. The system is subject to non-homogeneous Poisson shocks. Shocks can be partitioned into two types and the probabilities of the shock types depend on the number of shocks since the last replacement. Such a policy can be applied to the chemical industry (ref. Sheu et al., 2021). In the chemical industry, unit 1 (pneumatic pump) picks out cold water to regulate the temperature of unit 2 (metal container). The temperature of the tank may increase and the surface of the tank may be corroded, when the function of unit 1 is abnormal. The thickness of unit 2 will now decrease. We can consider the reduction in wall thickness as damage. Once the total reduction in wall thickness has reached a predefined value, unit 2 fails.

The proposed article is divided into the following sections. Section 5.2 describes the replacement policy considered here. Section 5.3 establishes the average cost rate. Section 5.4 deals with the optimization of the proposed policy. We give a computational example in Section 5.5 to demonstrate the above results. Finally, some concluding remarks are drawn in Section 5.6.

5.2 DESCRIPTION OF GENERAL REPLACEMENT POLICY

The system considered consists of two units (names unit 1 and unit 2) and is subject to non-homogeneous Poisson shocks process $\{N(t), t \geq 0\}$ with the intensity function $\lambda(t)$. Each shock results in one of the two categories denoted as type 1 and type 2 shocks. At a type 1 shock arrival, unit 1 will have a minor failure and require minimal repair. While at the type 2 shock arrival, the system fails and must be correctively replaced. The probability of type 2 shock depends on the number of shocks since the last replacement. Let $\overline{P}_j = \Pr(M > j)$ represent the probability that the first j shocks are type

1 shock, where M denotes the number of shocks until the first type 2 shock from the last replacement. Hence, it is natural to assume that $1 = \overline{P}_0 \geq \overline{P}_1 \geq \overline{P}_2 \geq \ldots$, and M is independent of the process $\{N(t), t \geq 0\}$.

Each minor failure of unit 1 also brings some damage to unit 2 and these damages will be additive to trigger a replacement operation. Replacements can be classified into two types: preventive replacement and corrective replacement, based on the damage level. If accumulative damage to unit 2 is exceeding a predefined level k but less than K, the replacement is referred to as the preventive replacement with a cost C_2. Otherwise, if the accumulative damage to unit 2 has reached a failure level K, the system fails, and then corrective replacement is carried out with a cost C_4. Let $H_j(w) = P(W_j \leq w)$ denote the distribution function of the damage amount W_j to unit 2 due to the jth minor failure of unit 1. We first define the notation "*" by the Stieltjes convolution. Therefore, we can get the accumulative damage to unit 2 up to the jth minor failure of unit 1 as

$$H^{(j)}(y) = P\left(\sum_{i=1}^{j} W_j \leq y\right) = P(Y_j \leq y) = H_1 * H_2 * \cdots * H_j(y) \tag{5.1}$$

for $j = 1, 2, 3, \cdots$ and $H^{(j)}(y) = 1$ for $j = 0$.

Besides, unit 2 with accumulative damage of y will have a minor failure with probability $q(y)$ at a unit 1 failure instant. This minor failure is fixed by a minimal repair with a cost $\beta(y)$. The replacement time is too small not worth mentioning. The system is replaced at failure. That is, the system is replaced when type 2 shock occurs or the accumulative damage to unit 2 has exceeded K, whichever comes first. To reduce the replacement cost after failure, the system is also replaced when the accumulative damage has reached k $(\leq K)$, or the system age reaches T, whichever comes first.

Let C_3 denote the replacement cost when type 2 shock occurs and cost C_1 represents the replacement cost when the system is replaced at age T. Assume that the cost of the minimal repair at the jth minor failure of unit 1 with age t is a non-decreasing function of age and the number of repairs, denoted by $g(R(t), r_j(t))$, where $R(t)$ is the t-dependent random part and $r_j(t)$ is the t and j dependent deterministic part. The expected cost of $g(R(t), r_j(t))$ is represented by $\alpha_j(t)$. Let S_j represent the arrival time of the j th shock for $j = 0, 1, 2, \cdots$, then the cost per unit time of maintenance of the system at time $t \in [(S_j, S_{j+1})$ is $m_j(t)$.

5.3 FORMULATION

According to the above scheme, we can derive the following corresponding probability of each replacement event in a renewal cycle. The probability of implementing preventive replacement when the system age reaches T is

$$\pi_T = \sum_{j=0}^{\infty} P(N(T) = j, M > j, Y_j < k) = \sum_{j=0}^{\infty} H^{(j)}(k)\overline{F}_j(T), \tag{5.2}$$

where $\overline{P}_j\{[\Lambda(T)]^j/j!\}\exp(-\Lambda(T)) \equiv \overline{F}_j(T)$ and $\Lambda(t) = \int_0^t \lambda(x)\mathrm{d}x$, the probability of implementing preventive replacement due to the accumulative damage to unit 2 exceeding k ($\leq K$) is

$$\pi_k = \sum_{j=0}^{\infty} P(Y_j < k \leq Y_{j+1} < K) \int_0^T P(M > j+1)P(S_{j+1} \in \mathrm{d}t)$$
$$= \sum_{j=0}^{\infty} \int_0^k [H_{j+1}(K-y) - H_{j+1}(k-y)]\mathrm{d}H^{(j)}(y) \int_0^T q_{j+1}\overline{F}_j(t)\lambda(t)\mathrm{d}t, \tag{5.3}$$

the probability of implementing the corrective replacement at the first type 2 shock is

$$\pi_2 = \sum_{j=0}^{\infty} P(Y_j < k) \int_0^T P(M = j+1)P(S_{j+1} \in \mathrm{d}t)$$
$$= \sum_{j=0}^{\infty} H^{(j)}(k) \int_0^T [1 - q_{j+1}]\overline{F}_j(t)\lambda(t)\mathrm{d}t, \tag{5.4}$$

where $q_{j+1} = \overline{P}_{j+1}/\overline{P}_j$, and the probability of carrying out corrective replacement due to the accumulative damage to unit 2 exceeding K is

$$\pi_K = \sum_{j=0}^{\infty} P(Y_j < k \leq K \leq Y_{j+1}) \int_0^T P(M > j+1)P(S_{j+1} \in \mathrm{d}t)$$
$$= \sum_{j=0}^{\infty} \int_0^k \overline{H}_{j+1}(K-y)\mathrm{d}H^{(j)}(y) \int_0^T q_{j+1}\overline{F}_j(t)\lambda(t)\mathrm{d}t, \tag{5.5}$$

where note that $\pi_T + \pi_k + \pi_2 + \pi_K = 1$.

For our replacement policy, the average length of a replacement cycle is

$$\sum_{j=0}^{\infty} TH^{(j)}(k)\overline{F}_j(T)$$
$$+ \sum_{j=0}^{\infty} \int_0^k [H_{j+1}(K-y) - H_{j+1}(k-y)]\mathrm{d}H^{(j)}(y) \int_0^T tq_{j+1}\overline{F}_j(t)\lambda(t)dt$$
$$+ \sum_{j=0}^{\infty} \int_0^k \overline{H}_{j+1}(K-y)\mathrm{d}H^{(j)}(y) \int_0^T tq_{j+1}\overline{F}_j(t)\lambda(t)\mathrm{d}t$$
$$+ \sum_{j=0}^{\infty} H^{(j)}(k) \int_0^T t[1 - q_{j+1}]\overline{F}_j(t)\lambda(t)\mathrm{d}t$$
$$= \sum_{j=0}^{\infty} H^{(j)}(k) \int_0^T \overline{F}_j(t)\mathrm{d}t.$$

The operational cost in a replacement cycle can be expressed as follows.

$$C_1 \sum_{j=0}^{\infty} I_{\{N(T)=j\}} I_{\{M>j\}} I_{\{Y_j<k\}} + C_2 \sum_{j=0}^{\infty} I_{\{S_{j+1}<T\}} I_{\{M>j+1\}} I_{\{Y_j<k\le Y_{j+1}<K\}}$$
$$+ C_3 \sum_{j=0}^{\infty} I_{\{S_{j+1}<T\}} I_{\{M=j+1\}} I_{\{Y_j<k\}}$$
$$+ C_4 \sum_{j=0}^{\infty} I_{\{S_{j+1}<T\}} I_{\{M>j+1\}} I_{\{Y_j<k\le K\le Y_{j+1}\}}$$
$$+ \sum_{j=1}^{\infty} g(R(S_j), r_j(S_j)) I_{\{S_j<T\}} I_{\{M>j\}} I_{\{Y_j<k\}}$$
$$+ \sum_{j=1}^{\infty} q(Y_j)\beta(Y_j) I_{\{S_j<T\}} I_{\{M>j\}} I_{\{Y_j<k\}}$$
$$+ \int_0^T m_{N(t)}(t) I_{\{M>N(t)\}} I_{\{Y_{N(t)}<k\}} \mathrm{d}t,$$

where $I_{\{A\}}$ is the indicator function of event A.

Thus, the average operational cost in a replacement cycle is obtained as

$$C_1\pi_T + C_2\pi_k + C_3\pi_2 + C_4\pi_K + \sum_{j=1}^{\infty} H^{(j)}(k) \int_0^T \alpha_j(t) q_j \overline{F}_{j-1}(t)\lambda(t)\mathrm{d}t$$
$$+ \sum_{j=1}^{\infty} \int_0^k q(y)\beta(y)\mathrm{d}H^{(j)}(y) \int_0^T q_j \overline{F}_{j-1}(t)\lambda(t)\mathrm{d}t$$
$$+ \sum_{j=0}^{\infty} H^{(j)}(k) \int_0^T m_j(t)\overline{F}_j(t)\mathrm{d}t$$
$$= C_1 + (C_2 - C_1)(1-\pi_T) + (C_3 - C_2)\pi_2 + (C_4 - C_2)\pi_K$$
$$+ \sum_{j=1}^{\infty} H^{(j)}(k) \int_0^T [\alpha_j(t) + \varphi_j(k)] q_j \overline{F}_{j-1}(t)\lambda(t)\mathrm{d}t$$
$$+ \sum_{j=0}^{\infty} H^{(j)}(k) \int_0^T m_j(t)\overline{F}_j(t)\mathrm{d}t,$$

where $\varphi_j(k) = \int_0^k q(y)\beta(y)\mathrm{d}H^{(j)}(y)/H^{(j)}(k)$.

Consequently, the average cost rate is given by

$$MCR(T,k) = \frac{\begin{array}{l} C_1 + (C_2 - C_1)(1-\pi_T) + (C_3 - C_2)\pi_2 + (C_4 - C_2)\pi_K \\ + \sum_{j=1}^{\infty} H^{(j)}(k) \int_0^T [\alpha_j(t) + \varphi_j(k)] q_j \overline{F}_{j-1}(t)\lambda(t)\mathrm{d}t \\ + \sum_{j=0}^{\infty} H^{(j)}(k) \int_0^T m_j(t)\overline{F}_j(t)\mathrm{d}t \end{array}}{\sum_{j=0}^{\infty} H^{(j)}(k) \int_0^T \overline{F}_j(t)\mathrm{d}t}. \tag{5.6}$$

Remark: The proposed policy can be regarded as a generalization of existing policies. This generalization makes the proposed policy more flexible in examining the effect of replacement policy parameters on the average cost rates.

5.4 OPTIMAL POLICY

Under the assumption that $\lambda(t)$ increases strictly with t from 0 to ∞, we derive the optimal policy (T^*, k^*), which minimizes $MCR(T, k)$ in (5.6). Setting the first derivative function of the average cost rate equal to zero, we obtain

$$\begin{aligned}
&V_T(T;k)\sum_{j=0}^{\infty} H^{(j)}(k)\int_0^T \overline{F}_j(t)\mathrm{d}t \\
&-(C_2-C_1)(1-\pi_T)-(C_3-C_2)\pi_2-(C_4-C_2)\pi_K \\
&-\sum_{j=0}^{\infty} H^{(j+1)}(k)\int_0^T [\alpha_{j+1}(t)+\varphi_{j+1}(k)]q_{j+1}\overline{F}_j(t)\lambda(t)\mathrm{d}t \\
&-\sum_{j=0}^{\infty} H^{(j)}(k)\int_0^T m_j(t)\overline{F}_j(t)\mathrm{d}t = C_1,
\end{aligned} \tag{5.7}$$

where

$$\begin{aligned}
V_T(T;k) = {} & (C_2-C_1)\lambda(T)\frac{\sum_{j=0}^{\infty}[H^{(j)}(k)\overline{F}_j(T)-H^{(j+1)}(k)q_{j+1}\overline{F}_j(T)]}{\sum_{j=0}^{\infty}H^{(j)}(k)\overline{F}_j(T)} \\
&+(C_3-C_2)\lambda(T)\frac{\sum_{j=0}^{\infty}H^{(j)}(k)[1-q_{j+1}]\overline{F}_j(T)}{\sum_{j=0}^{\infty}H^{(j)}(k)\overline{F}_j(T)} \\
&+(C_4-C_2)\lambda(T)\frac{\sum_{j=0}^{\infty}q_{j+1}\overline{F}_j(T)\int_0^k\overline{H}_{j+1}(K-y)\mathrm{d}H^{(j)}(y)}{\sum_{j=0}^{\infty}H^{(j)}(k)\overline{F}_j(T)} \\
&+\lambda(T)\frac{\sum_{j=0}^{\infty}H^{(j+1)}(k)[\alpha_{j+1}(T)+\varphi_{j+1}(k)]q_{j+1}\overline{F}_j(T)}{\sum_{j=0}^{\infty}H^{(j)}(k)\overline{F}_j(T)} \\
&+\frac{\sum_{j=0}^{\infty}H^{(j)}(k)m_j(T)\overline{F}_j(T)}{\sum_{j=0}^{\infty}H^{(j)}(k)\overline{F}_j(T)}.
\end{aligned}$$

Letting $Q_T(T;k)$ represent the left-hand side of the above equation,

$$\frac{\mathrm{d}Q_T(T;k)}{\mathrm{d}T}=\frac{\mathrm{d}V_T(T;k)}{\mathrm{d}T}\sum_{j=0}^{\infty}H^{(j)}(k)\int_0^T\overline{F}_j(t)\mathrm{d}t,$$

which follows that $Q_T(T;k)$ is strictly increasing if $\mathrm{d}V_T(T;k)/\mathrm{d}T|_{T=T^*}>0$. Therefore, if $Q_T(\infty;k)>C_1$, then T^* is a finite and unique solution satisfying (5.7).

Next, setting $\partial MCR(T,k)/\partial k = 0$ leads to

$$\begin{aligned}
&V_k(k;T)\sum_{j=0}^{\infty} H^{(j)}(k)\int_0^T \overline{F}_j(t)\mathrm{d}t \\
&-(C_2-C_1)(1-\pi_T)-(C_3-C_2)\pi_2-(C_4-C_2)\pi_K \\
&-\sum_{j=0}^{\infty} H^{(j+1)}(k)\int_0^T [\alpha_{j+1}(t)+\varphi_{j+1}(k)]q_{j+1}\overline{F}_j(t)\lambda(t)\mathrm{d}t \\
&-\sum_{j=0}^{\infty} H^{(j)}(k)\int_0^T m_j(t)\overline{F}_j(t)\mathrm{d}t = C_1,
\end{aligned} \tag{5.8}$$

where

$$\begin{aligned}
&V_k(k;T) \\
&=(\alpha_j(T)+q(k)\beta(k)+C_1-C_2)\frac{\sum_{j=0}^{\infty} h^{(j)}(k)\overline{F}_j(T)}{\sum_{j=0}^{\infty} h^{(j)}(k)\int_0^T F_j(t)\mathrm{d}t} \\
&\quad+(C_3-C_2)\frac{\sum_{j=0}^{\infty} h^{(j)}(k)\int_0^T [1-q_{j+1}]\overline{F}_j(t)\lambda(t)\mathrm{d}t}{\sum_{j=0}^{\infty} h^{(j)}(k)\int_0^T \overline{F}_j(t)\mathrm{d}t} \\
&\quad+(C_4-C_2)\frac{\sum_{j=0}^{\infty} \overline{H}_{j+1}(K-k)h^{(j)}(k)\int_0^T q_{j+1}\overline{F}_j(t)\lambda(t)\mathrm{d}t}{\sum_{j=0}^{\infty} h^{(j)}(k)\int_0^T \overline{F}_j(t)\mathrm{d}t} \\
&\quad+\frac{\sum_{j=1}^{\infty} h^{(j)}(k)\int_0^T [\alpha_j(t)\lambda(t)+q(k)\beta(k)\lambda(t)-\alpha_j'(t)+m_j(t)]\overline{F}_j(t)\mathrm{d}t}{\sum_{j=0}^{\infty} h^{(j)}(k)\int_0^T \overline{F}_j(t)\mathrm{d}t},
\end{aligned}$$

$$h^{(j)}(k)=\frac{\mathrm{d}H^{(j)}(k)}{\mathrm{d}k} \quad \text{and} \quad \alpha_j'(t)=\frac{\mathrm{d}\alpha_j(t)}{\mathrm{d}t}.$$

Letting $Q_k(k;T)$ represent the left-hand side of the above equation,

$$\begin{aligned}
Q_k(0;T) &= V_k(0;T)\int_0^T \exp(-\Lambda(t))\mathrm{d}t-(C_2-C_1)(1-\exp(-\Lambda(T))) \\
&\quad-(C_3-C_2)[1-\overline{P}_1]\exp(-\Lambda(T))-\int_0^T m_0(t)\exp(-\Lambda(t))\mathrm{d}t,
\end{aligned}$$

$$\begin{aligned}
Q_k(K;T) &= V_k(K;T)\sum_{j=0}^{\infty} H^{(j)}(K)\int_0^T \overline{F}_j(t)\mathrm{d}t \\
&\quad-(C_2-C_1)\left(1-\sum_{j=0}^{\infty} H^{(j)}(K)\overline{F}_j(T)\right)
\end{aligned}$$

$$
\begin{aligned}
&-(C_3-C_2)\sum_{j=0}^{\infty}H^{(j)}(K)\int_0^T[1-q_{j+1}]\overline{F}_j(t)\lambda(t)\mathrm{d}t\\
&-(C_4-C_2)\sum_{j=0}^{\infty}H^{(j+1)}(K)\int_0^T q_{j+1}\overline{F}_j(t)\lambda(t)\mathrm{d}t\\
&-\sum_{j=0}^{\infty}H^{(j+1)}(K)\int_0^T[\alpha_{j+1}(t)+\varphi_{j+1}(k)]q_{j+1}\overline{F}_j(t)\lambda(t)\mathrm{d}t\\
&-\sum_{j=0}^{\infty}H^{(j)}(K)\int_0^T m_j(t)\overline{F}_j(t)\mathrm{d}t,
\end{aligned}
$$

$$
\frac{\mathrm{d}Q_k(k;T)}{\mathrm{d}k}=\frac{\mathrm{d}V_k(k;T)}{\mathrm{d}T}\sum_{j=0}^{\infty}H^{(j)}(k)\int_0^T\overline{F}_j(t)\mathrm{d}t.
$$

Thus, if $V_k(k;T)$ increases strictly with k, $Q_k(k;T)$ is strictly increasing with k. If $Q_k(0;T)<R_1$ and $Q_k(K;T)<R_1$, k^* is finite and unique.

Therefore, the optimal replacement policy (T^*,k^*) must meet both (5.7) and (5.8). Usually, an explicit closed-form solution of these equations cannot be obtained. Hence, these optimal values will be determined numerically.

5.5 NUMERICAL EXAMPLE

Assume first that shocks occur according to the non-homogeneous Poisson process with $\lambda(t)=\eta t^{\eta-1}/\lambda^{\eta}$ and $H_j(w)=1-\mathrm{e}^{-w/\mu}$. The repair cost in practice is a random variable with distribution $A(t)$ and density function $a(t)$. Here repairing or replacing a failed system is based on the estimated repair cost R. Upon the ith shock, a replacement is carried out if $R>\delta_i c_\infty$; elsewise a repair is done, where c_∞ is a constant cost and δ_i is a fraction of c_∞ at the ith shock. The parameter δ_i meets the equation $q_i=\int_0^{\delta_i c_\infty}a(x)\mathrm{d}x$. Letting $g(R(t),r_i(t))=c_i$, then $\alpha_i(t)=E_{R(t)}[g(R(t),r_i(t))]=c_i$.

For convenience, the following cases are considered:

Case 1: $\overline{P}_i=q^i,\quad i=0,1,2,\ldots,\quad 0\le q\le 1,$

Case 2: $\overline{P}_i=q^{i^{0.5}},\quad i=0,1,2,\ldots,\quad 0\le q\le 1.$

First, we show the curves of $MCR(T,k)$ with respect to (T,k) in Figure 5.1, where the parameters are given as follows: $C_1=C_2=1000$, $C_3=C_4=1500$, $\overline{P}_i=q^i$, $q=0.8$, $m_j(t)=3t+5j$, $\beta(w)=100w$, $c_\infty=1000$, $\mu=1$, $\eta=2$, $\lambda=1$, $q(w)=1-\mathrm{e}^{-w}$, $K=10$, $R\sim N(100,20^2)$. With the information of Figure 5.1, we can get that there is the optimal solution (T^*,k^*) to minimize the average cost rate.

The optimal solutions and the average cost rates are shown in Table 5.1, where the parameters are below: $C_1=C_2=1000$, $C_3=C_4=1500$, $\beta(w)=25w$, $m_j(t)=3t+5j$, $c_\infty=1000$, $\mu=1$, $\eta=2$, $\lambda=1$, $q(w)=1-\mathrm{e}^{-w}$, $K=10$, $R\sim N(100,20^2)$.

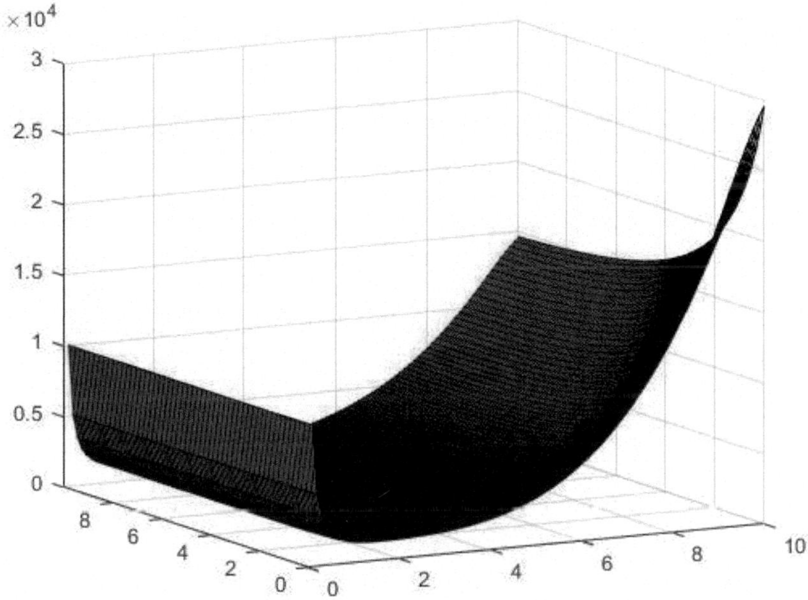

Figure 5.1 The effect of (T, k) on $MCR(T, k)$.

Based on the numerical results, we can obtain the following observations:

1) Table 5.1 reveal that as the value of q increases (that is, type 1 shock is easier to occur), the average cost rate decreases. This indicates that the average cost rate becomes larger as type 2 shock is easier to occur.

2) The average cost rate can be reduced when the probability of type 1 shock depends on the number of shocks.

3) As the probability of type 1 shock becomes larger, the optimal replacement age should be extended and the optimal preventive replacement damage level should be reduced to lessen the average cost rate.

Table 5.1
Optimal (T^*, k^*) and $MCR(T^*, k^*)$ for various q.

	$\overline{P}_i = q^i$			$\overline{P}_i = q^{i^{0.5}}$		
q	T^*	k^*	$MCR(T^*, k^*)$	T^*	k^*	$MCR(T^*, k^*)$
0.1	1.46	8.84	1636.68	1.51	7.47	1622.24
0.2	1.51	7.51	1560.16	1.62	6.86	1524.35
0.3	1.56	6.61	1481.29	1.74	6.39	1423.32
0.4	1.61	5.97	1399.66	1.85	6.00	1322.71
0.5	1.67	5.49	1314.83	1.94	5.66	1225.26
0.6	1.74	5.14	1226.33	2.01	5.37	1132.75
0.7	1.82	4.87	1133.58	2.06	5.11	1046.08
0.8	1.92	4.67	1035.90	2.10	4.87	965.60
0.9	2.03	4.53	932.54	2.14	4.65	891.31

5.6 CONCLUSIONS

This paper proposed a bivariate replacement policy for a two-unit system that is subject to non-homogeneous Poisson shocks. The policy considers that the system is replaced when the accumulative damage to unit 2 exceeds k $(\leq K)$, or when the accumulative damage to unit 2 exceeds K, or at type 2 shock, or at age T, whichever comes first. The average cost rate is established to determine the optimal preventive replacement schedule at the minimal average cost rate. It is shown the existence and uniqueness of the optimal preventive replacement policy.

REFERENCES

1. Barlow, R. E., & Hunter, L. C. (1960). Optimum preventive maintenance policies, *Operational Research*, 8, 90-100.

2. Chang, C. C., & Chen. (2019). Optimization of continuous and discrete scheduled times for a cumulative damage system with age-dependent maintenance, *Communications in Statistics - Theory and Methods*, 48, 4261-4277.

3. Cox, D. R. (1962). *Renewal Theory*, London: Methuen.

4. Dong, W., Liu, S., Cao, Y., & Javed, S. S. (2021). Scheduling optimal replacement policies for a stochastically deteriorating system subject to two types of shocks, ISA Transcations, 112, 292-301.

5. Eryilmaz, S., & Kan, C. (2019). Reliability and optimal replacement policy for an extreme shock model with a change point, *Reliability Engineering & System Safety*, 190, 106513.

6. Huang, X., Jin, S., He, X., & He, D. (2019). Reliability analysis of coherent systems subject to internal failures and external shocks, *Reliability Engineering & System Safety*, 181, 75-83.

7. Jafary, B., Nagaraju, V., & Fiondella, L. (2017). Impact of correlated component failure on preventive maintenance policies, *IEEE Transactions on Reliability*, 66, 575-586.

8. Kaio, N. (2018). A study on optimal preventive maintenance policies for cumulative damage models, In: Ram M., Davim J. (eds) Modeling and Simulation in *Industrial Engineering. Management and Industrial Engineering.* Springer, Cham.

9. Kitagawa, T., Yuge, T., & Yanagi, S. (2017). Maintenance modeling for a system equipped on ship, *IEICE Transactions on Fundamentals of Electronics, Communications and Computer Sciences* E100.A 629-638.

10. Lai, M. T., & Chen, C. H. (2017). Optimal replacement period with cumulative repair cost limit under a cumulative damage model, *International Journal of Industrial and Systems Engineering*, 26, 133-147.

11. Nakagawa, T., & Kowada, M. (1983). Analysis of a system with minimal repair and its application to replacement policy, *European Journal of Operational Research*, 12, 176-183.

12. Mizutani, S., & Nakagawa, T (2018). Maintenance overtime policy with cumulative damage, *International Journal of Mathematical, Engineering and Management Sciences*, 3, 123-135.

13. Sheu, S. H., Liu, T. H., Zhang, Z. G., Tsai, H. N., & Chen, J. C. (2016). Optimal two-threshold replacement policy in a cumulative damage model, *Annals of Operations Research*, 244, 23-47.

14. Sheu, S. H., Liu, T. H., & Zhang, Z. G. (2019). Extended optimal preventive replacement policies with random working cycle, *Reliability Engineering & System Safety*, 188, 398-415.

15. Sheu, S. H., Liu, T. H., Zhang, G., & Tsai, H. N. (2020). Optimum replacement policy for cumulative damage models based on multi-attributes, *Computers & Industrial Engineering*, 139, 106206.

16. Sheu, S. H., Liu, T. H., Zhang, Z. G., Zhao, X., & Chien, Y. H. (2021). A generalized age-dependent minimal repair with random working times, *Computers & Industrial Engineering*, 156, 107248.

17. Sheu, S. H., Liu, T. H., Sheu, W. T., Zhang, Z. G., & Ke, J. C. (2021). Optimal replacement policy with replacement last under cumulative damage models, *Reliability Engineering & System Safety*, 209, 107445.

18. Tsai, H. N., Sheu, S. H., & Zhang, Z. G. (2017). A trivariate optimal replacement policy for a deteriorating system based on cumulative damage and inspections, *Reliability Engineering and System Safety*, 160, 74-88.

19. Zarezadeh, S., & Asadi, M. (2019). Coherent systems subject to multiple shocks with applications to preventative maintenance, *Reliability Engineering & System Safety*, 185, 124-132.

20. Zhao, X., Cai, K., Wang, X., & Song, Y. (2018). Optimal replacement policies for a shock model with a change point, *Computers & Industrial Engineering*, 118, 383-393.

21. Zhao, X., Cai, J., Mizutani, S., & Nakagawa, T. (2020). Preventive replacement policies with time of operations, mission durations, minimal repairs and maintenance triggering approaches, *Journal of Manufacturing Systems*, https://doi.org/10.1016/j.jmsy.2020.04.003.

22. Zhao, X., Guo, X., & Wang, X. (2018). Reliability and maintenance policies for a two-stage shock model with self-healing mechanism, *Reliability Engineering & System Safety*, 172, 185-194.

23. Zhao X., & Nakagawa, T. (2017). *Advanced Maintenance Policies for Shock and Damage Models*, Springer, Cham.

Section II

Reliability Modeling & Application

6 Optimal Checking Policy for a Server System with a Cyber Attack

Mitsuhiro Imaizumi
Aichi Gakusen University, Japan

CONTENTS

6.1 INTRODUCTION

In this chapter, we survey a server system which is attacked from various computers at random. As Internet technology and cloud computing have developed, huge amounts of data have been generated and stored in servers. These data are collected and analyzed and their results are utilized in many areas of business. The demands for the improvement of the reliability and security of servers connected with the Internet have increased. Cyber attacks on the Internet have become a problem in recent years, and it has been becoming more sophisticated and complicated. In order to solve this problem, several schemes have been proposed. To prevent the server from cyber attack, configuring the proxy server or setting up a firewall and IDS (intrusion detection system) have been introduced. As one of the useful ways to detect cyber attacks, IDS has been widely used [1, 2]. There are two types of IDS. One is a host-based IDS and the other is a network-based IDS. A host-based IDS is an intrusion detection system that runs on a host and monitors the intruder attacks, activities, and malicious behavior. On the other hand, a network-based IDS is an intrusion detection system that detects malicious traffic on the network.

After a cyber attack has occurred, the downloader, which includes malware, enters a server, and other malware is downloaded through the Internet. After that, the RAT (Remote Administration Tool, Remote Access Trojan) that lets intruders remotely control the server through the Internet is downloaded repeatedly. Therefore, the risk of data leakage and destruction of vital data increase as time passes. Cyber attacks cause heavy damage to the system. It is important to detect a cyber attack in the early stage. A cyber attack is performed so that it circumvents server monitoring. Therefore, random checking by IDS is one of the effective ways to detect cyber attacks. On the other hand, when the number of checks increases, the total cost of the checking increases. In terms of monitoring, it is necessary to perform checking under appropriate management policy.

In terms of IDS, the accuracy level for the detection of a cyber attack is also important. If it is low, the overhead of system processing becomes larger. That is, if the IDS judges the occurrence of a cyber attack erroneously, and the server system executes processing based on incorrect judgement, the system load increases. There are two types of errors. One is a type I error and the other is a type II error. In terms of type I error, when a cyber attack has not occurred, it is misjudged that it has occurred. In terms of type II error, when a cyber attack has occurred, it is misjudged that it has not occurred.

The simulation about the policy for monitoring and detection of a cyber attack has already introduced [3, 4, 5], and there are some formalized stochastic models. We have proposed some stochastic models of a system with a cyber attack. By using Markov renewal processes [6, 7], we have discussed optimal monitoring policies [8, 9].

In this chapter, we pay attention to a server which is connected with the Internet, and treat five stochastic models. The server has a host-based IDS. We apply the maintenance theory [10, 11, 12] to a system with a cyber attack. In Model 1, we formulate a stochastic model in which the cyber attack is checked at periodic time considering the type II error. We assume that the cost between attack occurrence and its detection increases with time linearly. In Model 2, we modify Model 1 considering non-linear parameter. We assume that the cost between attack occurrence and its detection increases with time non-linearly. In Model 3, we modify Model 1 considering two types of checks, a simple check and a detailed check. The detailed check is performed every 2 checks. In Model 4, we modify Model 3. The detailed check is performed every 3 checks. In Model 5, we modify Model 1 considering random checking. We assume that a cyber attack is checked at both random and periodic times.

In each model, the expected costs until a cyber attack is detected are derived. Further, optimal policies which minimize them are discussed. Finally, numerical examples are given.

6.2 MODEL 1

This paper pays attention to only a server which is connected with the Internet. We have the following assumptions.

(1) After the server begins to operate, a cyber attack occurs according to a general distribution $F(t)$ with finite mean $1/\lambda$, and does not disappear. That is, $1/\lambda$ represents the mean time to cyber attack occurrence.

(2) A cyber attack is checked at times $T_k(k = 1, 2, \cdots)$, where $T_0 = 0$.

(3) If a cyber attack has occurred, it is detected by the i-th periodic check with probability $p_i (i = 1, 2, \cdots ; 0 < p_i \leq 1)$, and it is not detected with probability $1 - p_i$, which is called a type II error. If a cyber attack has not occurred, the occurrence of cyber attack is judged correctly with probability 1.

We obtain the expected cost until a cyber attack is detected and discuss an optimal policy which minimizes it. Let c_p be the cost of one check and c_l be the cost per unit of time for the time elapsed between attack occurrence and its detection.

When $p_i = p$, the expected number of checks is

$$\sum_{j=0}^{\infty}(1-p)^j p \sum_{k=0}^{\infty} \int_{T_k}^{T_{k+1}} (k+1+j) dF(t).$$

The mean time elapsed between attack occurrence and its detection is

$$\sum_{j=0}^{\infty}(1-p)^j p \sum_{k=0}^{\infty} \int_{T_k}^{T_{k+1}} (T_{k+1+j} - t) dF(t).$$

Therefore, the expected cost $C_1(T_1, T_2, \cdots)$ is

$$\begin{aligned} C_1(T_1, T_2, \cdots) &= \sum_{j=0}^{\infty}(1-p)^j p \sum_{k=0}^{\infty} \int_{T_k}^{T_{k+1}} [c_p(k+1+j) \\ &\quad + c_l(T_{k+1+j} - t)] dF(t) \\ &= \sum_{j=0}^{\infty}(1-p)^j p \Big\{ \sum_{k=0}^{\infty} [c_p + c_l(T_{k+1+j} - T_{k+j})] \overline{F}(T_k) \\ &\quad + j c_p + c_l T_j \Big\} - \frac{c_l}{\lambda}, \end{aligned} \tag{6.1}$$

where

$$\overline{\Phi}(t) \equiv 1 - \Phi(t).$$

In particular, when $T_k = kT(k = 1, 2, \cdots)$, that is, a cyber attack is checked at periodic times kT, the expected cost in (6.1) becomes

$$C_1(T) = \sum_{k=0}^{\infty}(c_p + c_l T)\overline{F}(kT) + \frac{1-p}{p}(c_p + c_l T) - \frac{c_l}{\lambda}. \tag{6.2}$$

Moreover, when $F(t) = 1 - e^{-\lambda t}$, that is, a cyber attack occurs according to an exponential distribution, the expected cost in (6.2) becomes

$$C_1(T) = c_p\left(\frac{1}{1-e^{-\lambda T}} + \frac{1-p}{p}\right) + c_l\left(\frac{1}{1-e^{-\lambda T}} + \frac{1-p}{p}\right)T - \frac{c_l}{\lambda}. \tag{6.3}$$

We seek an optimal time T^* which minimizes $C_1(T)$. Differentiating equation (6.3) with respect to T and setting it equal to zero, we have

$$\frac{1-p}{p}(1-e^{-\lambda T})^2 e^{\lambda T} + e^{\lambda T} - (1 + \lambda T) = \lambda\frac{c_p}{c_l}, \tag{6.4}$$

whose the left-hand side increases strictly from 0 to ∞. Thus, there exists a finite and unique optimal $T^*(0 < T^* < \infty)$ which satisfies equation (6.4). Clearly, the left-hand side decreases with p from ∞ to $e^{\lambda T} - (1 + \lambda T)$, and hence, T^* increases with p.

6.3 MODEL 2

We modify Model 1 considering non-linear parameter.

(1) The same as (1) of Model 1.

(2) A cyber attack is checked at periodic times $kT(k = 1, 2, \cdots)$ (periodic check).

(3) The same as (3) of Model 1.

Let c_p be the cost of the periodic check. The cost between attack occurrence and its detection increases with T non-linearly, i.e., $c_d e^{\alpha[(k+1)T-t]}(k = 0, 1, 2, \cdots)$. When $p_i = p(i = 1, 2, \cdots, n-1; n \geq 2)$ and $p_n = 1$, that is, when a cyber attack is detected the n-th periodic check with probability 1 after a cyber attack has occurred, the expected cost $C_2(T)$ until a cyber attack is detected is given by

$$\begin{aligned} C_2(T) &= \sum_{m=0}^{n-2}(1-p)^m p \sum_{k=0}^{\infty}\int_{kT}^{(k+1)T}\{(k+m+1)c_p \\ &+c_d e^{\alpha[(k+m+1)T-t]}\}dF(t) \\ &+(1-p)^{n-1}\sum_{k=0}^{\infty}\int_{kT}^{(k+1)T}\{(k+n)c_p + c_d e^{\alpha[(k+n)T-t]}\}dF(t). \end{aligned} \tag{6.5}$$

Especially, when $p_1 = p$ and $p_2 = 1$, that is, when a cyber attack is detected at the second periodic check with probability 1 after a cyber attack has occurred, the expected cost in (6.5) becomes

$$\begin{aligned} C_2(T) &= p\sum_{k=0}^{\infty}\int_{kT}^{(k+1)T}\{(k+1)c_p + c_d e^{\alpha[(k+1)T-t]}\}dF(t) \\ &+(1-p)\sum_{k=0}^{\infty}\int_{kT}^{(k+1)T}\{(k+2)c_p + c_d e^{\alpha[(k+2)T-t]}\}dF(t). \end{aligned} \tag{6.6}$$

When $F(t) = 1 - e^{-\lambda T}$ the expected cost in (6.6) becomes

$$C_2(T) = \frac{c_p}{1-e^{-\lambda T}} + (1-p)c_p + c_d\frac{\lambda}{\lambda+\alpha}\frac{e^{\alpha T}-e^{-\lambda T}}{1-e^{-\lambda T}}[p+(1-p)e^{\alpha T}]. \tag{6.7}$$

We seek an optimal T^* which minimizes $C_2(T)$. Differentiating (6.7) with respect to T and setting it equal to zero,

$$\begin{aligned} &\left\{e^{\alpha T}\left[\left(\frac{\alpha}{\lambda+\alpha}\right)e^{\lambda T}-1\right]+\frac{\lambda}{\lambda+\alpha}\right\}[p+(1-p)e^{\alpha T}] \\ &+(1-p)\left(\frac{\alpha}{\lambda+\alpha}\right)(e^{\alpha T}-e^{-\lambda T})(1-e^{-\lambda T}) = \frac{c_p}{c_d}, \end{aligned} \tag{6.8}$$

whose the left-hand side increases strictly from 0 to ∞. Thus, there exists a finite and unique optimal $T^*(0 < T^* < \infty)$ which satisfies equation (6.8).

6.4 MODEL 3

We modify Model 1 considering two types of checks. One is a simple check, the other is a detailed check.

(1) The same as (1) of Model 1.

(2) The simple check is performed at times T_{2k-1} $(k = 1, 2, \cdots)$. The detailed check is performed at times T_{2k} $(k = 1, 2, \cdots)$, where $T_0 = 0$. That is, it is performed every 2 checks.

(3) If a cyber attack has occurred, it is detected by the simple check with probability p_s $(0 < p_s \leq 1)$, and it is detected by the detailed check with probability p_d $(0 < p_d \leq 1; p_d \geq p_s)$. If a cyber attack has not occurred, the occurrence of a cyber attack is judged correctly with probability 1.

Let c_{ps} be the cost of one simple check, c_{pd} be the cost of one detailed check, and c_l be the cost per unit of time for the time elapsed between attack occurrence and its detection. Then, the expected cost $C_3(T_1, T_2, \cdots)$ until a cyber attack is detected is

$$
\begin{aligned}
&C_3(T_1, T_2, \cdots) \\
&= \sum_{j=0}^{\infty}[(1-p_s)(1-p_d)]^j \\
&\times\Big\{p_s\sum_{k=0}^{\infty}\int_{T_{2k}}^{T_{2k+1}}[c_{ps}(k+1+j)+c_l(T_{2k+1+2j}-t)+c_{pd}(k+j)]dF(t) \\
&+p_d\sum_{k=0}^{\infty}\int_{T_{2k+1}}^{T_{2k+2}}[c_{ps}(k+1+j)+c_l(T_{2k+2+2j}-t)+c_{pd}(k+1+j)]dF(t) \\
&+(1-p_s)p_d\sum_{k=0}^{\infty}\int_{T_{2k}}^{T_{2k+1}}[c_{ps}(k+1+j)+c_l(T_{2k+2+2j}-t) \\
&+c_{pd}(k+1+j)]dF(t) \\
&+(1-p_d)p_s\sum_{k=0}^{\infty}\int_{T_{2k+1}}^{T_{2k+2}}[c_{ps}(k+2+j)+c_l(T_{2k+3+2j}-t) \\
&+c_{pd}(k+1+j)]dF(t)\Big\}.
\end{aligned}
\tag{6.9}
$$

When $F(t) = 1 - e^{-\lambda t}$ and $T_k = kT$, (6.9) is

$$
\begin{aligned}
C_3(T) \quad = \quad & c_{ps}\Big[\frac{1}{1-e^{-2\lambda T}}+\frac{(1-p_d)p_s e^{-\lambda T}}{[1-(1-p_s)(1-p_d)](1+e^{-\lambda T})} \\
&+\frac{(1-p_s)(1-p_d)}{1-(1-p_s)(1-p_d)}\Big] \\
&+c_{pd}\Big[\frac{e^{-\lambda T}}{1-e^{-2\lambda T}}+\frac{(1-p_s)p_d}{[1-(1-p_s)(1-p_d)](1+e^{-\lambda T})} \\
&+\frac{(1-p_s)(1-p_d)}{1-(1-p_s)(1-p_d)}\Big] \\
&+c_l\Big\{\Big[\frac{1}{1-e^{-\lambda T}}+\frac{(1-p_s)p_d+(1-p_d)p_s e^{-\lambda T}}{[1-(1-p_s)(1-p_d)](1+e^{-\lambda T})}
\end{aligned}
$$

$$+\frac{2(1-p_s)(1-p_d)}{1-(1-p_s)(1-p_d)}\Big]T-\frac{1}{\lambda}\Big\}. \tag{6.10}$$

We seek an optimal time T^* which minimizes $C_3(T)$ when $c_{pd} = \gamma c_{ps}(\gamma \geq 1)$. Differentiating equation (6.10) with respect to T and setting it equal to zero, we have

$$\frac{\left\{\begin{array}{l}\dfrac{p_d-p_s}{[1-(1-p_s)(1-p_d)]}\left(\dfrac{1-e^{-\lambda T}}{1+e^{-\lambda T}}\right)^2\lambda T+e^{\lambda T}-(1+\lambda T)\\ +\Big[\dfrac{(1-p_s)p_d}{[1-(1-p_s)(1-p_d)]}\left(\dfrac{1}{1+e^{-\lambda T}}\right)\\ +\dfrac{(1-p_d)p_s}{[1-(1-p_s)(1-p_d)]}\left(\dfrac{e^{-\lambda T}}{1+e^{-\lambda T}}\right)\\ +\dfrac{2(1-p_s)(1-p_d)}{1-(1-p_s)(1-p_d)}\Big]e^{\lambda T}(1-e^{-\lambda T})^2\end{array}\right\}}{\left\{\begin{array}{l}\dfrac{2e^{-\lambda T}}{(1+e^{-\lambda T})^2}+\dfrac{(1-p_d)p_s}{[1-(1-p_s)(1-p_d)]}\left(\dfrac{1-e^{-\lambda T}}{1+e^{-\lambda T}}\right)^2\\ +\gamma\left[\dfrac{1+e^{-2\lambda T}}{(1+e^{-\lambda T})^2}-\dfrac{(1-p_s)p_d}{[1-(1-p_s)(1-p_d)]}\left(\dfrac{1-e^{-\lambda T}}{1+e^{-\lambda T}}\right)^2\right]\end{array}\right\}} = \lambda\frac{c_{ps}}{c_l}. \tag{6.11}$$

Denoting the left-hand side of (6.11) by $L_3(T)$, we have

$$\begin{aligned} L_3(0) &= 0, \\ L_3(\infty) &= \infty. \end{aligned}$$

When $L_3'(T) > 0$, $L_3(T)$ is strictly increasing in T from 0 to ∞. Thus, in this case, there exists a finite and unique optimal $T^*(0 < T^* < \infty)$ which satisfies equation (6.11).

When $\gamma = 1$, (6.11) is

$$\begin{aligned} &\frac{p_d-p_s}{[1-(1-p_s)(1-p_d)]}\left(\frac{1-e^{-\lambda T}}{1+e^{-\lambda T}}\right)^2\lambda T+e^{\lambda T}-(1+\lambda T)\\ &+\Big\{\frac{(1-p_s)p_d}{[1-(1-p_s)(1-p_d)]}\left(\frac{1}{1+e^{-\lambda T}}\right)\\ &+\frac{(1-p_d)p_s}{[1-(1-p_s)(1-p_d)]}\left(\frac{e^{-\lambda T}}{1+e^{-\lambda T}}\right)\\ &+\frac{2(1-p_s)(1-p_d)}{1-(1-p_s)(1-p_d)}\Big\}e^{\lambda T}(1-e^{-\lambda T})^2 = \lambda\frac{c_{ps}}{c_l}, \end{aligned} \tag{6.12}$$

whose the left-hand side increases strictly from 0 to ∞. Thus, there exists a finite and unique optimal $T^*(0 < T^* < \infty)$ which satisfies equation (6.12).

6.5 MODEL 4

We modify Model 3.

(1) The same as (1) of Model 1.

(2) The simple check is performed at times T_{3k-2}, T_{3k-1} $(k = 1, 2, \cdots)$. The detailed check is performed at times T_{3k} $(k = 1, 2, \cdots)$, where $T_0 = 0$. That is, it is performed every 3 checks.

(3) The same as (3) of Model 3.

Let c_{ps} be the cost of one simple check, c_{pd} be the cost of one detailed check and c_l be the cost per unit of time for the time elapsed between attack occurrence and its detection. Then, the expected cost $C_4(T_1, T_2, \cdots)$ until cyber attack is detected is

$$
\begin{aligned}
&C_4(T_1, T_2, \cdots) \\
&= \sum_{j=0}^{\infty} [(1-p_s)^2(1-p_d)]^j \\
&\quad \times \Big\{ p_s \sum_{k=0}^{\infty} \int_{T_{3k}}^{T_{3k+1}} [c_{ps}(2k+2j+1) + c_l(T_{3k+3j+1} - t) + c_{pd}(k+j)] dF(t) \\
&\quad + p_s \sum_{k=0}^{\infty} \int_{T_{3k+1}}^{T_{3k+2}} [c_{ps}(2k+2j+2) + c_l(T_{3k+3j+2} - t) + c_{pd}(k+j)] dF(t) \\
&\quad + p_d \sum_{k=0}^{\infty} \int_{T_{3k+2}}^{T_{3k+3}} [c_{ps}(2k+2j+2) + c_l(T_{3k+3j+3} - t) + c_{pd}(k+j+1)] dF(t) \\
&\quad + (1-p_s)p_s \sum_{k=0}^{\infty} \int_{T_{3k}}^{T_{3k+1}} [c_{ps}(2k+2j+2) + c_l(T_{3k+3j+2} - t) \\
&\quad + c_{pd}(k+j)] dF(t) \\
&\quad + (1-p_s)p_d \sum_{k=0}^{\infty} \int_{T_{3k+1}}^{T_{3k+2}} [c_{ps}(2k+2j+2) + c_l(T_{3k+3j+3} - t) \\
&\quad + c_{pd}(k+j+1)] dF(t) \\
&\quad + (1-p_d)p_s \sum_{k=0}^{\infty} \int_{T_{3k+2}}^{T_{3k+3}} [c_{ps}(2k+2j+3) + c_l(T_{3k+3j+4} - t) \\
&\quad + c_{pd}(k+j+1)] dF(t) \\
&\quad + (1-p_s)^2 p_d \sum_{k=0}^{\infty} \int_{T_{3k}}^{T_{3k+1}} [c_{ps}(2k+2j+2) + c_l(T_{3k+3j+3} - t) \\
&\quad + c_{pd}(k+j+1)] dF(t)
\end{aligned}
$$

$$+(1-p_s)(1-p_d)p_s\sum_{k=0}^{\infty}\int_{T_{3k+1}}^{T_{3k+2}}[c_{ps}(2k+2j+3)$$
$$+c_l(T_{3k+3j+4}-t)+c_{pd}(k+j+1)]dF(t)$$
$$+(1-p_s)(1-p_d)p_s\sum_{k=0}^{\infty}\int_{T_{3k+2}}^{T_{3k+3}}[c_{ps}(2k+2j+4)$$
$$+c_l(T_{3k+3j+5}-t)+c_{pd}(k+j+1)]dF(t)\Big\}. \tag{6.13}$$

When $F(t)=1-e^{-\lambda T}$ and $T_k=kT$, (6.13) is

$$\begin{aligned}
&C_4(T)\\
&=(1-e^{-\lambda T})\Big\{\frac{(2c_{ps}+c_{pd})B}{1-(1-p_s)^2(1-p_d)}\frac{e^{-3\lambda T}}{(1-e^{-3\lambda T})^2}\\
&+\Big[\frac{c_{ps}p_s+2c_{ps}(1-p_s)p_s+(2c_{ps}+c_{pd})(1-p_s)^2p_d}{1-(1-p_s)^2(1-p_d)}\\
&+\frac{2c_{ps}p_s+(2c_{ps}+c_{pd})(1-p_s)p_d+(3c_{ps}+c_{pd})(1-p_s)(1-p_d)p_s}{1-(1-p_s)^2(1-p_d)}e^{-\lambda T}\\
&+\frac{(2c_{ps}+c_{pd})p_d+(3c_{ps}+c_{pd})(1-p_d)p_s+(4c_{ps}+c_{pd})(1-p_s)(1-p_d)p_s}{1-(1-p_s)^2(1-p_d)}e^{-2\lambda T}\\
&+(2c_{ps}+c_{pd})AB\Big]\frac{1}{1-e^{-3\lambda T}}\Big\}\\
&+c_l\Big\{(1-e^{-\lambda T})\Big\{\frac{3B}{1-(1-p_s)^2(1-p_d)}\frac{e^{-3\lambda T}}{(1-e^{-3\lambda T})^2}\\
&+\Big[\frac{p_s+2(1-p_s)p_s+3(1-p_s)^2p_d}{1-(1-p_s)^2(1-p_d)}\\
&+\frac{2p_s+3(1-p_s)p_d+4(1-p_s)(1-p_d)p_s}{1-(1-p_s)^2(1-p_d)}e^{-\lambda T}\\
&+\frac{3p_d+4(1-p_d)p_s+5(1-p_s)(1-p_d)p_s}{1-(1-p_s)^2(1-p_d)}e^{-2\lambda T}+3AB\Big]\frac{1}{1-e^{-3\lambda T}}\Big\}T-\frac{1}{\lambda}\Big\},
\end{aligned}$$

where

$$\begin{aligned}
A &\equiv \frac{(1-p_s)^2(1-p_d)}{[1-(1-p_s)^2(1-p_d)]^2},\\
B &\equiv [p_s+(1-p_s)p_s+(1-p_s)^2p_d]\\
&\quad+[p_s+(1-p_s)p_d+(1-p_s)(1-p_d)p_s]e^{-\lambda T}\\
&\quad+[p_d+(1-p_d)p_s+(1-p_s)(1-p_d)p_s]e^{-2\lambda T}.
\end{aligned}$$

We can discuss the optimal policy which minimizes $C_4(T)$ by the similar method of Model 3. Especially, when $p_s=p_d=p$ and $c_{ps}=c_{pd}$, this corresponds to the expected cost of Model 1.

6.6 MODEL 5

We modify Model 1 considering random checking.

(1) The same as 1 of Model 1.

(2) A cyber attack is checked at successive times $Y_j (j = 1, 2, \cdots)$ (random check), where $Y_0 \equiv 0$, and at periodic times $kT (k = 1, 2, \cdots)$ (periodic check).

(3) If a cyber attack has occurred, it is detected by random check and periodic check with probability 1. If a cyber attack has not occurred, the occurrence of a cyber attack is judged correctly with probability 1.

(4) The distribution of $Y_j - Y_{j-1} (j = 1, 2, \cdots)$ has a general distribution $G(t)$ with finite mean $1/\theta (\theta > \lambda)$. The distribution of Y_j is represented by the j-th fold convolution $G^{(j)}(t)$ of $G(t)$ with itself, i.e., $G^{(j)}(t) \equiv G^{(j-1)}(t) * G(t), G(t) * G(t) \equiv \int_0^t G(t-u) dG(u), G^{(0)}(t) \equiv 1$, where the asterisk denotes the Stieltjes convolution.

The probability that a cyber attack is detected by periodic check is given by

$$\sum_{k=0}^{\infty} \int_{kT}^{(k+1)T} \sum_{j=0}^{\infty} \int_0^t \overline{G}[(k+1)T - x] dG^{(j)}(x) dF(t). \tag{6.14}$$

Moreover, the probability that a cyber attack is detected by random check is given by

$$\sum_{k=0}^{\infty} \int_{kT}^{(k+1)T} \sum_{j=0}^{\infty} \int_0^t \{G[(k+1)T - x] - G(t-x)\} dG^{(j)}(x) dF(t). \tag{6.15}$$

Let c_p be the cost of the periodic check, $c_r (c_r > c_p)$ be the cost of the random check. The cost between an attack occurrence and its detection increases with T non-linearly, i.e., $c_d e^{\alpha[(k+1)T-t]} (k = 0, 1, 2, \cdots)$, where $\theta > \alpha$ and $\theta - \alpha - \lambda > 0$. Then, the expected cost $C_5(T)$ until a cyber attack is detected is given by

$$\begin{aligned} C_5(T) = & \sum_{k=0}^{\infty} \int_{kT}^{(k+1)T} \sum_{j=0}^{\infty} \{(k+1)c_p + jc_r + c_d e^{\alpha[(k+1)T-t]}\} \\ & \times \left\{ \int_0^t \overline{G}[(k+1)T - x] dG^{(j)}(x) \right\} dF(t) \\ & + \sum_{k=0}^{\infty} \int_{kT}^{(k+1)T} \sum_{j=0}^{\infty} \int_0^t \int_{t-x}^{(k+1)T-x} [kc_p \end{aligned}$$

$$
\begin{aligned}
&+(j+1)c_r + c_d e^{\alpha(x+y-t)}]dG(y)dG^{(j)}(x)dF(t) \\
= \;& c_p \sum_{k=0}^{\infty} \overline{F}(kT) + c_r \sum_{j=0}^{\infty} j \int_0^{\infty} [G^{(j)}(t) - G^{(j+1)}(t)]dF(t) \\
&-(c_p - c_r) \sum_{k=0}^{\infty} \int_{kT}^{(k+1)T} \Big\{ G[(k+1)] - G(t) \\
&+ \int_0^t \{G[(k+1)T - x] - G(t-x)\} dM(x) \Big\} dF(t) \\
&+c_d + c_d \sum_{k=0}^{\infty} \int_{kT}^{(k+1)T} \Big[\int_t^{(k+1)T} \overline{G}(y)dy \\
&+ \int_0^t \int_{t-x}^{(k+1)T-x} \overline{G}(y) d e^{\alpha(x+y-t)} dM(x) \Big] dF(t), \qquad (6.16)
\end{aligned}
$$

where

$$M(x) \equiv \sum_{j=1}^{\infty} G^{(j)}(x)$$

represents the expected number of checks during $(0, x]$.

In particular, when $G(t) = 1 - e^{-\theta t}$ the expected cost in (6.16) becomes

$$
\begin{aligned}
C_5(T) \;=\;& \sum_{k=0}^{\infty} \overline{F}(kT) + c_r \frac{\theta}{\lambda} \\
&-(c_p - c_r) \sum_{k=0}^{\infty} \int_{kT}^{(k+1)T} \{1 - e^{-\theta[(k+1)T-t]}\} dF(t) \\
&+c_d + c_d \sum_{k=0}^{\infty} \int_{kT}^{(k+1)T} \frac{\alpha}{\theta - \alpha} \{1 - e^{-(\theta-\alpha)[(k+1)T - \alpha t]} e^{\theta t}\} dF(t).
\end{aligned} \qquad (6.17)
$$

Moreover, when $F(t) = 1 - e^{-\lambda t}$, the expected cost in (6.17) becomes

$$
\begin{aligned}
C_5(T) \;=\;& \frac{c_p}{1 - e^{-\lambda T}} + c_r \frac{\theta}{\lambda} - (c_p - c_r) \left(1 - \frac{\lambda}{\theta - \lambda} \frac{e^{-\lambda T} - e^{-\theta T}}{1 - e^{-\lambda T}} \right) \\
&+c_d \left\{ 1 + \frac{\alpha}{\theta - \alpha} \left[1 - \frac{\lambda}{(\theta - \alpha) - \lambda} \frac{e^{-\lambda T} - e^{-(\theta-\alpha)T}}{1 - e^{-\lambda T}} \right] \right\}.
\end{aligned} \qquad (6.18)
$$

We seek an optimal T^* which minimizes $C_5(T)$. Differentiating (6.18) with respect to T and setting it equal to zero,

$$\frac{\theta}{\theta-\lambda}[1-e^{-(\theta-\lambda)T}]-(1-e^{-\theta T})$$
$$+\frac{c_d}{c_r-c_p}\frac{\alpha}{\theta-\alpha}\left\{\frac{\theta-\alpha}{(\theta-\alpha)-\lambda}\{1-e^{-[(\theta-\alpha)-\lambda]}\}-[1-e^{-(\theta-\alpha)T}]\right\}=\frac{c_p}{c_r-c_p}. \tag{6.19}$$

Denoting the left-hand side (6.19) by $L_5(T)$

$$\begin{aligned} L_5(0) &= 0, \\ L_5(\infty) &= \frac{\lambda}{\theta-\lambda}+\frac{c_d}{c_r-c_p}\frac{\alpha}{\theta-\alpha}\frac{\lambda}{(\theta-\alpha)-\lambda}, \\ L_5'(T) &= \left[\theta e^{-\theta T}+\frac{c_d}{c_r-c_p}\alpha e^{-(\theta-\alpha)T}\right](e^{\lambda T}-1). \end{aligned}$$

Hence, $L_5(T)$ is strictly increasing in T from 0 to $L_5(\infty)$. Thus, we can characterize the optimal policy:

(1) If $L_5(\infty) > c_p/(c_r-c_p)$, then there exists a finite and unique optimal $T^*(0 < T^* < \infty)$ which satisfies (6.19).

(2) If $L_5(\infty) \leq c_p/(c_r-c_p)$, then $T^* = \infty$. In this case, we should only perform the random check.

6.7 NUMERICAL EXAMPLES

In terms of Models 1–5, we numerically compute T^*, which minimizes the expected cost. It is assumed that the mean time to cyber attack occurrence is $1/\lambda = 40$–400, the mean time to random check is $1/\theta = 10$, the probability that a cyber attack is detected by the periodic check is $p = 0.6$–1.0, the cost per unit of time for the time elapsed between occurrence of cyber attack and its detection is $c_l = 1$, the cost of the periodic check is $c_p = 1$, the cost of one simple check is $c_{ps} = 10$–50, the cost rate of one detailed check for the cost of one simple check is $\gamma = 1$–5, the cost of the random check $c_r = 5$–20, parameters of the cost between attack occurrence and its detection are $c_d = 1$–20 and $\alpha = 0.01$–0.04.

Table 6.1 gives the optimal checking time T^* which minimizes the expected cost $C_1(T)$ of Model 1. For example, when $1/\lambda = 200$, $p = 0.9$ and $c_p/c_l = 30$, the optimal $T^* = 93$. This indicates that T^* increases with $1/\lambda$, p and c_p/c_l. When p is small, T^* is small. In this case, we should perform the check at small intervals.

Table 6.2 gives the optimal checking time T^* which minimizes the expected cost $C_2(T)$ of Model 2 when $1/\lambda = 60$ and $\alpha = 0.01$. This indicates that T^* decreases with c_d, however, increases with p. When p is large and c_d is small, T^* is large.

Table 6.1
Optimal checking time T^* to minimize $C_1(T)$.

$1/\lambda$	p	c_p/c_l				
		10	20	30	40	50
100	0.7	32	44	54	61	68
	0.8	35	49	59	67	74
	0.9	38	53	64	72	80
	1	42	57	69	78	86
200	0.7	45	64	77	89	99
	0.8	50	70	85	97	108
	0.9	55	77	93	106	117
	1	60	83	100	114	127
300	0.7	56	78	95	110	122
	0.8	62	86	105	120	134
	0.9	68	95	115	131	146
	1	74	103	125	143	158
400	0.7	65	91	111	127	142
	0.8	72	100	122	140	156
	0.9	79	110	134	153	170
	1	86	120	146	166	185

Table 6.2
Optimal checking time T_2^* to minimize $C_2(T)$ when $1/\lambda = 60$ and $\alpha = 0.01$.

c_d	p				
	1.0	0.9	0.8	0.7	0.6
1	69	62	58	55	52
5	38	35	32	30	29
10	29	26	24	23	22
15	24	22	20	19	18

Table 6.3 gives the optimal checking time T^* which minimizes the expected cost $C_3(T)$ of Model 3 when $c_{ps} = 10$ and $c_l = 1$. This indicates that T^* increases with $1/\lambda$, p_s, p_d and γ. When γ is large, that is, when the cost rate of one detailed check for the cost of one simple check is large, T^* is large.

Table 6.4 gives the optimal checking time T^* which minimizes the expected cost $C_4(T)$ of Model 4 when $c_{ps} = 10$ and $c_l = 1$. This indicates that T^* increases with $1/\lambda$, p_s, p_d and γ. Compared to Table 6.3, T^* is small. Moreover, when γ is small, T^* is similar to that of Table 6.4.

Table 6.3
Optimal checking time T^* to minimize $C_3(T)$.

$1/\lambda$	p_d	p_s	$\gamma(=c_{pd}/c_{ps})$				
			1	2	3	4	5
100	0.8	0.6	31	38	44	49	53
		0.7	33	40	46	51	56
		0.8	35	43	49	55	60
	0.9	0.7	34	42	48	53	58
		0.8	36	44	50	56	61
		0.9	38	47	54	60	65
200	0.8	0.6	45	55	63	70	77
		0.7	47	58	66	74	81
		0.8	50	61	70	78	86
	0.9	0.7	50	60	69	77	84
		0.8	52	63	73	81	88
		0.9	55	67	77	86	94
300	0.8	0.6	56	68	78	87	95
		0.7	59	71	82	91	100
		0.8	62	75	87	97	106
	0.9	0.7	61	75	86	95	104
		0.8	64	78	90	100	109
		0.9	68	83	95	106	116
400	0.8	0.6	65	79	90	101	110
		0.7	68	83	95	106	116
		0.8	72	87	101	112	123
	0.9	0.7	71	87	100	111	121
		0.8	75	91	105	117	127
		0.9	79	96	110	123	135

Table 6.5 gives the optimal checking time T^* which minimizes the expected cost $C_5(T)$ of Model 5 when $1/\theta = 10$, $c_r = 10$ and $c_d = 10$. This indicates that T^* decreases with α, however, increases with $1/\lambda$. This table shows that when α is large, T^* is small. In this case, we should perform the periodic check at small intervals.

Table 6.6 gives the optimal checking time T^* which minimizes the expected cost $C_5(T)$ of Model 5 when $1/\theta = 10$, $\alpha = 0.01$ and $c_d = 10$. This indicates that T^* decreases with c_r, however, increases with $1/\lambda$. This table shows that when c_r is large compared to c_p, T^* is small. In this case, we should perform the periodic check at small intervals.

Table 6.4
Optimal checking time T^* to minimize $C_4(T)$.

$1/\lambda$	p_d	p_s	$\gamma(=c_{pd}/c_{ps})$				
			1	2	3	4	5
100	0.8	0.6	30	34	38	42	45
		0.7	32	37	41	45	49
		0.8	35	40	45	49	53
	0.9	0.7	33	38	42	46	49
		0.8	35	41	45	49	53
		0.9	38	44	49	53	57
200	0.8	0.6	43	50	55	60	65
		0.7	46	53	60	65	70
		0.8	50	58	64	70	76
	0.9	0.7	48	55	61	66	72
		0.8	51	59	65	71	77
		0.9	55	63	71	77	83
300	0.8	0.6	53	61	68	75	80
		0.7	57	66	74	80	87
		0.8	62	71	79	87	94
	0.9	0.7	59	68	75	82	89
		0.8	63	73	81	88	95
		0.9	68	78	87	95	103
400	0.8	0.6	62	71	79	87	93
		0.7	66	77	85	93	101
		0.8	72	82	92	101	109
	0.9	0.7	68	79	88	96	103
		0.8	73	84	94	103	111
		0.9	79	91	101	111	119

Table 6.5
Optimal checking time T^* to minimize $C_5(T)$ when $1/\theta = 10$, $c_r = 10$ and $c_d = 10$.

$1/\lambda$	α			
	0.01	0.02	0.03	0.04
40	13	12	11	10
50	15	14	13	12
60	18	16	15	13
70	22	19	17	15
80	26	22	19	17

Table 6.6
Optimal checking time T^* to minimize $C_5(T)$ when $1/\theta = 10$, $\alpha = 0.01$ and $c_d = 10$.

$1/\lambda$	c_r			
	5	10	15	20
40	23	13	10	8
50	32	15	11	9
60	54	18	13	11

6.8 CONCLUSIONS

We have formulated stochastic models for a server system with a cyber attack. We have assumed that a cyber attack is checked by IDS, and the cost between attack occurrence and its detection increases with time linearly and non-linearly. In this chapter, we have considered type II errors where IDS judges the occurrence of a cyber attack erroneously when it occurs. Moreover, we have considered two types of checks, a simple check and a detailed check. Further, we have analyzed the case that a cyber attack is checked at both random and periodic times. We have derived the expected costs until a cyber attack is detected and have discussed the optimal checking time which minimizes them. Finally, to understand the results easily, we have given numerical examples of each model and have evaluated them for various standard parameters. In some, parameters are estimated from actual data, and we could select the best policy. From the numerical examples, we have shown that the optimal checking time decreases with the cost of the random check and parameters of the cost between attack occurrence and its detection, however, increases with the mean time to cyber attack occurrence and the probability that a cyber attack is detected by the check.

It would be very important to evaluate and improve the reliability of a server system with a cyber attack. The results derived in this paper would be applied in practical fields by making some suitable modification and extensions. Further studies for such subject would be expected.

REFERENCES

1. Ohta, K., & Mansfield, G. (2000). Illegal access detection on the internet-present status and future directions, *Transactions of Institute of Electronics, Information and Communication Engineers*, J83-B(9), 1209-1216.

2. Miyamoto, T., Kojima, A., Izumi, M., & Fukunaga, K. (2004). Network traffic anomaly detection using SVM, *Transactions of Institute of Electronics, Information and Communication Engineers*, J87-B(4), 593-598.

3. Sawada, A., Takakura, H., & Okabe, Y. (2003). A support system for monitoring log information of IDS applied to large scaled open networks, *Transactions of Information, Processing Society of Japan*, 44(8), 1861-1871.

4. Kanaoka, A., & Okamoto, E. (2003). The problems and countermeasures of intrusion detection exchange format in Internet engineering task force, *Transactions of Information Processing Society of Japan*, 44(8), 1830-1837.

5. Ohya, H., Miyaji, R., Kawaguchi, N., Shigeno, H., & Okada, K. (2004). A technique to reduce false positives of network IDS with machine learning, *Transactions of Information Processing Society of Japan*, 45(8), 2105-2112.

6. Osaki, S. (1992). *Applied Stochastic System Modeling*, Springer-Verlag, Berlin.

7. Yasui, K., Nakagawa, T., & Sandoh, H. (2002). Reliability models in data communication systems, in *Stochastic Models in Reliability and Maintenance*, (Osaki, S. Ed.) Springer-Verlag, Berlin.

8. Imaizumi, M., & Kimura, M. (2017). A two-stage policy for a server system with illegal access, *Proceedings of 23rd International Society of Science and Applied Technologies International Conference on Reliability and Quality in Design*, 176-180.

9. Imaizumi, M., & Kimura, M. (2019). Reliability analysis of a system with multiple-stage policy considering type I and type II errors, *International Journal of Reliability, Quality and Safety Engineering*, 26(6), (1950028)1-13.

10. Balow, R. E., & F. Proschan, F. (1965). *Mathematical Theory of Reliability*, JohnWiley and Sons, New York.

11. Nakagawa, T. (2005). *Maintenance Theory of Reliability*, Springer-Verlag, London.

12. Nakagawa T. (2014). *Random Maintenance Policies*, Springer-Verlag, London.

7 Reliability Analysis of Congestion Control Scheme with Code Error Correction Methods

Mitsutaka Kimura
Aichi University, Japan

CONTENTS

7.1 INTRODUCTION

Recently, congestion control schemes have been widely used in a radio environment and video stream transmission. In general, the window flow control scheme has been used as a classical method and provided a practical improvement method [5, 6]: A server sends the number of packets equivalent to the window size to a client [5]. If a router discards some packets because of network congestion, it sends half of the first window size. Then there is an issue that the throughput decreases rapidly. In order to deal with the issue, the congestion control executes by extending packet transmission intervals, which is called HpFP (High-performance and Flexible Protocol) [8, 9, 10, 11].

On the other hand, code error correction methods have been important techniques in data transmission. In general, when some packet losses occur, the client makes a request for loss packets and the server retransmits the only loss packets, which is called Selective Repeat ARQ [1, 2]. But it is pointed out that some delays occur in transmission. In order to prevent such delays, the server transmits error correction packets as well as data packets and the client restores the loss packets with the error correction packets, which is called FEC (Forward Error Correction). Moreover, if the client cannot restore all loss packets, it requests the remaining loss packets for the server, and the server transmits only loss packets to it, which is called Hybrid ARQ [3, 4].

In Section 7.2, we consider a reliability model of congestion control based on packet transmission interval with FEC. We formulate a stochastic model of a communication system in which a web server transmits data packets to a client. The server transmits n data packets which corresponds to a window size and m redundant packets of FEC. If less than m packets have failed, the client restores all loss packets by the redundant packets. If more than m packets have failed, a packet transmission interval is prolonged and all data packets retransmit again. Using techniques of Markov renewal processes, we derive the mean time until the transmission of n packets succeeds and discuss an optimal policy to maximize the number of packets per unit of time until the transmission succeeds (the throughput).

In Section 7.3, we consider an extended model of congestion control based on packet transmission interval with Hybrid ARQ. That is, the server transmits n data packets which corresponds to a window size and m redundant packets. If less than m packets have failed, the client restores all loss packets by the redundant packets. If more than m packets have failed, the client restores m loss packets and requests the remaining loss packets. If such retransmissions have been done at k times, a packet transmission interval is prolonged. We assume a communication system model is identical to Section 7.2 and derive the mean time until the transmission of n packets succeeds and discuss an optimal window size and an optimal number of redundant packets to maximize the throughput.

7.2 CONGESTION CONTROL SCHEME WITH FEC

This section formulates a stochastic model of congestion control based on packet transmission interval with FEC for data transmission. That is, the server transmits $n(n = 1, 2, \cdots)$ data packets which corresponds to a window size and $m(m = 1, 2, \cdots, n)$ redundant packets of FEC. If less than m packets have failed, the client restores all loss packets by the error correction packet. If more than m packets have failed, a packet transmission interval is prolonged and n packets retransmit again. The mean time until packet transmissions succeed is derived analytically, and the window size n^* which maximizes the number of packets per unit of time until the transmission succeeds is discussed.

7.2.1 RELIABILITY QUANTITIES

We consider a communication system in which a web server transmits data packets to a client. First, when the client requests data packets to the server, the connection between them is made, and the server transmits both n data packets and m redundant packets. A sample operation of the above communication system when $n = 2$ and $m = 1$ is shown in Figure 7.1. When less than m packets have failed, the client restores all loss packets by using the redundant packets. If more than m packets have failed, a packet transmission interval is prolonged and n data packets retransmit again.

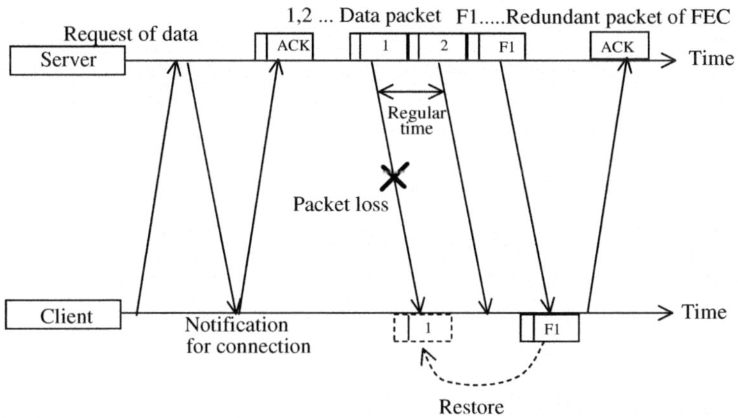

Figure 7.1 Operation of a communication system with the window size $n = 2$ and the number of redundant packets $m = 1$.

Then, we formulate the stochastic model as follows [11]:

(1) When a client requests data packets to a server, the connection between them is made according to a general distribution $A(t)$ with mean a.

(2) After the server establishes a connection, it successively transmits n data packets, which corresponds to a window size, and m redundant packets of FEC to the client.

   (i) The transmission interval of one packet requires the time according to a general distribution $B_1(t)$ with mean b_1.

   (ii) The server uses FEC as the protocol for the data transmission.

      (a) If less than m data packets transmission has failed, the client restores loss packets from the redundant packets. The time to restore a loss packet has a general distribution $C(t)$ with mean c, and the probability that a packet loss occurs is p_1 $(0 < p_1 < 1)$.

      (b) When the client has received n packets correctly, it returns ACK to the server.

(c) If more than m data packets transmission has failed, the client returns NAK to the server. The time to receive ACK or NAK has a general distribution $D(t)$ with mean d. The server interrupts the retransmission and retransmits $n+m$ packets whose time is longer than the regular time. That is, the transmission time needs more time according to a general distribution $B_2(t)$ with mean b_2 and $b_2 \geq b_1$, and the probability that a packet loss occurs is $p_2(p_2 \leq p_1 \leq 1)$.

(3) If more than m data packets transmission has failed again, the connection is made newly after constant time μ, where $G(t) \equiv 0$ for $t < \mu$ and 1 for $t \geq \mu$.
(4) If the server has received ACK of all packets n, the transmission succeeds.

Under the above assumptions, we define the following states of the system:

State 0: Connection between client and server starts at the time with $A(t)$.

State 1: Server transmits $n+m$ packets at the time with $B_1(t)$ to the client.

State 2: More than m data packets transmission has failed, and server transmits $n+m$ packets at the time with $B_2(t)$ again.

State F: More than m data packets transmission has failed again, and the connection is made newly between server and client after the time μ.

State S: n packets transmissions succeed.

The system states defined above form a Markov renewal process [13], where S is an absorbing state. A transition diagram between system states is shown in Figure 7.2.

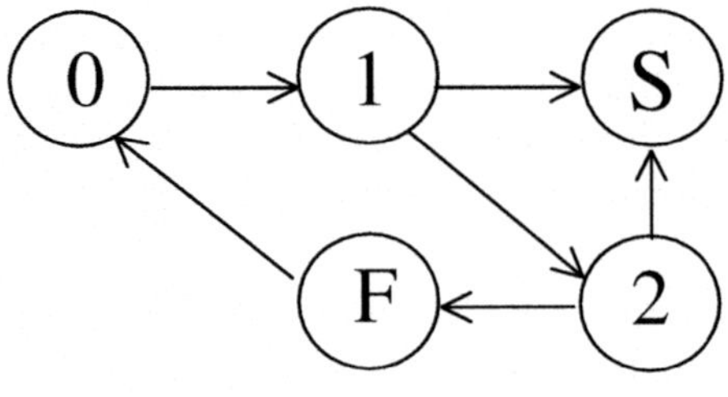

Figure 7.2 Transition diagram between system states.

The transition probabilities $Q_{i,\ell}(t)$ from State $i(i=1,2)$ to State $\ell(\ell = 2, F, S)$ are given by the following equations:

$$\begin{aligned}Q_{0,1}(t) &\equiv A(t), \quad Q_{F,0}(t) \equiv G(t),\\ Q_{i,S}(t) &\equiv \sum_{j=0}^{m}\binom{n}{j} p_i^j(1-p_i)^{n-j}B_i^{(n+m)}(t) * D(t) * C^{(j)}(t),\\ Q_{i,\ell}(t) &\equiv \sum_{j=m+1}^{n}\binom{n}{j} p_i^j(1-p_i)^{n-j}B_i^{(n+m)}(t) * D(t),\end{aligned}$$

where $\Phi^{(i)}(t)(i = 1, 2, \cdots)$ is the i-fold Stieltjes convolution of $\Phi(t)$, i.e., $\Phi^{(i)}(t) \equiv \int_0^t \Phi^{(i-1)}(t-u)d\Phi(u)(i = 1, 2, \cdots), \Phi^{(0)}(t) \equiv 1$ for $t \geq 0$, and the asterisk mark denotes the Stieltjes convolution with each other, i.e., $\Phi_1(t) * \Phi_2(t) \equiv \int_0^t \Phi_1(t-u)d\Phi_2(u)$ [12].

LS (Laplace-Stieltjes) transforms $Q_{i,\ell}^*(s)$ of transition probabilities $Q_{i,\ell}(t)$ are given by the following equations:

$$Q_{0,1}^*(s) \equiv A^*(s), \quad Q_{F,0}^*(s) \equiv G^*(s), \tag{7.1}$$

$$Q_{i,S}^*(s) \equiv \sum_{j=0}^{m}\binom{n}{j}[p_iC^*(s)]^j(1-p_i)^{n-j}B_i{}^*(s)^{n+m}D^*(s), \tag{7.2}$$

$$Q_{i,\ell}^*(s) \equiv \left[1-\sum_{j=0}^{m}\binom{n}{j}p_i^j(1-p_i)^{n-j}\right]B_i{}^*(s)^{n+m}D^*(s), \tag{7.3}$$

where $\Phi^*(s) \equiv \int_0^\infty e^{-st}d\Phi(t)$, i.e., $\int_0^\infty e^{-st}d[\Phi_1(t) * \Phi_2(t)] = \Phi_1^*(s)\Phi_2^*(s)$, and $Q_{i,S}^*(0) + Q_{i,\ell}^*(0) = 1$.

First, we derive the mean time $\ell_{0,S}(n)$ until the transmission of n packets succeeds. Letting $H_{0,S}^*(s)$ be LS transform of the first-passage distribution $H_{0,S}(t)$ from State 0 to State S,

$$\begin{aligned}H_{0,S}(t) &= \sum_{i=1}^{\infty}\left[Q_{0,1}(t) * Q_{1,2}(t) * Q_{2,F}(t) * Q_{F,0}(t)\right]^{(i-1)}\\ &\quad * Q_{0,1}(t) * \left[Q_{1,S}(t) + Q_{1,2}(t) * Q_{2,S}(t)\right],\end{aligned} \tag{7.4}$$

$$H_{0,S}^*(s) = \frac{Q_{0,1}^*(s)[Q_{1,S}^*(s) + Q_{1,2}^*(s)Q_{2,S}^*(s)]}{1 - Q_{0,1}^*(s)Q_{1,2}^*(s)Q_{2,F}^*(s)Q_{F,0}^*(s)}. \tag{7.5}$$

Hence, the mean time $\ell_{0,S}(n)$ is given by

$$\ell_{0,S}(n) \equiv \int_0^\infty t dH_{0,S}(t) = \lim_{s\to 0} -\frac{dH^*_{0,S}(s)}{ds}$$
$$= \frac{\begin{array}{l} a+d+\mu+(n+m)b_1+c\sum_{j=0}^{m}\binom{n}{j} jp_1^j(1-p_1)^{n-j} \\ +\sum_{j=m+1}^{n}\binom{n}{j} p_1^j(1-p_1)^{n-j}\Big[(n+m)b_2+d \\ \qquad +c\sum_{j=0}^{m}\binom{n}{j} jp_2^j(1-p_2)^{n-j}\Big]\end{array}}{1-\sum_{j=m+1}^{n}\binom{n}{j} p_1^j(1-p_1)^{n-j}\sum_{j=m+1}^{n}\binom{n}{j} p_2^j(1-p_2)^{n-j}} - \mu. \tag{7.6}$$

7.2.2 OPTIMAL POLICY

We propose an optimal policy which maximizes the number n of data packets per unit of time. That is, we calculate the throughput and derive an optimal window size to maximize it. We define throughput $E(n)$, which represents the rate of n packets to their mean transmission time given by

$$E(n) \equiv \frac{n}{\ell_{0,S}(n)}, \tag{7.7}$$

and seek optimal window size n^* to maximize $E(n)$ for given m. In order to simplify $E(n)$, we define $Z_1(n), X_i(n,m), Y_i(n,m) (i = 1,2)$ as follows:

$$Z_1(n) \equiv \ell_{0,S}(n)+\mu,$$
$$X_i(n,m) \equiv \sum_{j=m+1}^{n}\binom{n}{j} p_i^j(1-p_i)^{n-j},$$
$$Y_i(n,m) \equiv \sum_{j=0}^{m}\binom{n}{j} jp_i^j(1-p_i)^{n-j}.$$

Then, from (7.6),

$$Z_1(n) \equiv \frac{\begin{array}{c} a+d+\mu+(n+m)b_1+cY_1(n,m) \\ +X_1(n,m)\Big[(n+m)b_2+d+cY_2(n,m)\Big]\end{array}}{1-X_1(n,m)X_2(n,m)}.$$

We discuss optimal n^* to maximize $E(n)$. Putting formally that $A(n) \equiv 1/E(n)$, we seek n^* to minimize $A(n)$. Therefore, forming the inequality $A(n+1) - A(n) \geq 0$,

$$nZ_1(n+1) - (n+1)Z_1(n) + \mu \geq 0. \tag{7.8}$$

Denoting the left side of (7.8) by $L(n)$,

$$L(n+1) - L(n) = (n+1)[J_1(n+1) - J_1(n)],$$

where

$$J_1(n) \equiv Z_1(n+1) - Z_1(n).$$

If $J_1(n)$ is strictly increasing in n, $L(n)$ is strictly increasing in n from $L(1)$ to ∞, Therefore, we have the following optimal policy:

(i) If $L(1) < 0$ then there exists a finite and unique $n^*(> 1)$ which satisfies (7.8).

(ii) If $L(1) \geq 0$ then $n^* = 1$.

7.2.3 EXAMPLE 1

We compute numerically optimal window size n^* and optimal throughput $E(n^*)$. Suppose that the mean time b_1 until transmitting one packet is a unit of time. It is assumed that the mean time required for connection establishment is $a/b_1 = 10$, the mean time until transmitting one packet which is longer than the regular time is $b_2/b_1 = 2, 4$, the mean time for the server until receiving ACK or NAK is $d/b_1 = 2$, the probability that loss packets occur is $p_1 = 0.1, 0.2$ and $p_2 = 0.05, 0.1$, the number of redundant packets of FEC is $m = 8$, the mean time to restore one packet loss is $c/b_1 = 2, 4$ and the mean time until the connection is made again is $\mu/b_1 = 30$.

Table 7.1 gives optimal window size n^* to maximize the throughput $E(n)$, the mean time $\ell_{0,S}(n^*)$ and the throughput $E(n^*)$ by FEC.

For example, when $p_1 = 0.1, p_2 = 0.05, d/b_1 = 2$ and $b_2/b_1 = 2$, optimal window size $n^* = 45$, the mean time is $\ell_{0,S}(n^*) = 76.989$ and the throughput is $E(n^*) = 0.58450$. This indicates that n^* increase with c/b_1 but $E(n^*)$ and $\ell_{0,S}(n^*)$ decrease with it. $n^*, \ell_{0,S}(n^*)$ and $E(n^*)$ decrease with p_1, p_2 and b_2/b_1.

Table 7.2 gives optimal window size n^* to maximize the throughput $E(n)$, the mean time $\ell_{0,S}(n^*)$ and the throughput $E(n^*)$ by Selective Repeat ARQ when $d/c_1 = 2$ [9].

These indicate that $n^*, \ell_{0,S}(n^*)$ and $E(n^*)$ similarly show the above trends of Table 7.1. These values of Table 7.1 are larger than those of Table 7.2. Especially, the throughput $E(n^*)$ of Table 7.1 are larger than those of Table 7.2. That is, the throughput of FEC is much better than that of Selective Repeat ARQ.

Table 7.1
Optimal window size n^* to maximize $E(n)$ by FEC.

p_1	p_2	c/b_1	b_2/b_1	n^*	$\ell_{0,S}(n^*)$	$E(n^*)$
0.1	0.05	2	2	45	76.989	0.58450
			4	40	70.775	0.56517
		4	2	46	87.343	0.52666
			4	41	80.453	0.50962
	0.1	2	2	45	77.222	0.58273
			4	40	70.857	0.56452
		4	2	45	85.888	0.52394
			4	40	78.685	0.50836
0.2	0.1	2	2	26	59.723	0.43534
			4	22	52.967	0.41535
		4	2	27	71.867	0.37570
			4	23	64.064	0.35902
	0.2	2	2	25	57.907	0.43173
			4	22	53.082	0.41446
		4	2	25	67.463	0.37058
			4	22	61.676	0.35670

Table 7.2
Optimal window size n^* to maximize $E(n)$ by Selective Repeat ARQ when $d/b_1 = 2$.

p_1	p_2	b_2/b_1	n^*	$\ell_{0,S}(n^*)$	$E(n^*)$
0.1	0.05	2	5	27.214	0.18373
		4	4	25.746	0.15536
	0.1	2	4	26.070	0.15343
		4	3	22.662	0.13238
0.2	0.1	2	3	26.357	0.11382
		4	2	21.095	0.094810
	0.2	2	2	23.033	0.086832
		4	2	24.688	0.081013

7.3 CONGESTION CONTROL SCHEME WITH HYBRID ARQ

This section considers a stochastic model of congestion control based on packet transmission interval with Hybrid ARQ for data transmission. That is, if loss packets occur, the client restores loss packets by using the error correction packet. The client requests the remaining loss packets in which it cannot restore. If such retransmissions have been done at constant times, a packet transmission interval is prolonged. We assume the same communication system model as Section 7.2, and derive the mean time until packet transmissions succeed and discuss the window size which maximizes throughput $E(n)$.

7.3.1 RELIABILITY QUANTITIES

We consider a communication system in which a web server transmits data packets to a client: When $n = 2, k = 2$ and $m = 1$, the operation is shown in Figure 7.3. First, when the client requests data packets to the server, the connection between them is made. The server transmits n data packets and m redundant packets. When more than m packets have failed, the client restores m loss packets and requests the remaining loss packets. If such retransmissions have been done at $k(k = 1, 2, 3, \cdots)$ times, a packet transmission interval is prolonged. If less than m packets have failed, the client restores all loss packets by using the redundant packets. Then, we formulate the stochastic model in the same manner as Section 7.2. If the server has failed to retransmit at k times again, the connection is made newly after constant time μ, which has a degenerate distribution $G(t) \equiv 0$ for $t < \mu$ and 1 for $t \geq \mu$.

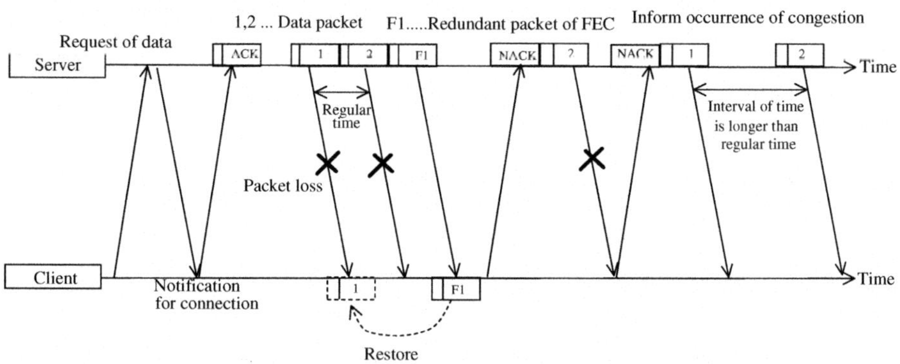

Figure 7.3 Operation of a communication system when the window size $n = 2$, the number of retransmissions $k = 2$ and the number of redundant packets $m = 1$.

We define the following states of a communication system:

State 0: Connection between client and server is made.

State 1: Server transmits $n+m$ packets according to the time $B_1(t)$ to client, or more than m data packets transmission has failed, and client restores m loss packets by using the redundant packets and requests the remaining loss packets to server.

State 2: When server fails to retransmit at k times, server transmits $n+m$ packets according to the time $B_2(t)$ again, or when more than m data packets transmission has failed, client restores m loss packets by using the redundant packets and requests the remaining loss packets to server.

State F: When server fails to retransmit at k times again, the connection is made newly after constant time μ.

State S: Transmission of all n packets succeeds.

The system states defined above form a Markov renewal process [13], where S is an absorbing state. A transition diagram between system states is shown in Figure 7.4.

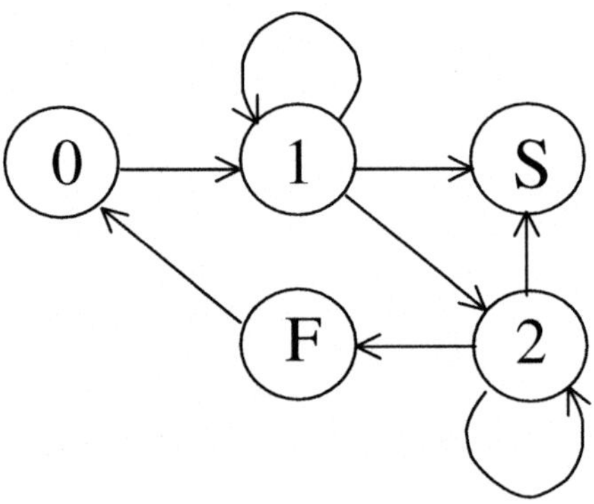

Figure 7.4 Transition diagram between communication system states.

The transition probabilities $Q_{i,\ell}(t)$ from State $i(i=0,1,2,F)$ to State $\ell(\ell=1,2,F,S)$ are given by the following equations:

$$
\begin{aligned}
&Q_{0,1}(t) \equiv A(t), \qquad Q_{F,0}(t) \equiv G(t), \\
&Q_{i,S}(t) \equiv (1-p_i)^n B_i^{(n+m)}(t) * D(t) \\
&+ \sum_{m_1=1}^{m} \binom{n}{m_1} p_i^{m_1}(1-p_i)^{n-m_1} B_i^{(n+m)}(t) * D(t) * C(t)^{(m_1)} \\
&+ \sum_{m_1=m+1}^{n} \binom{n}{m_1} p_i^{m_1}(1-p_i)^{n-m_1} B_i^{(n+m)}(t) * D(t) * C(t)^{(m)} \\
&*(1-p_i)^{m_1-m} B_i^{(m_1-m)}(t) * D(t) \\
&+ \sum_{m_1=m+1}^{n} \binom{n}{m_1} p_i^{m_1}(1-p_i)^{n-m_1} B_i^{(n+m)}(t) * D(t) * C(t)^{(m)}
\end{aligned}
$$

$$
\begin{aligned}
&* \sum_{m_2=1}^{m_1-m} \binom{m_1-m}{m_2} p_i^{m_2}(1-p_i)^{m_1-m-m_2} B_i^{(m_1-m)}(t) * D(t) \\
&*(1-p_i)^{m_2} B_i^{(m_2)}(t) * D(t) \\
&+\cdots
\end{aligned}
$$

$$
\begin{aligned}
&+ \sum_{m_1=m+1}^{n} \binom{n}{m_1} p_i^{m_1}(1-p_i)^{n-m_1} B_i^{(n+m)}(t) * D(t) * C(t)^{(m)} \\
&* \sum_{m_2=1}^{m_1-m} \binom{m_1-m}{m_2} p_i^{m_2}(1-p_i)^{m_1-m-m_2} B_i^{(m_1-m)}(t) * D(t) \\
&* \sum_{m_3=1}^{m_2} \binom{m_2}{m_3} p_i^{m_3}(1-p_i)^{m_2-m_3} B_i^{(m_2)}(t) * D(t) \\
&* \cdots \\
&* \sum_{m_{k-1}=1}^{m_{k-2}} \binom{m_{k-2}}{m_{k-1}} p_i^{m_{k-1}}(1-p_i)^{m_{k-2}-m_{k-1}} B_i^{(m_{k-2})}(t) * D(t) \\
&*(1-p_i)^{m_{k-1}} * B_i^{m_{k-1}}(t) * D(t) \qquad (i=1,2),
\end{aligned}
$$

$$
\begin{aligned}
Q_{i,j}(t) \equiv\ & \sum_{m_1=m+1}^{n} \binom{n}{m_1} p_i^{m_1}(1-p_i)^{n-m_1} B_i^{(n+m)}(t) * D(t) * C(t)^{(m)} \\
&* \sum_{m_2=1}^{m_1-m} \binom{m_1-m}{m_2} p_i^{m_2}(1-p_i)^{m_1-m-m_2} B_i^{(m_1-m)}(t) * D(t) \\
&* \cdots \\
&* \sum_{m_{k-1}=1}^{m_{k-2}} \binom{m_{k-2}}{m_{k-1}} p_i^{m_{k-1}}(1-p_i)^{m_{k-2}-m_{k-1}} B_i^{(m_{k-2})}(t) * D(t) \\
&*[1-(1-p_i)^{m_{k-1}}] B_i^{(m_{k-1})}(t) * D(t) \qquad (i=1,2,\ j=2,F).
\end{aligned}
$$

LS transforms $Q_{i,\ell}^*(s)$ of transition probabilities $Q_{i,\ell}(t)$ are given by the following equations:

$$Q_{0,1}^*(s) \equiv A^*(s), \qquad Q_{F,0}^*(s) \equiv G^*(s),$$

$$Q_{i,S}^*(s) \equiv \sum_{j=0}^{m} \binom{n}{j} [p_i C^*(s)]^j (1-p_i)^{n-j} B_i{}^*(s)^{n+m} D^*(s)$$

$$+\sum_{\ell=2}^{k} \sum_{j=m+1}^{n} \binom{n}{j} [p_i B_i{}^*(s)]^j (1-p_i)^{n-m} B_i{}^*(s)^n C^*(s)^m$$

$$\times D^*(s)^\ell \left\{ \left[\sum_{r=0}^{\ell-2} [p_i B_i{}^*(s)]^r\right]^{j-m} - \left[\sum_{r=0}^{\ell-3} [p_i B_i{}^*(s)]^r\right]^{j-m} \right\}$$

$$(i=1,2),$$

$$Q_{i,\ell}^*(s) \equiv \sum_{j=m+1}^{n} \binom{n}{j} (p_i B_i{}^*(s))^j (1-p_i)^{n-m} B_i{}^*(s)^n C^*(s)^m$$

$$\times D^*(s)^k \left\{ \left[\sum_{r=0}^{k-3} [p_i B_i{}^*(s)]^r + \frac{[p_i B_i{}^*(s)]^{k-2}}{1-p_i}\right]^{j-m} - \left[\sum_{r=0}^{k-2} [p_i B_i{}^*(s)]^r\right]^{j-m} \right\}$$

$$(i=1,2,\ell=2,F),$$

where $Q_{i,S}^*(0) + Q_{i,\ell}^*(0) = 1$.

We derive the mean time $\ell_{0,S}(n)$ until the transmissions of n packets succeeds. Letting $H_{0,S}^*(s)$ be LS transform of the first-passage distribution $H_{0,S}(t)$ from State 0 to State S,

$$H_{0,S}^*(s) = \frac{Q_{0,1}^*(s)[Q_{1,S}^*(s) + Q_{1,2}^*(s)Q_{2,S}^*(s)]}{1 - Q_{0,1}^*(s)Q_{1,2}^*(s)Q_{2,F}^*(s)Q_{F,0}^*(s)}. \tag{7.9}$$

Hence, the mean time $\ell_{0,S}(n,m,k)$ is given by

$$\begin{aligned} \ell_{0,S}(n,m,k) &\equiv \int_0^\infty t dH_{0,S}(t) = \lim_{s\to 0} -\frac{dH_{0,S}^*(s)}{ds} \\ &= \frac{a + W_1(n,m,k) + X_1(n,m,k)W_2(n,m,k) + \mu}{1 - X_1(n,m,k)X_2(n,m,k)} - \mu \end{aligned} \tag{7.10}$$

where

$$X_i(n,m,k) \equiv \sum_{j=m+1}^{n} \binom{n}{j} p_i^j (1-p_i)^{n-j}[1-(1-p_i^{k-1})^{j-m}],$$

$$W_i(n,m,k) \equiv \sum_{j=0}^{m} \binom{n}{j} p_i^j (1-p_i)^{n-j}[(n+m)b_i + jc + d]$$

$$+\sum_{j=m+1}^{n}\binom{n}{j}p_i^j(1-p_i)^{n-j}$$

$$\times\left[(n+j)b_i+mc+kd+d\sum_{r=0}^{k-3}(1-p_i^{r+1})^{j-m}+b_i(j-m)\frac{p_i(1-p_i^{k-2})}{1-p_i}\right],$$

$$(k=2,3,\cdots).$$

7.3.2 OPTIMAL POLICY

We define throughput $E(n,m,k)$ given by

$$E(n,m,k)\equiv\frac{n}{\ell_{0,S}(n,m,k)},\qquad(7.11)$$

and seek optimal window size n^* to maximize $E(n,m,k)$. Then, from (7.10),

$$Z(n,m,k)\equiv\frac{a+W_1(n,m,k)+X_1(n,m,k)W_2(n,m,k)+\mu}{1-X_1(n,m,k)X_2(n,m,k)}.$$

We discuss optimal n^* to maximize $E(n,m,k)$ for given m and k. Putting formally that $A(n,m,k)\equiv 1/E(n,m,k)$, we seek n^* to minimize $A(n,m,k)$. Therefore, forming the inequality $A(n+1,m,k)-A(n,m,k)\geq 0$,

$$nZ(n+1,m,k)-(n+1)Z(n,m,k)+\mu\geq 0\quad(k=2,3,\cdots).\qquad(7.12)$$

Denoting the left side of (7.12) by $L(n,m,k)$,

$$L(n+1,m,k)-L(n,m,k)=(n+1)[J(n+1,m,k)-J(n,m,k)],$$

where

$$J(n,m,k)\equiv Z(n+1,m,k)-Z(n,m,k).$$

If $J(n,m,k)$ is strictly increasing in n, $L(n,m,k)$ is strictly increasing in n from $L(1,m,k)$ to ∞, Thus, there exists a finite and unique $n^*(>1)$ which satisfies (7.12).

7.3.3 EXAMPLE 2

We compute numerically optimal window size n^* and optimal throughput $E(n^*,m,k)$. Suppose that the mean time b_1 until transmitting one packet is a unit of time. It is assumed that the mean time required for connection establishment is $a/b_1=10$, the number of retransmission for detecting network congestion is $k=3,4$, the mean time until transmitting one packet which is longer than the regular time is $b_2/b_1 = 2, 4$, the mean time for the server until receiving ACK or NAK is $d/b_1 = 2, 4$, the probability that loss packets occur is $p_1 = 0.1$ and $p_2 = 0.05$, the number of redundant packets of FEC is $m = 1 \sim 30$, the mean time to restore one packet loss is $c/b_1 = 2$ and the mean time until the connection is made again is $\mu/b_1 = 30$.

Table 7.3 gives optimal window size n^* to maximize the throughput $E(n, m, k)$, the mean time $\ell_{0,S}(n^*, m, k)$ and the throughput $E(n^*, m, k)$ by Hybrid ARQ when $c/b_1 = 2$ and $m = 8$. For example, when $a/b_1 = 10, p_1 = 0.1, p_2 = 0.05, \mu/b_1 = 30, d/b_1 = 2, b_2/b_1 = 2, c/b_1 = 2$ and $m = 8$, optimal window size $n^* = 115$, the mean time is $\ell_{0,S}(n^*, m, k) = 168.21$ and the throughput is $E(n^*, m, k) = 0.68368$. These indicate that $n^*, \ell_{0,S}(n^*, m, k)$ and $E(n^*, m, k)$ similarly show the above trends of Table 7.1. By comparison with Table 7.1, and these values of Table 7.3 are larger than those of Table 7.1. That is, the throughput $E(n^*)$ of Hybrid ARQ are much better than those of FEC.

Table 7.3
Optimal window size n^* to maximize $E(n, m, k)$ by Hybrid ARQ.

k	p_1	p_2	d/b_1	b_2/b_1	n^*	$\ell_{0,S}(n^*, m, k)$	$E(n^*, m, k)$
3	0.1	0.05	2	2	115	168.21	0.68368
				4	84	126.65	0.66327
			4	2	123	186.55	0.65935
				4	81	126.17	0.64199
		0.1	2	2	107	157.19	0.68069
				4	83	125.33	0.66224
			4	2	105	160.11	0.65582
				4	80	124.79	0.64109
	0.2	0.1	2	2	41	83.072	0.49355
				4	35	73.390	0.47691
			4	2	39	82.852	0.47072
				4	34	74.350	0.45730
		0.2	2	2	39	79.565	0.49016
				4	34	71.530	0.47533
			4	2	37	79.082	0.46787
				4	33	72.404	0.45578

Figure 7.5 draws the optimal throughput $E(n^*, m, k)$ for m when $k = 2, a/b_1 = 10, b_1/b_2 = 2, d/b_1 = 2, p_1 = 0.1, p_2 = 0.05, \mu/b_1 = 30$ and $c/b_1 = 2$. From this figure, $E(n^*, m, 2)$ increase with m and converges to $E(n^*, m^*, 2)$ when $m^* = 25$. That is, it should be enough that the number of redundant packets of FEC is 25 when the optimal window size n^* is 169.

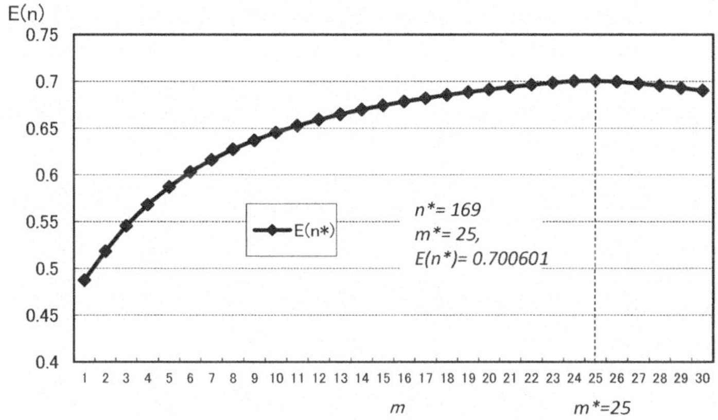

Figure 7.5 Optimal throughput $E(n^*, m, k)$ for m when $k = 2, a/b_1 = 10, b_1/b_2 = 2, d/b_1 = 2, p_1 = 0.1, p_2 = 0.05, \mu/b_1 = 30$ and $c/b_1 = 2$.

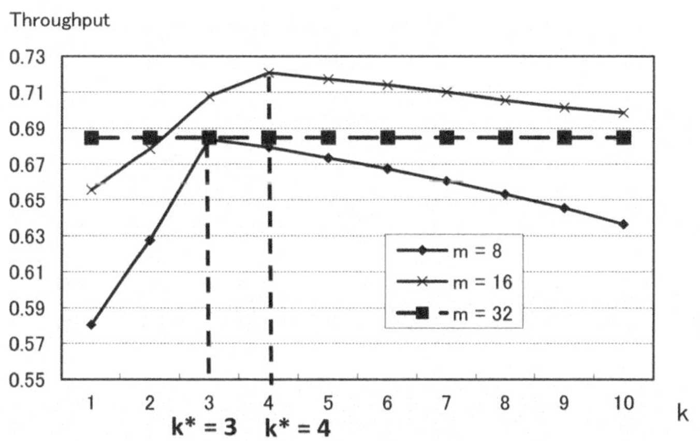

Figure 7.6 Optimal throughput $E(n^*)$ for k and m when $a/b_1 = 10, b_1/b_2 = 2, d/b_1 = 2, p_1 = 0.1, p_2 = 0.05, \mu/b_1 = 30$ and $c/b_1 = 2$.

Figure 7.6 draws the optimal throughput $E(n^*, m, k)$ for k and m when $a/b_1 = 10, b_1/b_2 = 2, d/b_1 = 2, p_1 = 0.1, p_2 = 0.05, \mu/b_1 = 30$ and $c/b_1 = 2$. From this figure, $E(n^*, m, k)$ increase with k and converges to $E(n^*, 16, k^*)$ when $k^* = 4$ and $m = 16$. That is, it should be enough that the number of retransmissions for detecting network congestion is 4 when the optimal window size n^* is 164.

7.4 CONCLUSIONS

This chapter has analytically discussed two models of congestion control based on packet transmission interval in Sections 7.2 and 7.3. Moreover, we have derived the mean time until the transmissions of window size succeeds by using Markov renewal processes, and have discussed the optimal policies to maximize the throughput.

From numerical examples, we have shown that the optimal window size and throughput decrease with the mean time until transmitting one packet which is longer than the regular time. That is, we should transmit at double one packet interval when the probability that loss packets occur is large. The optimal throughput of FEC is much better than that of Selective Repeat ARQ. Furthermore, the optimal throughput of Hybrid ARQ are much better than those of FEC.

REFERENCES

1. Anagnostou, M., & Protonotarios, E. (1986) Performance Analysis of the Selective Repeat ARQ Protocol, *IEEE Transactions on Communications, 34(2),* 127-135.

2. Shikama, T., Watanabe, T., & Mizuno, T. (2005) Delay analysis of the selective-repeat ARQ with the per flow resequencing, *IEEE International Conference on Communications, 1,* 26-32.

3. Zhao, B., & Valenti, M.C. (2005). Practical relay networks: a generalization of hybrid-ARQ, *IEEE Journal on Selected Areas in Communications, 23(1),* 7-18.

4. Tanaka, H., & Takahashi, Y. (2004). Modeling and Performance Analysis of Realtime Hybrid FEC/ARQ, *IEICE Technical Report. Information Networks, 104(121),* 13-18.

5. Jacobson, V. (1988). Congestion avoidance and control, *Computer Communication Review, 18(4),* 314-329.

6. Stevens, W. (1997). TCP slow start, congestion avoidance, fast retransmit and fast recovery algorithms, *RFC2001.*

7. Floyd, S. (1994). TCP and explicit congestion notification, *ACM Comput. Commun. Rev. 24(5),* 10-23.

8. Murata, T., Mizuhara, T., Takagi, A., Fukushima, K., Yamamoto, K., Nagaya, Y., Muranaga, K., & Kimura, E. (2016). HpFP: A new protocol for LFNs with packet-loss based on UDP: A basic concept and detailed design of the protocol, *IEICE Technical Report, 115(484), IN2015-109,* 7-12.

9. Kimura, M., Imaizumi, M., & Nakagawa, T. (2017). Reliability Modelling of Window Flow Control Scheme for a Communication System based on Packet Transmission Interval, *Proceedings of 23th ISSAT International Conference on Reliability and Quality in Design,* 171-175.

10. Kimura, M., Imaizumi, M., & Nakagawa, T. (2018). Reliability Modelling of Window Flow Control Scheme based on Packet Transmission Interval with Explicit Congestion Notification, *Proceedings of 24th ISSAT International Conference on Reliability and Quality in Design*, 72-76.

11. Kimura, M., Imaizumi, M., & Nakagawa, T. (2019). Optimal Policy of Window Flow Control Based on Packet Transmission Interval with Explicit Congestion Notification, *International Journal of Reliability, Quality and Safety Engineering, 26(5),* 1950024-1 - 1950024-14, (14pages), DOI:10.1142/S0218539319500244.

12. Osaki, S. (1992). *Applied Stochastic System Modeling*, Springer-Verlag, Berlin.

13. Nakagawa, T. (2011). *Stochastic Processes with Applications to Reliability Theory*, Springer, London.

Section III

Warranty Analysis Manufacturing

8 The Optimal Design of Consecutive-k Systems

Hisashi Yamamoto
Tokyo Metropolitan University, Japan

Tomoaki Akiba
Chiba Institute of Technology, Japan

Taishin Nakamura
Tokai University, Japan

Lei Zhou
Yamaguchi University, Japan.

CONTENTS

8.1 INTRODUCTION

The achievements of science and technology have permeated our world in the form of systems. The various benefits accompanying these systems have allowed us to enjoy an affluent life we may not have had otherwise. However, with further advancements in science and technology, the systems that support our society and daily lives are becoming increasingly complex and

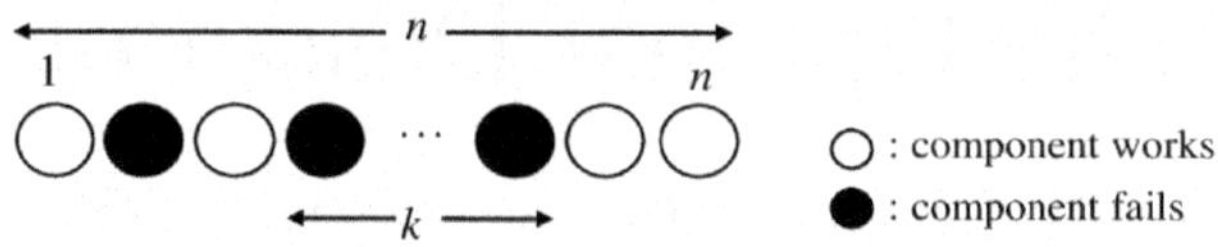

Figure 8.1 Example of a failed consecutive-k-out-of-n:F system.

extensive. Impairments in the functions and safety of these systems, due to abrupt failures or malfunctions, would significantly impact our lives and society. Several news articles on system failures at banks and security companies, and in transportation systems, among others, come up when searching for "system failures" online. There is an increasing importance of improving and maintaining the reliability of the "systems" around us.

Herein, we focus on a consecutive-k system, wherein the system fails (operates) when a part of it fails (operates), and provide an overview of the research (particularly the research of the authors) surrounding its design and maintenance.

This chapter is organized as follows.

"Consecutive-k system" is an all-inclusive term for the systems described above, and many types of systems are called as such. In Section 8.2, we introduce a few typical consecutive-k systems along with an overview of each. We consider three problems in consecutive-k systems. Studies aiming to obtain system reliabilities, when component reliabilities are provided, were conducted as far back as the 1960s. However, the computational load increases exponentially with the number of components. Therefore, as the first problem, in Section 8.3, we discuss the efficient calculation of system reliability and several lower and upper bounds. Additionally, we consider how to approximate values from the limit theorems. As the second problem, in Section 8.4, we consider the component assignment problem (CAP), wherein we obtain the optimal arrangement with the maximum system reliability when all of the component reliabilities are not necessarily the same. The third problem is a maintenance problem. In Section 8.5, we discuss how to obtain an optimal maintenance policy that minimizes the expected costs for one-dimensional consecutive-k systems. Lastly, a conclusion for this chapter is presented in Section 8.6.

8.2 CONSECUTIVE-K SYSTEMS

The study of consecutive-k systems began when Kontoleon [21] proposed a linear consecutive-k-out-of-n:F system. As shown in Figure 8.1, this system, consisting of n components arranged in a single line, fails when k or more consecutive components fail. Such a system can be applied to evaluate street lighting systems, pipeline systems, sensor systems, etc.

Table 8.1
Classifying the consecutive-k systems by system shape.

System Form		Connecting Components	Applications	Figure	Remarks
1-D	linear	adjacent	automatic flap gate turnstile system, street lighting system, pump of pipeline system	Figure 1	
	circular	adjacent	network by token ring		Connecting both ends of the linear system
2-D	linear (rectangular)	lattice	display system, lighting system of auditorium, fire (or water supply) sprinkler system	Figure 2(a)	Structure in which components are connected in a lattice
		triangular lattice		Figure 10 in Akiba et al. [4]	Structure in which components are connected triangle by 3 components
	circular (cylindlical)	lattice	sensor system of reactor	Figure 2(b)	Connecting both sides of the rectangular system. Similar type of radiation circle in Figure 2(c)
		triangular lattice		Figure 2 in Akiba et al. [11]	Connecting both sides of a rectangular (triangular lattice) system
	toroidal	lattice	processor architecture	Figure 2(d)	Connecting both sides, top and bottom, of a rectangular system
	polyhedral	triangular lattice	satellite layout of global positioning system, etc.	Figure 5 in Akiba et al. [11]	Connecting any sides of a triangular lattice system
3-D	rectangular solid	lattice	multilayer ceramic capacitor, etc.	Figure 11 in Akiba et al. [4]	Structure in which linear connected lattice systems are stacked multiple layers

Following Kontoleon's paper, many systems falling into the category of consecutive-k systems have been proposed. Consecutive-k systems are classified using four characteristics: 1) system shape, 2) shape of the set of failed components leading to system failure (hereafter, "fault shape"), 3) number of component or system states (operating, failure, or another intermediate state), and 4) dependence of the system component states on the surrounding component states.

Table 8.1 lists the results of classifying consecutive-k systems according to the system shape. A linear consecutive-k-out-of-n:F system with both ends connected is called a "circular consecutive-k-out-of-n:F system." If these two systems are regarded as one-dimensional systems, two-dimensional systems extending them to two dimensions also exist. Such systems are classified as "linear connected (rectangular)," "circular connected (cylindrical or radiation circle)," and "toroidal," as shown in Figure 8.2.

Table 8.2 lists the results of classifying consecutive-k systems by the fault shape. Each consecutive-k system can be expressed as a combination of the aforementioned shape of the system and the fault shape. The various

Table 8.2
Classifying the consecutive-k systems by fault shape.

Shape		Set of Failed Components	Figure	Remarks
1-D	consecutive-k	at least k consecutive components fail	Figure 1	
	k-within-r	at least k components fail in r consecutive components		
	m-consecutive-k	at least m-overlapping runs of k consecutive failed components in system		
	(n,f,k)	at least k consecutive components fail or at least f components fail in system		
	$\langle n,f,k \rangle$	at least k consecutive components fail and at least f components fail in system		
2-D	(r,s)	all components in the rectangle with r rows and s columns fail	Figure 2 (a)	
	k-within-(r,s)	at least k components fail in the rectangle with r rows and s columns	Figure 9 in Akiba et al. [4]	
	(r_1,s_1)-or-(r_2,s_2)-...	r_1 rows and s_1 columns fail or r_2 rows and s_2 columns fail or...	Figure 5 in Akiba et al. [4]	(1,2)-or-(2,1) is an example
	(r_1,s_1)-and-(r_2,s_2)-...	r_1 rows and s_1 columns fail and r_2 rows and s_2 columns fail and...	Figure 5 in Yamamoto and Akiba [7]	The shape of triangular lattice is an example
	$\cup X_\theta$	all components in the fault shape X_1 or X_2...	Figure 7 in Akiba et al. [4]	The shape of triangular lattice is an example
3-D	(r_1,r_2,r_3)	all components fail in $r_1 \times r_2 \times r_3$ rectanguler solid		
	k-within-(r_1,r_2,r_3)	at least k components fail in $r_1 \times r_2 \times r_3$ rectanguler solid	Figure 11 in Akiba et al. [4]	

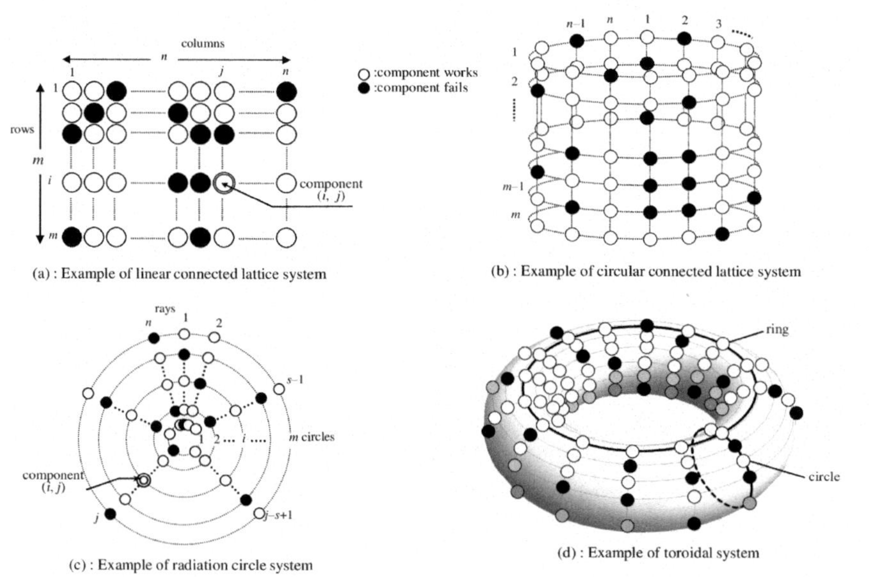

Figure 8.2 Examples of two-dimensional consecutive-k-out-of-n:F systems.

shapes of the systems and failure (operating) shapes of one-dimensional systems are extensively described in the survey papers listed below. Chap. 1 of Chang *et al.* [9], Chap. 9 of Kuo and Zuo [24], and Chap. 2 of Akiba *et al.* [3].

We now introduce the various shapes of the systems and fault shapes of two-dimensional systems. A two-dimensional consecutive-k system is called a "linear connected-**X**-out-of-(m, n):F lattice system," where (m, n) implies that the components of the system are arranged in a grid, forming a rectangle with m rows and n columns. **X** denotes the fault shape leading to system failure. Salvia and Lasher [36] proposed a linear connected-(r, s)-out-of-(m, n):F lattice system, where **X** is (r, s), i.e., the failure condition is that all components in the rectangle with r rows and s columns fail. Table 8.2 lists other fault shapes (conditions) for two-dimensional systems. Boehme *et al.* [6] defined the several complex fault shapes of two-dimensional systems called "connected-**X**." The two-dimensional systems above were summarized by Yamamoto and Akiba [43] and Cui and Dong [10].

Furthermore, the following papers focused on three-dimensional systems. Godbole *et al.* [16], Boushaba and Ghoraf [7], Akiba *et al.* [5], and Gharib *et al.* [15], etc.

Multi-state systems also exist within consecutive-k systems. In a multi-state system, which has been studied since the 1970s, both the system and components are allowed to be in $M+1$ possible states ranging from $0, 1, \ldots, M$. Here, M is a positive integer representing a system or component in a perfectly working state, while zero denotes a complete failure state. A reliability model for a multi-state system increases flexibility in modeling equipment conditions. Definitions of the multi-state consecutive-k-out-of-n:F and G systems were proposed by Huang *et al.* [19].

8.3 RELIABILITIES OF CONSECUTIVE-K SYSTEMS

8.3.1 SYSTEM RELIABILITY

Two main efficient approaches to obtaining the reliabilities of consecutive-k systems exist. The first is a recursive equation approach (REA), wherein the system reliability is expressed using a recursive equation, namely the system reliability for a large system is expressed using the reliabilities of smaller systems (wherein the states of certain components are determined). The other is the finite Markov chain imbedding approach (FMCIA), wherein certain states of a system are regarded as the states of a Markov process, the time of this process denotes the system size, and the system reliability is calculated by obtaining the transition probability of tracing paths in this process.

(1) Recursive Equation Approach (REA)

Hwang [20] first proposed a recursive equation for calculating the reliability of a consecutive-k-out-of-n:F system.

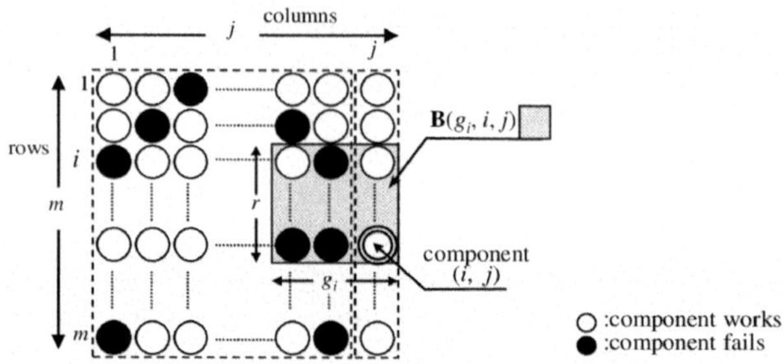

Figure 8.3 $R(r, s, m, j; \mathbf{G})$.

For two-dimensional systems, Yamamoto and Miyakawa [49] proposed a recursive equation for calculating the reliability of a linear connected-(r, s)-out-of-(m, n):F lattice system in the independent but not identically distributed (INID) case[1]. We define the following notations for describing this method.

p_{ij}: the reliability of a component (i, j).

δ_{ij}: index function that takes 1 when components $(i - r + 1, j), (i - r + 2, j), \ldots, (i, j)$ all fail and 0 otherwise.

$\mathbf{G}$: $(m - r + 1)$-dimensional vector, $(g_1, g_2, \ldots, g_{m-r+1})$.

$B(g_i, i, j)$: event that all components fail in an $r \times g_i$ rectangle with component (i, j) at its lower right corner.

$R(r, s, m, j; \mathbf{G})$: reliability of a linear connected-(r, s)-out-of-(m, j):F lattice system given that $B(g_i, i, j)$ does not occur for $i = r, r + 1, \ldots, m$, as shown in Figure 8.3, for each j.

Here, $\mathbf{G}$ is a vector representing the state of the system.

Using the notation above, we obtain

$$R(r, s, m, j; \mathbf{G}) =$$
$$\begin{cases} \sum_{\mathbf{x}_n} R(r, s, m, j - 1; \mathbf{G}') \prod_{i=1}^{m} \left(p_{ij}^{1-x_{ij}} (1 - p_{ij})^{x_{ij}} \right), & \text{if } i \geq \min_{\mathbf{G}} g_i \geq 1, \\ 1, & \text{if } \min_{\mathbf{G}} g_i \geq i \leq 1, \\ 0, & \text{if } \min_{\mathbf{G}} g_i = 0, \end{cases} \tag{8.1}$$

where $\mathbf{G}'$ is a $(m - r + 1)$-dimensional vector $(g'_1, g'_2, \ldots, g'_{m-r+1})$, and

$$g'_i = \begin{cases} g_i - 1, & \text{if } \delta_{ij} = 1, \\ s, & \text{otherwise.} \end{cases} \tag{8.2}$$

Additionally, $\mathbf{x}_j = (x_1 j, x_2 j, \ldots, x_{mj})$, a 0-1 binary vector representing the state of the column j, is given by δ_{ij}s. The following equation then holds.

$$R^L(r, s, m, n; [p_{ij}]) = R(r, s, m, n; \mathbf{s}), \tag{8.3}$$

where $\mathbf{s}$ is an $(m - r + 1) \times 1$ vector with all s's.

[1]The component lifetimes are independent but not identically distributed.

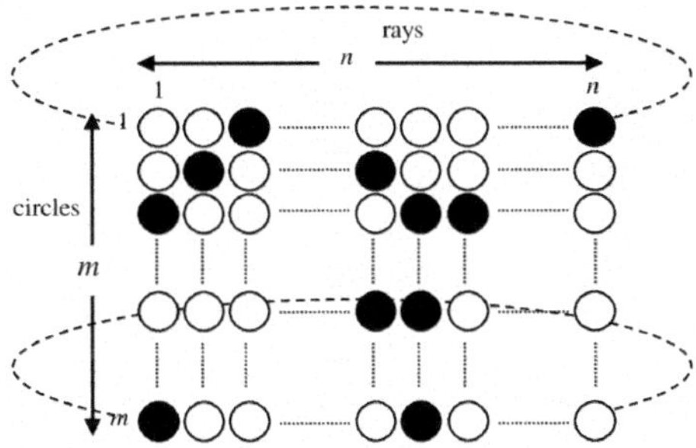

Figure 8.4 Conversion from a circular to a linear system.

The total time required to evaluate the system reliability using the REA above is $O(s^{(m-r)}m^2rn)$. Namely, its order is exponential for m but linear for r and n. Therefore, the computation time likely increases rapidly with m. We investigated the computation time for the linear connected-(2, 2)-out-of-(6, 6):F lattice system using C language with Visual Studio 2019. This was executed on a computer with Intel Xeon E3 3.0 GHz CPU, 32.0 GB memory, and a Windows 10 OS. Subsequently, Yamamoto and Miyakawa's REA took approximately 1 ms compared to the 7693 s (approximately 128 min) the enumeration method took.

Yamamoto [50] proposed the REA for obtaining the reliability of a linear connected-(r_1, s_1)-or-(r_2, s_2)-...-or-(r_k, s_k)-out-of-(m, n):F lattice system in the INID case by extending the concept above. Akiba and Yamamoto [4] proposed an REA for obtaining the reliability of a linear and circular connected k-within-rectangular connected-(r, s)-out-of-(m, n):F lattice system in the INID case using the same concept as that of Yamamoto and Miyakawa's algorithm. Huang *et al.* [19] proposed an exact algorithm to evaluate the system state distribution of a decreasing multi-state consecutive-k-out-of-n:F system. Zuo *et al.* [59] evaluated the system state distribution of a decreasing multi-state consecutive-k-out-of-n:G system. Furthermore, Yamamoto *et al.* [52] and Akiba *et al.* [2] proposed an REA for obtaining the system state distribution of increasing multi-state consecutive-k-out-of-n:F and G systems.

Yamamoto and Miyakawa [41] proposed the REA for circular systems in the INID case with the same concept as that of Yamamoto and Miyakawa's algorithm. The circular system becomes a linear connected one when the system is cut in the ray side direction. Thus, Yamamoto and Akiba [46] improved on Yamamoto and Miyakawa's REA.

First, we cut the circular system between rays 1 and n and treat it as a linear connected-(r, s)-out-of-(m, n):F lattice system, as shown in Figure 8.4. This cut system differs, however, from an ordinary linear connected-(r, s)-out-of-(m, n):F lattice system because, in addition to ordinary failure conditions, the system will also fail if all the components within an $r \times s$ grid on $2(s-1)$ rays consisting of rays "$n-s+2$ to n" and "rays 1 to $s-1$" fail. Namely, this

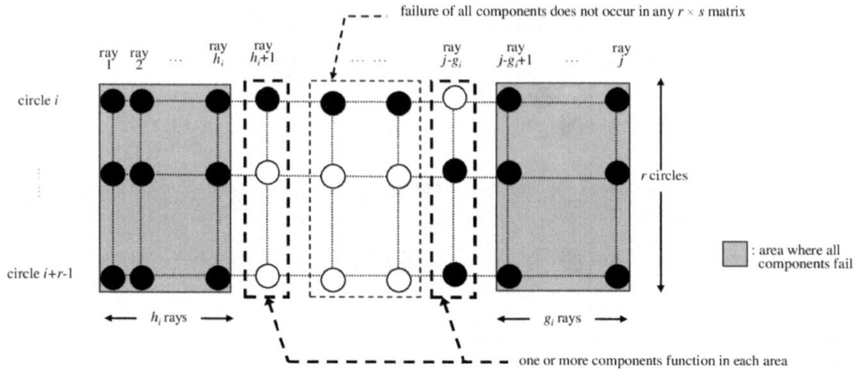

Figure 8.5 $B(h_i, g_i, i, j)$.

system becomes a linear connected-(r, s)-out-of-(m, n):F lattice system with component states on both ends provided, which is called a "special" linear connected-(r, s)-out-of-(m, n):F lattice system. By acquiring the reliability of this special system we can obtain the reliability of a circular connected-(r, s)-out-of-(m, n):F lattice system. By tracing this idea, the system reliability can be expressed as in Equation (8.5) with the following notations.

For $i = 1, 2, \ldots, m - r + 1$ and $j = 1, 2, \ldots, n$,

$B(h_i, g_i, i, j)$: event that a linear connected-(r, s)-out-of-$(r, j - h_i - g_i - 2)$:F lattice system functions with components $(i, h_i+2), (i+r-1, h_i+2), (i+r-1, j-g_i-1)$ and $(i, j-g_i-1)$ as the corners, and all components fail in an $r \times h_i$ rectangle and $r \times g_i$ rectangle at both ends, as shown in Figure 8.5.

$\mathbf{H}$: $(m - r + 1)$-dimensional nonnegative integer vector $(h_1, h_2, \ldots, h_{m-r+1})$ for $h_i = 0, 1, \ldots, s - 1$.

$\mathbf{G}$: $(m - r + 1)$-dimensional nonnegative integer vector $(g_1, g_2, \ldots, g_{m-r+1})$ for $g_i = 0, 1, \ldots, s - 1$.

$\mathbf{X}^{(i)}$: i-dimensional 0-1 binary vector $(x_1, x_2, \ldots, x_i)$, where $x_i = 1$ denotes all components $(i, j), (i + 1, j), \ldots, (i + r - 1, j)$ fail, $x_i = 0$ otherwise.

$F_j(\mathbf{X}^{(i)})$: probability that the status of ray j can be expressed $\mathbf{X}^{(i)}$.

$R(r, s, m, j; \mathbf{H}; \mathbf{G})$: reliability of a linear connected-(r, s)-out-of-(m, j):F lattice system given that $B(h_i, g_i, i, j)$ does not occur for each j.

Here, $R(r, s, m, j; \mathbf{H}; \mathbf{G})$ can be determined using a recursive equation for j given by the following.

$$R(r,s,m,j;\mathbf{H};\mathbf{G}) = \begin{cases} F_j(\mathbf{X}^{(m-r+1)}) \displaystyle\sum_{\mathbf{G}^{(1)}\in\Theta(\mathbf{H};\mathbf{G})} R(r,s,m,j-1;\mathbf{H};\mathbf{G}^{(1)}), \\ \quad \text{if } (g_i = j \text{ and } 1 \le j \le h_i) \text{ or } (g_i = 0 \text{ and } j = h_i + 1) \\ \quad \text{or } (h_i + 2 \le j) \text{ for any } i, \\ 0, \qquad\qquad\qquad\qquad \text{otherwise,} \end{cases} \tag{8.4}$$

where

$$\Theta(\mathbf{H};\mathbf{G}) = \left\{ \mathbf{G}^{(1)} \left| \begin{array}{ll} g_i^{(1)} = g_i - 1, & \text{if } (g_i = j \text{ and } 1 \le j \le h_i) \\ & \text{ or } (g_i > 0 \text{ and } h_i + 2 \le j) \\ g_i^{(1)} = j - 1, & \text{if } g_i = 0 \text{ and } j = h_i + 1 \\ g_i^{(1)} = 0, 1, \ldots, s-1, & \text{if } g_i = 0 \text{ and } h_i + 2 \le j \end{array} \right. \right\},$$

and

$$x_i = \begin{cases} 1, & \text{if } g_i > 0, \\ 0, & \text{if } g_i = 0, \end{cases}$$

for $i = 1, 2, \ldots, m - r + 1$. As the boundary condition, for $j = 0$,

$$R(r,s,m,j;\mathbf{H};\mathbf{G}) = \begin{cases} 1, & \text{if } g_1 = g_2 = \cdots = g_{m-r+1} = 0, \\ 0, & \text{otherwise.} \end{cases}$$

For $j = s, s+1, \ldots, n$, the reliability of a circular connected-(r,s)-out-of-(m,j):F lattice system is expressed by

$$R^C(r,s,m,j;[p_{ij}]) = \sum_{\mathbf{H}\in\phi} \sum_{\mathbf{G}\in A(j;\mathbf{H})} R(r,s,m,j;\mathbf{H};\mathbf{G}), \tag{8.5}$$

where

$$\phi = \left\{ \mathbf{H} \left| \prod_{u=1}^{m-r+1} \chi(h_u < s) = 1, A(j;\mathbf{H}) = \left\{ \mathbf{G} \left| \prod_{u=1}^{m-r+1} \chi(h_u + g_u < s) = 1 \right. \right\} \right. \right\},$$

for $\mathbf{H} \in \phi$, and $\chi(A)$ indicates an index function taking the value 1 when A is true and the value 0 otherwise.

Yamamoto and Akiba's [46] REA mentioned above for obtaining the reliability of a circular connected-(r,s)-out-of-(m,j):F lattice system includes recursive ideas for computing the state probability of each row (column) by efficiently checking the component status, for which $B(h_i, g_i, i, j)$ events do not occur.

Yamamoto and Akiba [44] proposed an REA for obtaining the reliability of a circular k-within-connected-(r,s)-out-of-(m,n):F lattice system in the INID case with a similar design, improving on their previous REA [4].

The reliability of a toroidal connected-(r, s)-out-of-(m, n):F lattice system can also be obtained by an REA. This is because it can be considered a special circular system if we cut the system between circles 1 and n [30]. Nakamura *et al.* [33] also proposed an REA for obtaining the reliability of a toroidal connected-$(1, 2)$-or-$(2, 1)$-out-of-(m, n):F lattice system with a similar design. Yamamoto *et al.* [47] proposed an REA for obtaining the reliability of an adjacent triangle-(m, n):F triangular lattice system in the INID case.

(2) Finite Markov Chain Imbedding Approach (FMCIA)

As mentioned above, the FMCIA constituted the other approach. Fu [12] and also Fu and Hu [13] first used the FMCIA to compute the reliability of a consecutive-k-out-of-n:F system. They presented the following $(k-1) \times (k-1)$ transition probability matrix, for $i = 1, 2, \ldots, n$.

$$A(i) = \begin{pmatrix} p_i & q_i & 0 & & 0 \\ p_i & 0 & q_i & \cdots & 0 \\ p_i & 0 & 0 & & 0 \\ & \vdots & & \ddots & \vdots \\ p_i & 0 & 0 & & q_i \\ 0 & 0 & 0 & & 1 \end{pmatrix}_{(k-1)\times(k-1)} . \tag{8.6}$$

We can then compute the reliability of the linear consecutive-k-out-of-n:F system using the transition probability matrix $A(i)$ as follows.

$$R^L(k, n; [p_n]) = \pi_0 \prod_{i=1}^{n} A(i)\mathbf{u}^{\mathsf{T}}, \tag{8.7}$$

where $\pi_0 = (1, 0, \ldots, 0)_{1\times(k-1)}$, $\mathbf{u}^{\mathsf{T}}$ is the transpose of the row vector $\mathbf{u} = (1, \ldots, 1, 0)_{1\times(k-1)}$.

Following which, Fu and Koutras [14] named this method the "finite Markov chain imbedding approach" (FMCIA). Zhao and Cui [53] introduced the accelerated scan finite Markov chain imbedding (AS-FMCI), which simplifies the computation for system reliability because it decreases the number of multiplied matrices. Additionally, Yamamoto and Akiba [42] and also Zhao *et al.* [54] proposed an FMCIA for computing the reliability of a linear connected-(r, s)-out-of-(m, n):F lattice system, independently. Furthermore, Nakamura *et al.* [31, 32] presented the FMCIAs to compute the reliabilities of circular and toroidal connected-$(1, 2)$-or-$(2, 1)$-out-of-(m, n):F lattice systems.

(3) Comparison of REA and FMCIA

We believe that REA and FMCIA are essentially the same. If system reliability is expressed as recursive equations, it can be expressed in a finite Markov chain, and vice versa. However, on implementing the two approaches on a computer, the efficiency of the computation differs with the expression of the system reliability, indicating a difference between the two approaches.

Table 8.3
Comparison between computation times of REA and FMCIA for a linear connected-$(1, 2)$-or-$(2, 1)$-out-of-$(20, n)$:F system in INID case (sec.).

n	REA	FMCIA
10	37.597	39.689
50	49.205	58.860
100	64.725	80.040

Table 8.4
Comparison between computation times of REA and FMCIA for a linear connected-$(1, 2)$-or-$(2, 1)$-out-of-$(17, n)$:F system in IID case (sec.).

n	REA	FMCIA
500	13.263	15.974
1000	24.791	17.384
2000	47.741	18.520

Nakamura [27] experimentally compared REA and FMCIA in terms of computation time by calculating the reliability of a linear connected-$(1, 2)$-or-$(2, 1)$-out-of-(m, n):F lattice system. A few of the results of this comparison are presented here. Each approach was programmed in MATLAB R2018a and executed on a computer with Intel Core i5 3.20 GHz CPU, 8.0 GB memory, and a Microsoft Windows 10 OS. Table 8.3 lists a comparison between the computation times in the INID case. The REA is evidently more efficient in all the cases within the scope of the experiment. Table 8.4 lists a comparison between the computation times in the independent and identically distributed (IID) case[2]. For relatively few columns and many rows, the REA is more efficient, whereas the FMCIA outperforms the REA for many columns. From these results, we can conclude that the superiority or inferiority of both approaches depends on the systems whose reliabilities are being computed. Although the REA and the FMCIA may appear different, they can essentially be the same. This is because the system reliability expressed as recursive equations can also be expressed in a finite Markov chain and vice versa. Therefore, we can select an expression for efficiently computing the system reliability depending on the system whose reliability is being computed.

[2]The component lifetimes are independent and identically distributed.

8.3.2 APPROXIMATION METHODS FOR SYSTEM RELIABILITY

Despite the methods proposed in Subsection 8.3.1, the problem of computing the system reliability is challenging for large systems, as mentioned above. Therefore, there have been many proposals for approximation methods wherein the value obtained slightly differs from the reliability; however, it can be obtained in a reasonable amount of time. The approximation methods can be roughly classified into two types: (1) upper and lower bounds, and (2) approximations based on a limit theorem. The upper (lower) bounds of the system reliability discussed in this subsection constitute values that have been theoretically proven to be greater (less) than the reliability. The limit theorem implies that the system reliability converges to a value except 0 and 1 when the system size (m and (or) n) is increased and the failure probability of a component simultaneously approaches zero.

(1) Upper and Lower Bounds

The values of the upper (lower) bounds of the system reliability have been proven to be theoretically greater (less) than the reliability. For instance, for a fictitious system that does not fail regardless of whether a few of the failure conditions of the target system hold, the system reliability becomes the upper bound of the target system. However, the reliability of a system wherein one failure condition is assumed to be independent of another becomes the lower bound of the target system. Based on this idea, Yamamoto and Miyakawa [49] proposed upper and lower bounds for the reliability of a rectangular connected-(r, s)-out-of-(m, n):F lattice system, as shown in the following equations. Let $R^s(r, j)$ be the reliability of a rectangular connected-(r, s)-out-of-(m, s):F lattice system consisting of components from column j to $j + s - 1$ in the target system. Additionally, let $R(j)$ be the reliability of a consecutive-r-out-of-m:F system of column j in the target system. Figure 8.6 shows the images of the notations $R^s(r, j)$s and $R(j)$, and those of the product of $R^s(r, j)$s for obtaining upper and lower bounds for reliability UB_1^L, UB_2^L and LB_1^L.

$$UB_1^L = \prod_{j=1}^{\lfloor n/s \rfloor} R^s(r, j \times s + 1),$$

$$UB_2^L = R^s(r, 1) \prod_{j=2}^{n-s} R(j)(1 - R^s(r, j + 1)),$$

$$LB_1^L = \prod_{j=1}^{n-s+1} R^s(r, j).$$

Table 8.5 lists the upper and lower bounds for the reliability of a linear connected-$(2, 2)$-out-of-$(20, 20)$:F lattice system in the IID case. For a connected-(r, s)-out-of-(m, s):F lattice system, several designs regarding the

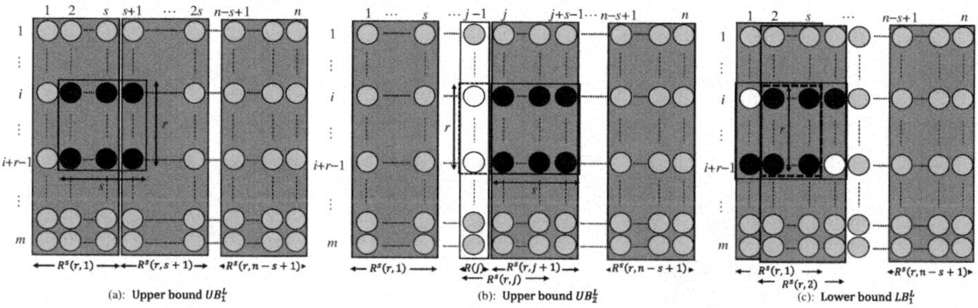

Figure 8.6 Images of the notations $R^s(r, j)$s and $R(j)$, and the product of $R^s(r, j)$s.

Table 8.5
Comparison of upper and lower bounds.

	Upper bounds		Lower bound
p	UB^L_1	UB^L_2	LB^L_1
0.30	0.00000	0.01998	0.00000
0.35	0.00000	0.05550	0.00000
0.40	0.00000	0.11923	0.00000
0.45	0.00000	0.20612	0.00000
0.50	0.00003	0.29420	0.00000
0.55	0.00106	0.35886	0.00000
0.60	0.01287	0.39713	0.00026
0.65	0.07412	0.43165	0.00713
0.70	0.23865	0.49226	0.06572
0.75	0.49445	0.59804	0.26232
0.80	0.74562	0.74215	0.57251
0.85	0.91006	0.88272	0.83605
0.90	0.98135	0.97015	0.96487
0.95	0.99882	0.99785	0.99775

upper and lower bounds of reliability have been proposed since 1990 [36]. Boutsikas and Koutras [8] also proposed upper and lower bound designs and compared them with previous ones.

For a linear (circular) k-within-connected-(r, s)-out-of-(m, n):F lattice system, the upper and lower bounds of reliability are given by Yamamoto and Akiba [45] as follows: For $i = 1, 2, \ldots, m$ and $j = 1, 2, \ldots, n$, we define the following sets.

$CS_1(i, j)$: set of all the components in an $r \times s$ matrix with component (i, j) as its right lower corner.

$CG_1(i, j)$: set of all the components in an $(r - 1) \times s$ matrix with component $(i - 1, j)$ as its right lower corner.

$CG_2(i, j)$: set of all components in an $r \times (s - 1)$ matrix with component $(i, j - 1)$ as its right lower corner.

$CS_2(i, j)$: set of components $(i, j - s + 1), (i, j - s), \ldots, (i, j)$.

$CS_3(i, j)$: set of components $(i - r + 1, j), (i - r, j), \ldots, (i, j)$.

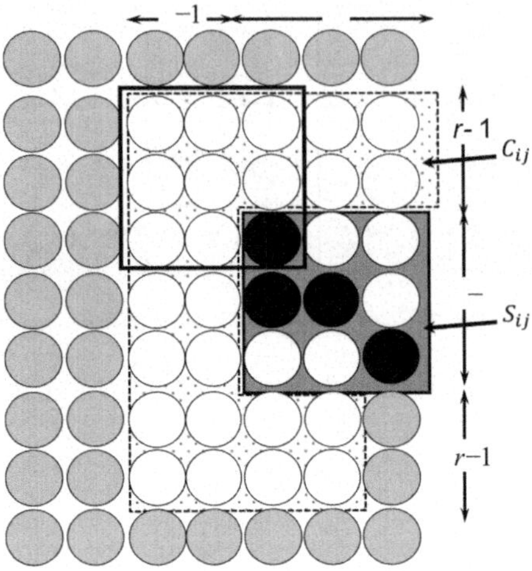

Figure 8.7 Areas of events C_{ij} and S_{ij}.

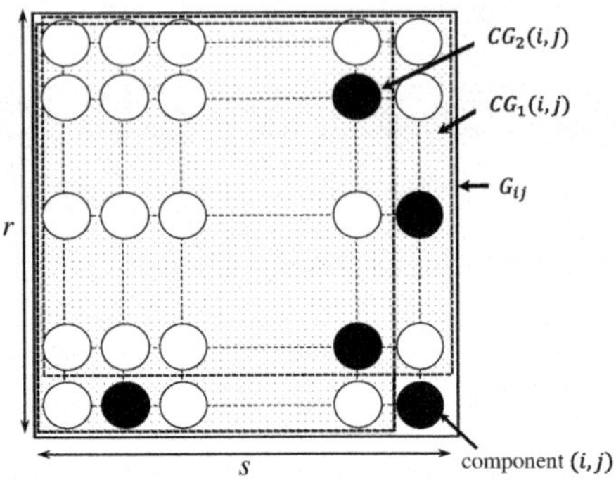

Figure 8.8 Area of event G_{ij}.

To express theorems and equations simply, we consider virtual components with a component reliability of 1 outside the system.

Furthermore, we introduce the following events, which are related to the above sets. For $i = 1, 2, \dots, m$ and $j = 1, 2, \dots, n$,

S_{ij}: event that "k or more components fail in $CS_1(i,j)$," "at least one component fails in $CS_2(i,j)$," and "at least one component fails in $CS_3(i,j)$."

Moreover, for $i = r, r+1, \dots, m$ and $j = s, s+1, \dots, n$, we define

C_{ij}: event that all the components function in the following set.

$$\bigcup_{(u,v)\in\{CS_1(u,v)|CS_1(i,j)\neq\phi,(u,v)<(i,j)\}} CS_1(u,v),$$

where $(u,v) < (i,j)$ indicates that $v < j$ or $v = j$ and $u < i$, (see Figure 8.7).

G_{ij}: the whole event, for $i = r$ and $j = s$, the event that less than k components fail in $CG_1(i,j)$, for $i \neq r$ and $j = s$, the event that less than k components fail in $CG_2(i,j)$, for $i = r$ and $j \neq s$, the event that "less than k components fail in $CG_1(i,j)$" and "less than k components fail in set $CG_2(i,j)$," for $i \neq r$ and $j \neq s$, (see Figure 8.8).

E_{ij}: event that "k or more components fail in $CS_1(i,j)$" and "event G_{ij} occurs."

Using the notations above, the upper (UB) and lower (LB) bounds for the reliability of a linear k-within-connected-(r,s)-out-of-(m,n):F lattice system are given as follows.

$$LB = \prod_{i=1}^{m} \prod_{j=1}^{n} (1 - \Pr\{S_{ij}\}),$$

$$UB = \prod_{i=r}^{m} \prod_{j=s}^{n} \left[1 - \Pr\{C_{ij}\} \frac{\Pr\{E_{ij}\}}{\Pr\{G_{ij}\}} \right].$$

For details regarding the lower bounds for the reliability of two-dimensional consecutive-k systems, please refer to Boutsikas and Koutras [8], Chap. 10 of Kuo and Zuo [24], and Chap. 3 of Akiba *et al.* [3].

(2) Limit Theorems

The limit theorem provides evidence that system reliability can be approximated as a value except 0 and 1 when m and n are large and the failure probability of a component is nearly zero. Furthermore, this theorem elucidates the relationship between system parameters (m, n, r, s) and component reliability to achieve the required system reliability.

A limit theorem for a linear connected-(r, s)-out-of-(m, n):F lattice system was presented by Fu and Koutras [14]. For a mathematical description, we redefined the component reliability p_n as a function of n. Letting $f(\cdot)$ be a monotone increasing function, we assumed $m = f(n)$ as the relationship between m and n (the notation and assumptions are used throughout this subsection when considering the limit theorems). Let $R^L(r, s, m, n; p_n)$ be the reliability of a linear connected-(r, s)-out-of-(m, n):F lattice system with component reliability p_n in the IID case and λ be a positive constant value. If $\lim_{n\to\infty} mn(1 - p_n)^{rs} = \lambda$, then

$$\lim_{n\to\infty} R^L(r, s, m, n; p_n) = \exp[-\lambda]. \tag{8.8}$$

This limit theorem can be proved easily using Corollary 2.3 in Fu and Koutras [14]. For a circular system, the following relationship hold.

$$R^L(r, s, m, n; p_n) \le R^C(r, s, m, n; p_n) \le R^L(r, s, m, n + s - 1; p_n). \tag{8.9}$$

Therefore, from Equations (8.8) and (8.9), we obtain

$$R^L(r, s, m, n; p_n) \approx \exp[-mn(1 - p)^{rs}], \tag{8.10}$$

$$R^C(r, s, m, n; p_n) \approx \exp[-mn(1 - p)^{rs}]. \tag{8.11}$$

Yamamoto [41] considered a limit theorem for a linear or circular connected-$\bigcup_{\theta=1}^{k} \mathbf{X}_\theta$-out-of-$(m, n)$:F lattice system, where $\mathbf{X}_\theta$ represents one of any connected failure patterns for $\theta = 1, 2, \ldots, k$. For instance, if $\mathbf{X}_1$ represents a failure pattern, as shown in Figure 8.9, then a linear connected-$\mathbf{X}_1$-out-of-$(7, 7)$:F system fails when failed components form a failure pattern $\mathbf{X}_1$ as shown in Figure 8.9(a). Such a system fails only on the condition

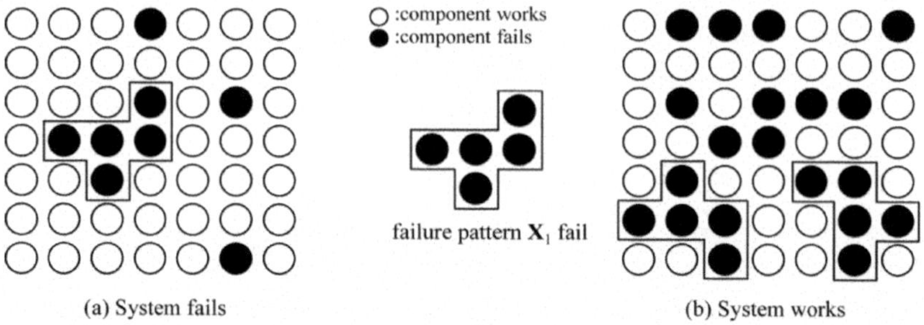

Figure 8.9 Linear connected-$\mathbf{X}_1$-out-of-$(7,7)$:F system.

that failed components form one of the $\mathbf{X}_\theta$'s in the system. For $\phi = L, C$, let $R^\phi(\bigcup_{\theta=1}^k \mathbf{X}_\theta, (m,n); p_n)$ be the reliability of a connected-$\bigcup_{\theta=1}^k \mathbf{X}_\theta$-out-of-$(m,n)$:F lattice system with component reliability p_n in the IID case. Notably, the system is linear when $\phi = L$, and the system is circular when $\phi = C$.

Let N be the number of failure patterns, where a failure pattern $\mathbf{X}_\theta$ consists of exactly M components, namely, $\#\{\theta | \#\mathbf{X}_\theta = M\}$, where $M = \min_{1\leq\theta\leq k}(\#\mathbf{X}_\theta)$. Notably, $\#\mathbf{X}_\theta$ denoted the number of components forming the failure pattern $\mathbf{X}_\theta$. Let λ be a positive constant value. If $\lim_{n\to\infty} mn(1-p_n)^M = \lambda$, then

$$\lim_{n\to\infty} R^\phi\left(\bigcup_{\theta=1}^k \mathbf{X}_\theta, (m,n); p_n\right) = \exp[-N\lambda], \tag{8.12}$$

for $\phi = L, C$.

Yamamoto and Akiba [46] proposed the limit theorems for linear or circular k-within-connected-(r,s)-out-of-(m,n):F lattice systems in the IID case. Let λ be a positive constant value. If $\lim_{n\to\infty} mn(1-p_n)^k = \lambda$, then

$$\lim_{n\to\infty} R^L(k,r,s,m,n;p_n) = \exp[-N(k)\lambda], \tag{8.13}$$

and

$$\lim_{n\to\infty} R^C(k,r,s,m,n;p_n) = \exp[-N(k)\lambda], \tag{8.14}$$

where

$$N(k) = \binom{rs}{k} - \binom{(r-1)s}{k} - \binom{r(s-1)}{k} + \binom{(r-1)(s-1)}{k}.$$

If $\lim_{n\to\infty} n(1-p_n)^k = \lambda$, then

$$\lim_{n\to\infty} R^L(k,r,s,m,n;p_n) = \exp[-(N_r(k) + (m-r)N(k))\lambda], \tag{8.15}$$

where

$$N_r(k) = \binom{rs}{k} - \binom{r(s-1)}{k}.$$

8.4 COMPONENT ASSIGNMENT PROBLEM (CAP)

There exists a problem, called the component assignment problem (CAP), of finding the component arrangement (optimal arrangement) that maximizes the system reliability when changing the positions of a few (or all) components does not affect system functioning. Here, we consider a situation where all of the component reliabilities are not necessarily the same because such component arrangements would not affect the reliability of the system otherwise.

This problem can be solved by finding the system reliability for all the component arrangements. However, finding the system reliabilities for one arrangement is laborious, as described earlier. To overcome this problem, eliminating component arrangements that are evidently not optimal, which are those that do not satisfy the conditions required, is necessary when searching for the optimal arrangement. This also aids in avoiding the time-consuming computational load of obtaining the system reliability as much as possible. Hence, we need to evaluate only the arrangements that satisfy the necessary conditions. These measures facilitate in reducing the computational effort of finding the optimal arrangement. Furthermore, the magnitude of component reliabilities determines whether the conditions are satisfied despite the exact component reliabilities being unknown. Consequently, any computations of the system reliability are not required when checking whether the conditions are satisfied. Therefore, the necessary conditions are beneficial for obtaining the optimal arrangement efficiently.

On the other hand, many studies have sought a pseudo-optimal solution instead of an exact one.

8.4.1 EFFICIENT ALGORITHM FOR OBTAINING THE OPTIMAL ARRANGEMENT

In this subsection, we first discuss the optimal assignment problem for one-dimensional systems, mainly linear consecutive-k-out-of-n:F ones, and then two-dimensional systems, mainly linear connected-(r, s)-out-of-(m, n):F lattice ones.

(1) Linear Consecutive-k-out-of-n:F System

We explain an efficient algorithm for solving the CAP for a linear consecutive-k-out-of-n:F system. The following notations are defined. For $i = 1, 2, \ldots, n$,

q_i: failure probability of component i, where $q_1 \geq q_2 \geq \cdots \geq q_n$. p_i $(= 1 - q_i)$.

Q: $Q \equiv (q_1, q_2, \ldots, q_n)$.

π: permutation consisting of the numbers 1 through n, where $\pi \equiv (\pi(1), \ldots, \pi(n))$.

S_n: set of all permutations of π.

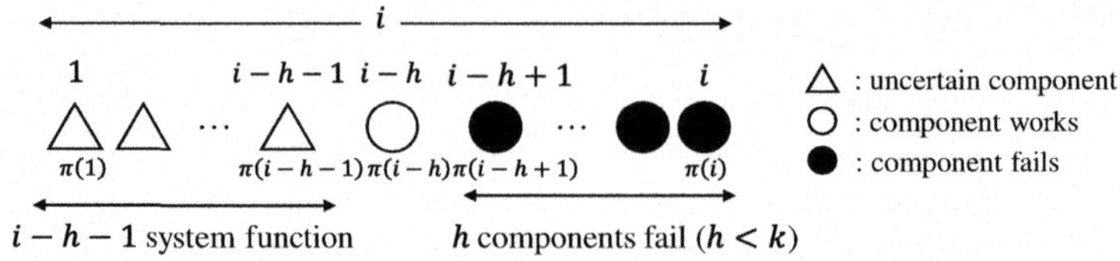

Figure 8.10 Image of $R(n, h; \pi, \mathbf{Q})$ by Hwang's algorithm [20].

For given failure probabilities $\mathbf{Q}$ and components assigned according to the permutation π, the system configuration is uniquely determined. Therefore, we call the permutation π the arrangement π.

The basic concept behind the algorithm is based on the branch-and-bound (BB) method, and the properties of the theoretical support for this algorithm are described below. From Theorem 2 of Kuo *et al.* [23], the necessary conditions for π to be an optimal configuration when $n \geq 2k$ are

(1) $\pi(1) < \pi(2) < \cdots < \pi(k)$ and

(2) $\pi(n-k+1) < \pi(n-k+2) < \cdots < \pi(n)$.

The system reliability, $R(n; \pi, \mathbf{Q})$, for $i = 1, 2, \ldots, n$, is

$$R(i; \pi, \mathbf{Q}) = \sum_{h=1}^{k-1} R(i, h; \pi, \mathbf{Q}), \tag{8.16}$$

where $R(n, h; \pi, \mathbf{Q})$ represents the state probability of the system, as shown in Figure 8.10 (Hwang's formula [20]). Using $R(n, h; \pi, \mathbf{Q})$ can be beneficial when computing the reliabilities of many linear consecutive-k-out-of-n:F systems, such as in optimal assignment problems. For instance, $R(4, h; (1, 2, 3, 4).\mathbf{Q})$ is used to obtain the system reliabilities of the arrangements $(1, 2, 3, 4, 5)$ and $(1, 2, 3, 4, 8)$, Therefore, by storing this value, the other reliabilities can be calculated easily after calculating one reliability.

An outline of this algorithm, which uses the method above for reliability and conditions necessary for the optimal arrangement, is described as follows. Figure 8.11 depicts a part of a search tree for the arrangements of a linear consecutive-3-out-of-8:F system. On the search tree, there are only arrangements with $\pi(1) < \pi(n)$, because the system reliabilities are identical to the ones with a reversed arrangement. On following this search tree with the depth-first search (DFS), all the arrangements can be obtained. Furthermore, the BB method is used in this case, and unnecessary searches can be eliminated using the following properties of Hanafusa and Yamamoto [17].

(a) The optimal arrangements must satisfy the necessary conditions (1) and (2) above.

(b) The system reliability is non-increasing for increasing n.

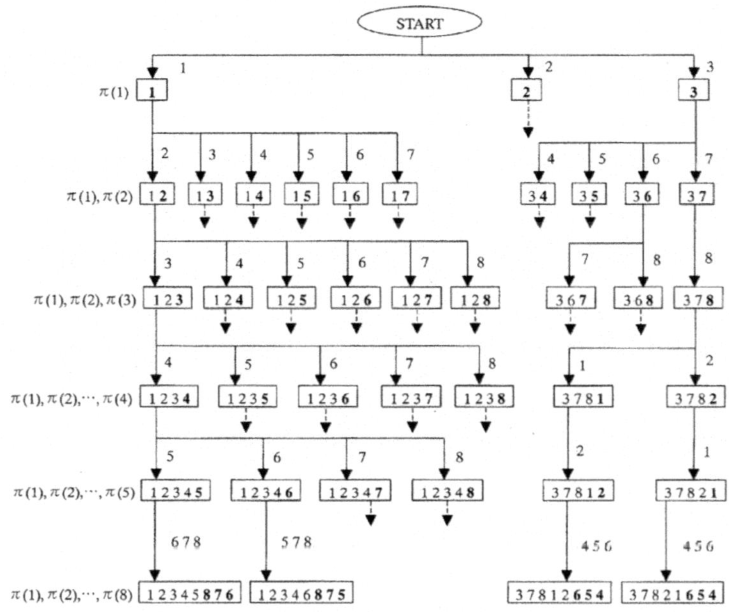

Figure 8.11 Part of a search tree of the arrangements by BB ($k = 3, n = 8$) [17].

Table 8.6
Comparison between computation times (sec.) [17].

	k =3		k =4		k =5		Enumeration
n	BB	MM	BB	MM	BB	MM	
11	0.05	0.60	0.00	0.05	0.00	0.00	158.60
12	0.61	7.64	0.06	0.55	0.00	0.00	2261.00
13	9.10	101.34	1.05	7.64	0.06	0.33	(approx.) 10 hour
14	66.62	1438.11	9.33	108.81	1.10	5.20	(approx.) 7 days
15	771.81	(approx.) 6 hour	83.32	1661.04	9.39	79.64	(approx.) 122 days
16	5051.00	(approx.) 5 days	136.15	27047.06	31.84	1291.63	(approx.) 6 years
17	N/A	(approx.) 100 days	2046.02	(approx.) 4 days	1216.71	22317.49	(approx.) 131 years

We solved a few CAPs to evaluate the algorithm, and the computation times are listed in Table 8.6. In the table, BB denotes the algorithm described above, MM denotes the (DFS) algorithm when implementing the properties above, except property (b). Enumeration indicates the algorithm wherein all the component arrangements are listed, and the system reliabilities are calculated separately. These results were computed using a Pentium III 500MHz CPU (in 1998). The results indicate that the BB algorithm is efficient, but the computation times continued to increase exponentially with system size n.

Based on the concept above, Shingyochi and Yamamoto [37] proposed an efficient algorithm for obtaining the optimal component arrangement for a circular consecutive-k-out-of-n:F system. Additionally, Yamamoto *et al.* [48] also proposed an efficient algorithm for the same purpose as above for a liner consecutive-k-out-of-r-from-n:F system. This algorithm employed a similar BB algorithm by using a recursive equation [25] to obtain the reliability of the system. Furthermore, Akiba *et al.* [1] proposed an algorithm for obtaining the optimal assignment of a multi-state consecutive-k-out-of-n:F system.

Recently, Zhu *et al.* [58] extended the original CAP to a multi-type one for linear consecutive-k-out-of-n:F systems. This problem is to find a component arrangement with the maximum system reliability under the assumption that each component should only be assigned to a position belonging to the same type of component. They developed Birnbaum importance-based local search method and genetic algorithm to solve this multi-type CAP.

(2) Linear Connected-(r, s)-out-of-(m, n):F Lattice System

Herein, we discuss the CAP for a linear connected-(r, s)-out-of-(m, n):F lattice system. Koutras *et al.* [22] established a necessary condition that the optimal arrangement must satisfy. While finding the optimal arrangement, a component arrangement that does not satisfy the necessary condition will not be considered optimal, as described in the subsection above. Hence, we only need to evaluate the optimal arrangements.

Omura *et al.* [34] developed an algorithm for finding the optimal arrangement available for $r = m$ and $s = 2$. It utilizes the invariant optimal arrangement, which only depends on the order of the values of component reliabilities, of a linear consecutive-k-out-of-n:F system [26] to produce fewer candidates for the optimal arrangement.

For $r = m - 1$ and $2s > n$, Omura *et al.* [35] developed a BB algorithm to find the optimal arrangement, which incorporates pruning based on the necessary condition and pruning for eliminating the equivalent arrangement.

Additionally, Nakamura *et al.* [29] proposed a BB algorithm specifically for $r = m - 1$ and $s = n - 1$. It has the feature of simplifying the CAP of a linear connected-$(m - 1, n - 1)$-out-of-(m, n):F lattice system to the problem of finding the optimal arrangement of a simple system. The basic procedure of this algorithm is as follows. As a preprocessing step, the $(m - 2)(n - 2)$ most reliable components are first assigned to particular positions of the system based on the necessary condition [22]. Consequently, the system obtained by eliminating the components can be regarded as a circular sub-system 8.12. Additionally, the least reliable component is assigned to the position corresponding to the 1st row and the 1st column in the system (see Figure 8.12). The remaining $2(m + n) - 5$ components are subsequently assigned to the remaining positions, and we apply the BB method to the problem of finding the optimal arrangement of a circular sub-system. To efficiently compute the reliability of a circular sub-system, Nakamura *et al.* [29] also proposed a recursive method.

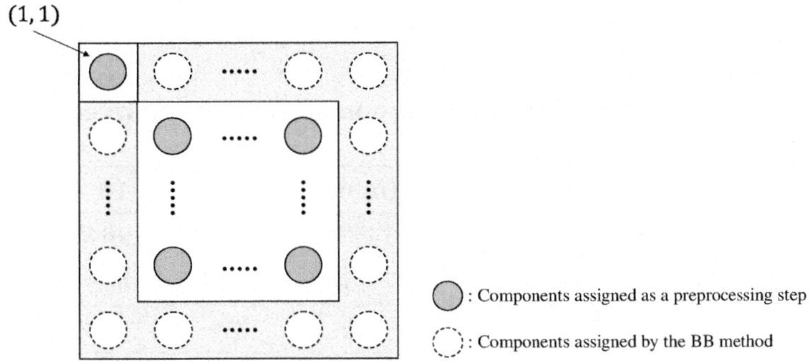

Figure 8.12 Circular sub-system.

The above algorithms are available for special cases. Nakamura and Yamamoto [28] developed an algorithm for finding the optimal arrangement of general connected-(r, s)-out-of-(m, n):F lattice systems. The main principle of this algorithm is incorporating conditions for pruning in the standard BB framework, producing fewer candidates for the optimal arrangement. There are three types of pruning: (a) pruning based on the necessary condition [22] that the optimal arrangement must satisfy, (b) pruning for eliminating the equivalent arrangement, and (c) pruning based on system reliability. A condition derived by Yamamoto *et al.* [51] is used to eliminate the equivalent arrangement. Notably, although the condition they provided was for the facility layout problem, it is also applicable to the CAP in a linear connected-(r, s)-out-of-(m, n):F lattice system. The system reliability is computed using the recursive equation proposed by Yamamoto and Miyakawa [49]. Moreover, to avoid redundant computations, this algorithm memorizes the specific values and utilizes them, enhancing its performance. Here, we present the results of comparing the algorithms proposed by Nakamura *et al.* [29] and Nakamura and Yamamoto [28], which are found in Nakamura and Yamamoto [28]. Table 8.7 lists the computation times required for each algorithm to find the optimal arrangement. Table 8.7 shows that the former [29] could obtain the optimal arrangement faster than the latter [28]. Thus, if a linear connected-(r, s)-out-of-(m, n):F lattice system satisfies $r = m-1$ and $s = n-1$, Nakamura *et al.*'s algorithm [29] is concluded to outperform Nakamura and Yamamoto's algorithm [28] in terms of the computation time. Accordingly, Nakamura *et al.*'s algorithm [29], which depends on the unique structure of a system, can considerably reduce the computational efforts in finding the optimal arrangement. However, for $r \neq m-1$ or $s \neq n-1$, Nakamura *et al.*'s algorithm [29] is inapplicable for finding the optimal arrangement. Nakamura and Yamamoto's algorithm [28] can find the optimal arrangement of a general linear connected-(r, s)-out-of-(m, n):F lattice system.

Table 8.7
Comparison between the algorithms proposed by Nakamura et al. [29] and Nakamura and Yamamoto [28].

(m, n)	(5, 5)	(5, 6)	(6, 5)	(6, 6)
(A) algorithm [28]	40.84	1144.77	1532.19	19645.68
(B) algorithm [29]	1.87	74.52	42.37	598.86
(B)/(A)	5%	7%	3%	3%
System reliability	0.998	0.992	0.992	0.997

8.4.2 ALGORITHMS FOR OBTAINING PSEUDO-OPTIMAL ARRANGEMENT

Despite using the above algorithms, solving the CAP for one-dimensional consecutive-k systems with a large n proves challenging. Hanafusa and Yamamoto [18] proposed a method that uses a genetic algorithm (GA) to obtain a pseudo-optimal solution for a linear consecutive-k-out-of-n:F system. The proposed method only searches for solutions that satisfy the necessary condition for the optimal solutions. Shingyochi and Yamamoto [38] improved this pseudo-optimal solution by proposing the allocation of components with low failure probabilities at equal or nearly equal intervals because such arrangements appear to have relatively high system reliabilities. Additionally, Shingyochi and Yamamoto [39] and Shingyochi *et al.* [40] proposed a method for obtaining a pseudo-optimal solution that uses a simulated annealing (SA) algorithm for linear (circular) consecutive-k-out-of-n:F systems.

8.5 MAINTENANCE PROBLEMS

In this section, we review a few maintenance problems in one-dimensional consecutive-k systems.

8.5.1 MAINTENANCE PROBLEMS IN LINEAR CONSECUTIVE-K-OUT-OF-N:F SYSTEM

Endharta *et al.* [11] discussed a condition-based maintenance (CBM) problem for linear consecutive-k-out-of-n:F systems. In CBM, the lifetime of a system is monitored, and preventive maintenance (PM) is performed at a time T_{PM} after a certain condition occurs. Endharta *et al.* [11] proposed a policy where PM is performed at a time T_{PM} after at least one minimal cut set comprising only one working component is generated. If the system fails prior to reaching the PM time after the condition is satisfied, the system undergoes corrective maintenance (CM).

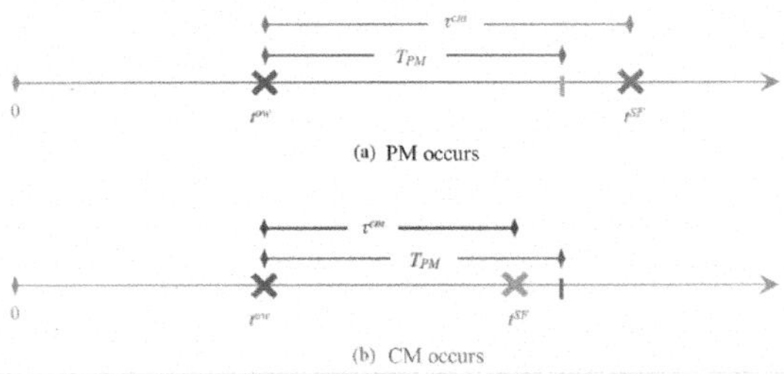

Figure 8.13 Illustration of the condition-based maintenance policy.

The CBM policy is illustrated in Figure 8.13. The time at which there is at least one minimal cut set consisting of only one working component is represented by t^{ow}, the time of system failure is represented by t^{SF}, and $\tau^{cm} = t^{SF} - t^{ow}$.

In this maintenance policy, the costs for CM, PM at time T_{PM}, and component replacement are c_{CM}, c_{PM}, and c_R, respectively. Therefore, the expected cost rate under the CBM model is

$$EC(T_{PM}) = \frac{c_{CM}F(T_{PM}) + c_{PM}(1 - F(T_{PM})) + c_R E[N(T_{PM})]}{E[T(T_{PM})]}, \quad (8.17)$$

where $F(T_{PM})$ denotes the system failure probability, $E[N(T_{PM})]$ denotes the expected number of component failures at maintenance time, and $E[T(T_{PM})]$ represents the expected replacement time. The optimal preventive maintenance interval (T_{PM}) is obtained by minimizing the expected cost rate. This policy was compared with existing maintenance policies, such as CM and age-dependent PM policies. The results are extensively described in Endharta *et al.* [11].

8.5.2 MAINTENANCE PROBLEMS IN LINEAR CONSECUTIVE-K-OUT-OF-N:G SYSTEMS

Zhou *et al.* [56] discussed the age-based replacement policy for a linear consecutive-k-out-of-n:G system, under which a system would be replaced at a planned time t or at the time of system failure, whichever occurs first. Figure 8.14 depicts the process of age replacement in one renewal cycle.

Let us define c_1 as the acquisition cost for each component and c_R as the replacement cost for a failed system. The expected cost rate is then

$$C_1(t; k, n) = \frac{nc_1 + c_R[1 - R(k, n, t)]}{\int_0^t R(k, n, x)dx}, \quad (8.18)$$

where $R(k, n, t)$ denotes the system reliability. The optimal replacement time is obtained by minimizing the expected cost rate. The result of the optimal maintenance policy is given below [56].

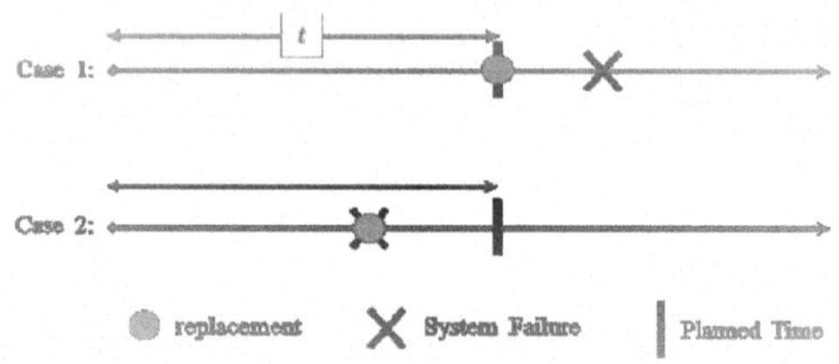

Figure 8.14 Illustration of the age-based maintenance policy (Case 1: PM occurs, Case 2: CM occurs).

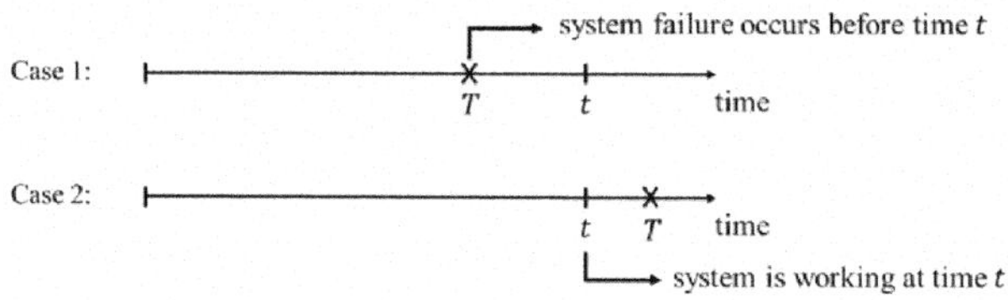

Figure 8.15 Case analysis for the expected number of failed components.

Proposition [56] For a linear consecutive-k-out-of-n:G system with IID components having an increasing failure rate (IFR), a unique t_1^* exists that can minimize the expected cost rate $C_1(t; k, n)$.

The other results obtained under the condition that components follow Weibull lifetime distributions are extensively discussed in Zhou *et al.* [56].

Economically, it is possible to solely replace failed components. Zhou *et al.* [57] provided the expected number of failed components in a coherent system. As shown in Figure 8.15, there are two cases. Let us define T as the lifetime of the system. Using the results presented by Zhou *et al.* [57], we obtain the expected results for the two cases of the linear consecutive-k-out-of-n:G system. Here, $E[X(T)|T \leq t]$ and $E[S(t)|T > t]$ denote the expected number of failed components under the condition that the system has failed prior to time t and under the condition that the system is working at time t, respectively.

Zhou [55] proposed that by using the same cost factors c_1 and c_R, the expected cost rate becomes

$$C_2(t; k, n) = \frac{Q_1(t) \Pr\{T > t\} + Q_2(t) \Pr\{T \leq t\}}{\int_0^t R(k, n, x) dx}, \tag{8.19}$$

where $Q_1(t) = c_1 E[S(t)|T > t]$ and $Q_2(t) = c_R + c_1 E[X(T)|T \leq t]$. The optimal replacement time is obtained by minimizing $C_2(t; k, n)$.

Furthermore, Table 8.8 [55] illustrates the optimal replacement time and a comparison between the minimum values of the expected cost rates $C_1(t_1^*; k, n)$ and $C_2(t_2^*; k, n)$. Let us assume that all components have an identical and

Table 8.8

Comparisons between $C_1(t_1^*; k, n)$ and $C_2(t_2^*; k, n)$ for a consecutive-k-out-of-n:G system for component lifetime distribution $F(t) = 1-\exp(-\lambda t)$ [55].

λ	k	n	$c_1 = 5, c_R = 50$				
			t_1^*	$C_1(t_1^*; k, n)$	t_2^*	$C_2(t_2^*; k, n)$	$\frac{C_2(t_2^*; k,n)}{C_1(t_1^*; k,n)}$
0.5	3	10	2.297	63.292	0.107	24.657	0.390
		12	2.516	63.842	0.274	28.749	0.450
		15	2.951	65.996	0.521	34.342	0.520
	4	12	3.446	95.188	0.032	29.884	0.314
		15	4.072	96.543	0.150	36.665	0.380
		20	∞	101.954	0.382	46.814	0.459
	5	15	∞	132.212	0.022	37.397	0.283
		20	∞	137.205	0.150	48.814	0.356
		25	∞	145.163	0.296	59.342	0.409

independent exponential lifetime distribution of rate λ. ∞ implies that the optimal replacement time is the moment of system failure. From the values of the ratio $C_2(t_2^*; k, n)/C_1(t_1^*; k, n)$, we observe that $C_2(t_2^*; k, n)$ outperforms $C_1(t_1^*; k, n)$.

8.6 CONCLUSIONS

In this chapter, we reviewed the main problems of the "consecutive-k systems," whose study began over 40 years ago. Many topics related to them have been studied, and papers have been published. We described a part of the results of many studies. First, we provided an overview of the consecutive-k system and introduced typical ones. We then reviewed three main topics pertaining to consecutive-k systems: a reliability evaluation method, the CAP, and maintenance problems in certain consecutive-k systems.

When component reliabilities are provided, proposing an efficient method for obtaining system reliability is important. However, the computational load increases exponentially with the number of components. To solve this problem, there are two types of approaches: REA and FMCIA. Moreover, we can obtain approximate values (lower and upper bounds) for the reliability in a reasonable time. We described a few lower and upper bounds and also approximate values from the limit theorems.

The CAP constitutes the other main topic discussed, which can be solved by obtaining the system reliabilities in all the component arrangements and finding their maximum values. However, finding the system reliability for even one arrangement is laborious. To overcome this problem, we introduced the BB method wherein only optimal component arrangements are considered.

The third topic discussed was the maintenance problem. Several studies were conducted on the maintenance of consecutive-k systems. We introduced a CBM problem for a linear consecutive-k-out-of-n:F system and the age-based replacement problem for the linear consecutive-k-out-of-n:G system.

We consider consecutive-k systems as quite simple, owing to which significant work has been conducted on them. Namely, consecutive-k systems have very minimal cuts (paths) that exist in an orderly sequence. Using these properties, we can construct many theories for these systems, and we should attempt to apply the many results obtained for them to general network type systems in the future.

ACKNOWLEDGEMENTS

We express our gratitude to Prof. Koji Shingyochi and Dr. Natsumi Takahashi for providing significant amounts of material for writing this paper. This work was supported by JSPS KAKENHI, Grant Numbers JP19K04908, JP20K04964, and JP21K14370.

REFERENCES

1. Akiba, T., Komuro, T., Yamamoto, H., & Nagatsuka, H. (2012). Fast algorithm for optimal arrangement problems of multi-state consecutive-k-out-of-n:F system. *Journal of Japan Industry Management Association*, 62(6), 247–255.

2. Akiba, T., Nagatsuka, H., Yamamoto, H., & Zuo, M. J. (2006). Recursive formulas for the reliability of multi-state consecutive-k-out-of-n:F system. *Journal of Japan Industrial Management Association*, 56(6), 439–446.

3. Akiba, T., Nakamura, T., Xiao, X., & Yamamoto, H. (2019). Evaluation methods for reliability of consecutive-k systems. In M. Ram & T. Dohi (Eds.), *Systems Engineering: Reliability Analysis Using k-out-of-n Structures* chapter 1, (pp. 1–24). CRC Press.

4. Akiba, T. & Yamamoto, H. (2001). Reliability of a 2-dimensional k-within-consecutive-$r \times s$-out-of-$m \times n$:F system. *Naval Research Logistics*, 48(7), 625–637.

5. Akiba, T., Yamamoto, H., & Kainuma, Y. (2004). Reliability of a 3-dimensional adjacent triangle:F triangular lattice system. In *Proceedings of the 5th Asia Pacific Industrial Engineering and Management Systems* (pp. 1–10).

6. Boehme, T. K., Kossow, A., & Preuss, W. (1992). A generalization of consecutive-k-out-of-n:F systems. *IEEE Transactions on Reliability*, 41(3), 451–457.

7. Boushaba, M. & Ghoraf, N. (2002). A 3-dimensional consecutive-k-out-n:F models. *International Journal of Reliability, Quality and Safety Engineering*, 9(2), 193–198.

8. Boutsikas, M. V. & Koutras, M. V. (2000). Generalized reliability bounds for coherent structures. *Journal of Applied Probability*, 37(3), 778–794.

9. Chang, G. J., Cui, L., & Hwang, F. K. (2000). *Reliabilities of Consecutive-k Systems*, volume 4. Kluwer Academic Publishers.

10. Cui, L. & Dong, Q. (2019). Consecutive k and related models—a survey. In *International Conference of Celebrating Professor Jinhua Cao's 80th Birthday* (pp. 3–18).: Springer.

11. Endharta, A. J. & Yun, W. Y. (2016). Preventive maintenance policy for linear consecutive-k-out-of-n:F system. *Journal of the Operations Research Society of Japan*, 59(4), 334–346.

12. Fu, J. C. (1986). Reliability of consecutive-k-out-of-n:F systems with $(k-1)$-step Markov dependence. *IEEE Transactions on Reliability*, 35(5), 602–606.

13. Fu, J. C. & Hu, B. (1987). On reliability of a large consecutive-k-out-of-n:F system with $(k-1)$-step Markov dependence. *IEEE Transactions on Reliability*, 36(1), 75 – 77.

14. Fu, J. C. & Koutras, M. V. (1994). Poisson approximations for 2-dimensional patterns. *Annals of the Institute of Statistical Mathematics*, 46(1), 179–192.

15. Gharib, M., El-Sayed, E., & Nashwan, I. (2010). Reliability of simple 3-dimensional consecutive-k-out-of-n:F systems. *Journal of Mathematics and Statistics*, 6(3), 261–264.

16. Godbole, A. P., Potter, L. K., & Sklar, J. K. (1998). Improved upper bounds for the reliability of d-dimensional consecutive-k-out-of-n:F systems. *Naval Research Logistics (NRL)*, 45(2), 219–230.

17. Hanafusa, T. & Yamamoto, H. (2000a). Branch and bound algorithm for the optimal assignment problem in linear consecutive-k-out-of-n:F system. *Reliability Engineering Association of Japan (in Japanese)*, 22(7), 641–661.

18. Hanafusa, T. & Yamamoto, H. (2000b). Genetic algorithm for optimal assignment problem in linear consecutive-k-out-of-n:F system. *Journal of Japan Industrial Management Association (in Japanese)*, 51(5), 293–301.

19. Huang, J., Zuo, M. J., & Fang, Z. (2003). Multi-state consecutive-k-out-of-n systems. *IIE Transactions*, 35(6), 527–534.

20. Hwang, F. K. (1982). Fast solutions for consecutive-k-out-of-n:F system. *IEEE Transactions on Reliability*, 31(5), 447–448.

21. Kontoleon, J. M. (1980). Reliability determination of a r-successive-out-of-n:F system. *IEEE Transactions on Reliability*, 29(5), 437.

22. Koutras, M. V., Papadopoulos, G. K., & Papastavridis, S. G. (1994). Note: Pairwise rearrangements in reliability structures. *Naval Research Logistics*, 41(5), 683–687.

23. Kuo, W., Zhang, W., & Zuo, M. (1990). A consecutive-k-out-of-n:G system: the mirror image of a consecutive-k-out-of-n:F system. *IEEE Transactions on Reliability*, 39(2), 244–253.

24. Kuo, W. & Zuo, M. J. (2003). *Optimal Reliability Modeling: Principles and Applications.* John Wiley and Sons.

25. Malinowski, J. & Preuss, W. (1995). A recursive algorithm evaluating the exact reliability of a consecutive k-within-m-out-of-n:F system. *Microelectronics Reliability*, 35(12), 1461–1465.

26. Malon, D. M. (1984). Optimal consecutive-2-out-of-n:F component sequencing. *IEEE Transactions on Reliability*, 33(5), 414–418.

27. Nakamura, T. (2020). *Study on Reliability Evaluation and Optimal Design of Connected-$\mathbf{X}$-out-of-(m, n):F Lattice Systems.* PhD thesis, Tokyo Metropolitan University. Japan.

28. Nakamura, T. & Yamamoto, H. (2020). Algorithm for solving optimal arrangement problem in connected-(r, s)-out-of-(m, n):F lattice system. *IEEE Transactions on Reliability*, 69(2), 497–509.

29. Nakamura, T., Yamamoto, H., & Akiba, T. (2018). Fast algorithm for optimal arrangement in connected-$(m-1, n-1)$-out-of-(m, n):F lattice system. *IEICE Transactions on Fundamentals of Electronics, Communications and Computer Sciences*, 101(12), 2446–2453.

30. Nakamura, T., Yamamoto, H., & Akiba, T. (2020). Reliability of a toroidal connected-(r, s)-out-of-(m, n):F lattice system. *Proceedings of the Institution of Mechanical Engineers, Part O: Journal of Risk and Reliability.*

31. Nakamura, T., Yamamoto, H., Shinzato, T., Xiao, X., & Akiba, T. (2017a). Proposal of calculation method for reliability of toroidal connected-$(1, 2)$-or-$(2, 1)$-out-of-(m, n):F lattice system with Markov chain. In S. Nakamura, C. H. Qian, & T. Nakagawa (Eds.), *Reliability Modeling with Computer and Maintenance Applications* chapter 7, (pp. 139–153). World Scientific.

32. Nakamura, T., Yamamoto, H., Shinzato, T., Xiao, X., & Akiba, T. (2017b). Reliability of a circular connected-$(1, 2)$-or-$(2, 1)$-out-of-(m, n):F lattice system with identical components. *IEICE Transactions on Fundamentals of Electronics, Communications and Computer Sciences*, 100(4), 1029–1036.

33. Nakamura, T., Yamamoto, H., Takashi, S., & Xiao, Xiao Akiba, T. (2016). Reliability of a toroidal connected-$(1, 2)$-or-$(2, 1)$-out-of-(m, n):F lattice system. In *Proceedings of the 7th Asia-Pacific International Symposium on Advanced Reliability and Maintenance Modeling* (pp. 399–406).

34. Omura, T., Akiba, T., Yamamoto, H., & Xiao, X. (2015a). Algorithm for obtaining optimal arrangement of a connected-(r, s)-out-of-(m, n):F system — the case of $m = r$ and $s = 2$ —. *IEICE Transactions on Fundamentals of Electronics, Communications and Computer Sciences*, E98-A(10), 2018–2024.

35. Omura, T., Yamamoto, H., Xiao, X., & Akiba, T. (2015b). Algorithm for obtaining optimal arrangement of a connected-(r, s)-out-of-(m, n):F system — the case of $m = r + 1$ and $2s > n$ —. In *Proceedings of the 16th Asia Pacific Industrial Engineering and Management Systems Conference* (pp. 652–659).

36. Salvia, A. A. & Lasher, W. C. (1990). 2-dimensional consecutive-k-out-of-n:F models. *IEEE Transactions on Reliability*, 39(3), 382–385.

37. Shingyochi, K. & Yamamoto, H. (2008). An efficient algorithm for solving optimal component arrangement problem in circular consecutive-k-out-of-n:F system. *Journal of Japan Industrial Management Association*, 58(6), 423–429.

38. Shingyochi, K. & Yamamoto, H. (2009). Efficient genetic algorithm for optimal arrangement in a linear consecutive-k-out-of-n:F system. *IEICE Transactions on Fundamentals of Electronics, Communications and Computer Sciences*, E92-A(7), 1578–1584.

39. Shingyochi, K. & Yamamoto, H. (2010). Comparative study of simulated annealing algorithms for optimal arrangement problems in a linear consecutive-k-out-of-n:F system. In *International Conference on Advanced Computer Science and Information Technology* (pp. 157–168).: Springer.

40. Shingyochi, K., Yamamoto, H., & Yamachi, H. (2012). Comparative study of several simulated annealing algorithms for optimal arrangement problems in a circular consecutive-k-out-of-n:F system. *Quality Technology & Quantitative Management*, 9(3), 295–303.

41. Yamamoto, H. (1996). Reliability of a connected-(r_1, s_1)-or-(r_2, s_2)-or-...-or-(r_k, s_k)-out-of-(m, n):F lattice system. *Microelectronics Reliability*, 36(2), 151–168.

42. Yamamoto, H. & Akiba, T. (1998). Efficient algorithm for the reliability of 2-dimensional consecutive-k-out-of-n:F system using finite Markov chain. *Reliability Engineering Association of Japan (in Japanese)*, 20(4), 256–257.

43. Yamamoto, H. & Akiba, T. (2003). Survey of reliability studies of multi-dimensional consecutive-k-out-of-n:F systems. *Reliability Engineering Association of Japan*, 25(8), 783–796.

44. Yamamoto, H. & Akiba, T. (2004). Efficient algorithm for the reliability of a 2-dimensional cylindrical k-within-consecutive-(r, s)-out-of-(m, n):F system. *IEICE Transactions on Fundamentals of Electronics, Communications and Computer Sciences*, 87(5), 1251–1257.

45. Yamamoto, H. & Akiba, T. (2005a). Evaluating methods for the reliability of a large 2-dimensional rectangular k-within-consecutive-(r, s)-out-of-(m, n):F system. *Naval Research Logistics (NRL)*, 52(3), 243–252.

46. Yamamoto, H. & Akiba, T. (2005b). A recursive algorithm for the reliability of a circular connected-(r, s)-out-of-(m, n):F lattice system. *Computers and Industrial Engineering*, 49(1), 21–34.

47. Yamamoto, H., Akiba, T., & Wakagi, Y. (2006a). Efficient algorithm for the reliability of an adjacent triangle:F triangular lattice system. *Reliability Engineering Association of Japan*, 28(1), 63–76.

48. Yamamoto, H., Akiba, T., & Yun, W. Y. (2006b). An efficient branch and bound algorithm for the optimal arrangement problem in the linear consecutive-k-out-of-r-from-n:F system. *Journal of Japan Industrial Management Association*, 56(6), 429–438.

49. Yamamoto, H. & Miyakawa, M. (1995). Reliability of a linear connected-(r, s)-out-of-(m, n):F lattice system. *IEEE Transactions on Reliability*, 44(2), 333–336.

50. Yamamoto, H. & Miyakawa, M. (1996). Reliability of a circular connected-(r, s)-out-of-(m, n):F lattice system. *Journal of the Operations Research Society of Japan*, 39(3), 389–406.

51. Yamamoto, H., Suzuki, A., & Nakamura, T. (2004). A branch and bound algorithm for solving facility layout problems using symmetries. *Journal of Japan Industrial Management Association (in Japanese)*, 55(5), 293–301.

52. Yamamoto, H., Zuo, M. J., Akiba, T., & Tian, Z. (2006c). Recursive formulas for the reliability of multi-state consecutive-k-out-of-n:G systems. *IEEE Transactions on Reliability*, 55(1), 98–104.

53. Zhao, X. & Cui, L. (2009). On the accelerated scan finite Markov chain imbedding approach. *IEEE Transactions on Reliability*, 58(2), 383–388.

54. Zhao, X., Cui, L., Zhao, W., & Liu, F. (2011). Exact reliability of a linear connected-(r, s)-out-of-(m, n):F system. *IEEE Transactions on Reliability*, 60(3), 689–698.

55. Zhou, L. (2020). *Study on Design and Maintenance Policies for Linear Consecutive-k-out-of-n:G Systems.* PhD thesis, Tokyo Metropolitan University. Japan.

56. Zhou, L., Yamamoto, H., Nakamura, T., & Xiao, X. (2020a). Optimization problems for consecutive-k-out-of-n:G systems. *IEICE Transactions on Fundamentals of Electronics, Communications and Computer Sciences*, E103-A(5), 741–748.

57. Zhou, L., Yamamoto, H., & Xiao, X. (2020b). Number of failed components in coherent systems. *Journal of Japan Industrial Management Association*, 71(2E), 92–98.

58. Zhu, X., Fu, Y., Yuan, T., & Wu, X. (2017). Birnbaum importance based heuristics for multi-type component assignment problems. *Reliability Engineering and System Safety*, 165, 209–221.

59. Zuo, M. J., Fang, Z., Huang, J., & Xu, X. (2003). Performance evaluation of decreasing multi-state consecutive-k-out-of-n:G systems. *International Journal of Reliability, Quality and Safety Engineering*, 10(03), 345–358.

9 Optimal Social Infrastructure Maintenance Models

Kodo Ito
Tottori University, Japan

Akihiro Yamane
Tottori University, Japan

Yoshiyuki Higuchi
Fukushima University, Japan

CONTENTS

9.1 INTRODUCTION

In 2023, 39% of road bridges and 27% of tunnels will be 50 years old in Japan, and large-scale maintenance of them has become a big national concern of the Japanese government, because such social infrastructures are fundamental and indispensable for our lives and economic activities [1]. Annual inspections and repairs of roads are performed by local governments which have serious financial difficulties because of the low fertility rate, the aging working population, and the regional economic downturn. So, efficient maintenance methods and plans which can innovatively solve the above issues are always needed.

Lots of studies concerning degradation prediction methods of roads and their optimal maintenance policies have been performed because infrastructures deteriorate with time stochastically and maintenance of them is a big issue. In case of highway rehabilitation, Friesz *et al.* and Markow *et al.* considered optimal rehabilitation frequencies of a single highway pavement [2, 3]. Tsunokawa *et al.* considered the optimal rehabilitation problem for multiple rehabilitations using the discontinuous pavement condition which is approximated by a linear deterioration rate continuous function [4]. Li *et al.* extended this consideration using the Markov decision process (MDP) [5]. Ouyang *et al.* considered optimal rehabilitation scheduling of multiple highway pavements as a mixed-integer nonlinear program and derived optimal rehabilitation timing for a single pavement analytically [6, 7]. Kobayshi *et al.* expressed the deterioration process of the pavement by the combined MDP model of the deterioration process of the road surface integrity and the deterioration process of the load capacity of the pavement, and proposed the optimal repair and update model to minimize the expected LCC [8]. In case of bridge and tunnel maintenance, Kaito *et al.* pointed out that Life Cycle Cost can separate average costs and relative costs in the case of bridge maintenance [9]. Sakai *et al.* proposed a model to predict deterioration of tunnel structures using a Poisson process [10]. Obama *et al.* proposed a methodology for designing optimal scrapping and maintenance policies for infrastructures that are to be demolished due to demand decline or deterioration progression [11]. Tsuda *et al.* proposed a methodology to estimate the Markovian transition probability model to forecast the deterioration process of bridge components [12]. Authors have also considered several optimal maintenance policies for infrastructures using a cumulative damage model and Markov chain model [13, 14, 15].

Comparing social infrastructure maintenance with maintenance of general mechanical systems, it has the following characteristics: 1) Some preventive maintenance might be delayed sometimes because the annual budget of local government has a maximum. 2) Maintenance costs change with their damage levels greatly. 3) Damage from natural disasters must be considered. We consider eight optimal maintenance models of social infrastructures considering the above characteristics applying a cumulative damage model. We obtain expected cost rates and discuss optimal damage levels which minimize them.

9.2 BASIC MODELS

First, we denote the fundamental model as follows [16]:

1) Social infrastructures such as roads and bridges are degraded with time by traffic, and its amount of damage is measured only at time $jT\,(j = 1, 2, \cdots)$ for a specified $T > 0$. Each amount W_j of damage between periodic times has an identical distribution $G(x) \equiv \mathrm{P_r}\{W_j \leq x\}$ with mean time $1/\mu$.

2) When the total damage exceeds a threshold level Z_K at time jT, the corrective maintenance (CM) is done at time jT, and when it is between Z_1 $(Z_1 < Z_K)$ and Z_K, the preventive maintenance (PM) is done at time jT.

3) CM cost is c_K and PM cost is $c_1(< c_K)$.

The expected cost rate is [17, p.45]

$$TC_0(Z_1) = \frac{c_K - (c_K - c_1)\left[G(Z_K) - \int_0^{Z_1} \overline{G}(Z_K - x)\mathrm{d}M_G(x)\right]}{1 + M_G(Z_1)}, \tag{9.1}$$

where $M_G(x) \equiv \sum_{j=1}^{\infty} G^{(j)}(x)$, $\phi^{(j)}(x)$ $(j = 1, 2, \cdots)$ denotes the j-fold Stieltjes convolution of distribution $\phi(x)$, and $\phi^{(0)}(x) \equiv 1$ for $x \geq 0$. Differentiating $C_0(Z_1)$ with respect to Z_1 and putting it equal to zero,

$$\int_{Z_K - Z_1}^{Z_K} [1 + M_G(Z_K - x)]\,\mathrm{d}G(x) = \frac{c_1}{c_K - c_1}, \tag{9.2}$$

whose left-hand side is strictly increasing from 0 to $M_G(Z_K)$. Thus, if $M_G(Z_K) > c_1/(c_K - c_1)$, then there exists a finite and unique Z_1^* $(0 < Z_1^* < Z_K)$ that satisfies (9.2), and its resulting cost rate is

$$TC_0(Z_1^*) = (c_K - c_1)\overline{G}(Z_K - Z_1^*). \tag{9.3}$$

Conversely, if $M_G(Z_K) \leq c_1/(c_K - c_1)$, then $Z_1^* = Z_K$, *i.e.*, the unit should be replaced only at failure, and the expected cost rate is

$$TC_0(Z_K) = \frac{c_K}{1 + M_G(Z_K)}. \tag{9.4}$$

We proposed four models which are expanded ones of the above fundamental model. Delayed maintenance is considered in Model 1. Multiple PM costs are introduced in Model 2, and multiple PM and additional replacement costs are introduced in Model 3. The delayed maintenance of multiple PM costs is considered in Model 4.

9.2.1 MODEL 1

We make assumptions 1) and 3) and rewrite 2) as follows:

2') When the total damage exceeds a threshold level Z_K at time jT, CM is done at time jT, and when it is between Z_1 and Z_K, PM is done at time $(j + 1)T$.

The probability that the total damage is between Z_1 and Z_K at time $(j+1)T$ and at time $(j + 2)T$, given that it is below Z_1 at time jT, is

$$\sum_{j=0}^{\infty} \int_0^{Z_1} \left[\int_0^{Z_K - x} G(Z_K - x - y)\mathrm{d}G(y)\right] \mathrm{d}G^{(j)}(x)$$
$$-\sum_{j=0}^{\infty} \int_0^{Z_1} \left[\int_0^{Z_1 - x} G(Z_K - x - y)\mathrm{d}G(y)\right] \mathrm{d}G^{(j)}(x)$$

$$= \sum_{j=0}^{\infty} \int_0^{Z_1} \left[\int_{Z_1-x}^{Z_K-x} G(Z_K - x - y) \mathrm{d}G(y) \right] \mathrm{d}G^{(j)}(x), \tag{9.5}$$

and the probability that it is between Z_1 and Z_K at time $(j+1)T$, given that it is below Z_1 at time jT, and exceeds Z_K at time $(j+2)T$ is

$$\sum_{j=0}^{\infty} \int_0^{Z_1} \left[\int_{Z_1-x}^{Z_K-x} \overline{G}(Z_K - x - y) \mathrm{d}G(y) \right] \mathrm{d}G^{(j)}(x) . \tag{9.6}$$

Thus, the mean time to maintenance is

$$\begin{aligned} & \sum_{j=0}^{\infty} (j+1)T \int_0^{Z_1} \overline{G}(Z_K - x) \mathrm{d}G^{(j)}(x) \\ & + \sum_{j=0}^{\infty} (j+2)T \int_0^{Z_1} \left[G(Z_K - x) - G(Z_1 - x) \right] \mathrm{d}G^{(j)}(x) \\ = {} & T \left[1 + G(Z_K) + \int_0^{Z_1} G(Z_K - x) \mathrm{d}M_G(x) \right], \end{aligned} \tag{9.7}$$

and the total expected cost until maintenance is

$$\begin{aligned} & c_K \sum_{j=0}^{\infty} \int_0^{Z_1} \overline{G}(Z_K - x) \mathrm{d}G^{(j)}(x) \\ & + c_1 \sum_{j=0}^{\infty} \int_0^{Z_1} \left[\int_{Z_1-x}^{Z_K-x} G(Z_K - x - y) \mathrm{d}G(y) \right] \mathrm{d}G^{(j)}(x) \\ & + c_K \sum_{j=0}^{\infty} \int_0^{Z_1} \left[\int_{Z_1-x}^{Z_K-x} \overline{G}(Z_K - x - y) \mathrm{d}G(y) \right] \mathrm{d}G^{(j)}(x) \\ = {} & c_K - (c_K - c_1) \left\{ \int_{Z_1}^{Z_K} G(Z_K - y) \mathrm{d}G(y) \right. \\ & \left. + \int_0^{Z_1} \left[\int_{Z_1-x}^{Z_K-x} G(Z_K - x - y) \mathrm{d}G(y) \right] \mathrm{d}M_G(x) \right\} . \end{aligned} \tag{9.8}$$

Thus, the expected cost rate is, from (9.7) and (9.8) [17, p.57],

$$TC_1(Z_1) = \frac{ \begin{array}{c} c_K - (c_K - c_1) \left\{ \int_{Z_1}^{Z_K} G(Z_K - y) \mathrm{d}G(y) \right. \\ \left. + \int_0^{Z_1} \left[\int_{Z_1-x}^{Z_K-x} G(Z_K - x - y) \mathrm{d}G(y) \right] \mathrm{d}M_G(x) \right\} \end{array} }{ 1 + G(Z_K) + \int_0^{Z_1} G(Z_K - x) \mathrm{d}M_G(x) } . \tag{9.9}$$

Assuming $G(x) = 1 - \exp(-\mu x)$, (9.9) is rewritten as,

$$TC_1(Z_1) = \frac{ c_K - (c_K - c_1) \left\{ 1 - [1 + \mu(Z_K - Z_1)] \mathrm{e}^{-\mu(Z_K - Z_1)} \right\} }{ 1 + \mu Z_1 + 1 - \mathrm{e}^{-\mu(Z_K - Z_1)} } . \tag{9.10}$$

We find optimal Z_1^* to minimize $C_1(Z_1^*)$. Differentiating $C_1(Z_1)$ with respect to Z_1 and putting it equal to zero,

$$\mathrm{e}^{-\mu(Z_K-Z_1)}\left[\frac{\mu(1+\mu Z_1)(Z_K-Z_1)}{1-\mathrm{e}^{-\mu(Z_K-Z_1)}}-1\right]=\frac{c_1}{c_K-c_1}. \tag{9.11}$$

Letting $Q_1(Z_1)$ denote the left-hand side of (9.11), $Q_1(Z_1)$ is strictly increasing in Z_1, and

$$Q_1(0)=\frac{\mu Z_K-\left(1-\mathrm{e}^{-\mu Z_K}\right)}{\mathrm{e}^{\mu Z_K}-1}, \quad Q_1(Z_K)=\mu Z_K.$$

Thus, if

$$\frac{\mu Z_K-\left(1-\mathrm{e}^{-\mu Z_K}\right)}{\mathrm{e}^{\mu Z_K}-1}<\frac{c_1}{c_K-c_1}<\mu Z_K,$$

then there exists a finite and unique Z_1^* $(0<Z_1^*<Z_K)$ which satisfies (9.11).

9.2.2 MODEL 2

We make assumption 1) of Model 1 and rewrite assumptions 2) and 3) as follows:

2") Degradation states are denoted as discrete degradation levels Z_i $(i=1,2,\cdots,K-1,K)$, where $Z_1<Z_2<\cdots<Z_K$ for $K>2$. When the total damage exceeds a threshold level Z_K at time jT, CM is done at time jT, and when it is between Z_1 and Z_K, PM is done at time jT.

3') CM cost is c_K and PM cost is c_i $(<c_{i+1})$ when the total damage is between Z_i and Z_{i+1} $(i=1,2,\cdots,K-1)$.

The probability that CM is done at time $(j+1)T$ is

$$\mathrm{P}_K = \sum_{j=0}^{\infty}\int_0^{Z_1}\overline{G}(Z_K-x)\mathrm{d}G^{(j)}(x), \tag{9.12}$$

and the probabilities that PM is done at time $(j+1)T$ are

$$\mathrm{P}_i=\sum_{j=0}^{\infty}\int_0^{Z_1}\left[G(Z_{i+1}-x)-G(Z_i-x)\right]\mathrm{d}G^{(j)}(x) \quad (i=1,2,\cdots,K-1). \tag{9.13}$$

The mean time to maintenance is

$$\sum_{j=0}^{\infty}(j+1)T\int_0^{Z_1}\overline{G}(Z_K-x)\mathrm{d}G^{(j)}(x)$$

$$
\begin{aligned}
&+\sum_{j=0}^{\infty}(j+1)T\sum_{i=1}^{K-1}\int_0^{Z_1}[G(Z_{i+1}-x)-G(Z_i-x)]\,\mathrm{d}G^{(j)}(x) \\
&= T\left[1+M_G(Z_1)\right]. \qquad (9.14)
\end{aligned}
$$

The total expected cost until maintenance is

$$
\begin{aligned}
\sum_{i=1}^{K}c_i\mathrm{P}_i &= c_K-\sum_{i=1}^{K-1}(c_K-c_i)\bigg\{G(Z_{i+1})-G(Z_i) \\
&\quad+\int_0^{Z_1}[G(Z_{i+1}-x)-G(Z_i-x)]\,\mathrm{d}M_G(x)\bigg\}. \qquad (9.15)
\end{aligned}
$$

Thus, the expected cost rate is, from (9.14) and (9.15) [17, p.59],

$$
TC_2(Z_1)=\frac{\begin{array}{c}c_K-\sum_{i=1}^{K-1}(c_K-c_i)\Big\{G(Z_{i+1})-G(Z_i) \\ +\int_0^{Z_1}[G(Z_{i+1}-x)-G(Z_i-x)]\,\mathrm{d}M_G(x)\Big\}\end{array}}{1+M_G(Z_1)}. \qquad (9.16)
$$

Because PM is done when the degradation level is greater than Z_1, $TC_2(Z_1)$ is expressed as a function of Z_1. We find optimal Z_1^* to minimize $C_2(Z_1)$. Differentiating $C_2(Z_1)$ with respect to Z_1 and putting it equal to zero,

$$
\sum_{i=2}^{K}(c_i-c_{i-1})\int_{Z_i-Z_1}^{Z_i}[1+M_G(Z_i-x)]\,\mathrm{d}G(x)=c_1\,. \qquad (9.17)
$$

Denoting the left-hand side of equation (9.17) by $Q_2(Z_1)$,

$$
\begin{aligned}
Q_2(0) &= 0, \\
Q_2(Z_2) &= (c_2-c_1)M_G(Z_2)+\sum_{i=3}^{K}(c_i-c_{i-1})\int_{Z_i-Z_2}^{Z_i}[1+M_G(Z_i-x)]\mathrm{d}G(x)\,.
\end{aligned}
$$

Thus, if $Q_2(Z_2)>c_1$, then there exists a finite and unique Z_1^* $(0<Z_1^*<Z_2)$ which satisfies (9.17).

When $K=2$, (9.17) is

$$
\int_{Z_2-Z_1}^{Z_2}[1+M_G(Z_2-x)]\,\mathrm{d}G(x)=\frac{c_1}{c_2-c_1}, \qquad (9.18)
$$

whose left-hand side increases strictly with Z_1 from 0 to $M_G(Z_2)$. Thus, if $M_G(Z_2)>c_1/(c_2-c_1)$, then there exists a finite and unique Z_1^* which satisfies (9.18). It can be easily shown that Z_1^* in (9.17) decreases with K to Z_1 in (9.18).

When $c_K > c_{K-1} = c_{K-2} = \cdots = c_2 > c_1$, the expected cost rate in (9.16) is

$$TC_2(Z_1) = \frac{\begin{array}{c} c_1 + (c_K - c_2)\left[\overline{G}(Z_K) - \int_0^{Z_1} \overline{G}(Z_K - x)\mathrm{d}M_G(x)\right] \\ +(c_2 - c_1)\left[\overline{G}(Z_2) - \int_0^{Z_1} \overline{G}(Z_2 - x)\mathrm{d}M_G(x)\right] \end{array}}{1 + M_G(Z_1)} . \tag{9.19}$$

(9.17) is

$$\begin{aligned} &(c_K - c_2)\int_{Z_K - Z_1}^{Z_K} [1 + M_G(Z_K - x)]\,\mathrm{d}G(x) \\ &+(c_2 - c_1)\int_{Z_2 - Z_1}^{Z_2} [1 + M_G(Z_2 - x)]\,\mathrm{d}G(x) = c_1 , \end{aligned} \tag{9.20}$$

whose left-hand side $Q_2(Z_1)$ increases strictly with Z_1 from 0 to

$$\begin{aligned} Q_2(Z_2) &= (c_K - c_2)\int_{Z_K - Z_2}^{Z_K} [1 + M_G(Z_K - x)]\,\mathrm{d}G(x) \\ &\quad +(c_2 - c_1)\int_0^{Z_2} [1 + M_G(Z_2 - x)]\,\mathrm{d}G(x) \\ &= (c_K - c_2)\int_{Z_K - Z_2}^{Z_K} [1 + M_G(Z_K - x)]\,\mathrm{d}G(x) + (c_2 - c_1)M_G(Z_2) . \end{aligned}$$

Thus, if $Q_2(Z_2) > c_1$, then there exists a finite and unique Z_1^* $(0 < Z_1^* < Z_2)$ which satisfies (9.20). It can be easily shown that Z_1^* in (9.17) is smaller than Z_1^* in (9.20).

When $G(x) = 1 - \exp(-\mu x)$, the expected cost rate in (9.16) is

$$TC_2(Z_1) = \frac{c_1 + \sum_{j=2}^{K}(c_j - c_{j-1})\mathrm{e}^{-\mu(Z_j - Z_1)}}{1 + \mu Z_1} , \tag{9.21}$$

and (9.17) is

$$\sum_{j=2}^{K}(c_j - c_{j-1})\mu Z_1 \mathrm{e}^{-\mu(Z_j - Z_1)} = c_1 , \tag{9.22}$$

whose left-hand side $Q_2(Z_1)$ increases strictly with Z_1 from 0 to

$$Q_2(Z_2) = \sum_{j=2}^{K}(c_j - c_{j-1})\mu Z_2 \mathrm{e}^{-\mu(Z_j - Z_2)} .$$

Thus, if $Q_2(Z_2) > c_1$, then there exists a finite and unique Z_1^* $(0 < Z_1^* < Z_2)$ which satisfies (9.22).

In particular, when $K = 2$, (9.22) is

$$\mu Z_1 e^{-\mu(Z_2 - Z_1)} = \frac{c_1}{c_2 - c_1}. \tag{9.23}$$

If $\mu Z_2 > c_1/(c_2 - c_1)$, then there exists a finite and unique Z_1^* $(0 < Z_1^* < Z_2)$ which satisfies (9.23).

9.3 MODEL 3

We make assumptions 1) of Model 1 and 2") of Model 2 and rewrite assumption 3) as follows:

3") CM cost is $c_K + c_0(Z_K)$ and PM cost is $c_i + c_0(x)$ when the total damage is between Z_i and Z_{i+1} $(i = 1, 2, \cdots, K-1)$, where $c_0(x)$ is an additional replacement cost that is variable for the total damage x with $c_0(0) = 0$.

The expected cost until maintenance is

$$\begin{aligned}
&[c_K + c_0(Z_K)] \sum_{j=0}^{\infty} \int_0^{Z_1} \overline{G}(Z_K - x) \mathrm{d}G^{(j)}(x) \\
&+ \sum_{i=1}^{K-1} \sum_{j=0}^{\infty} \int_0^{Z_1} \left\{ \int_{Z_i - x}^{Z_{i+1} - x} [c_i + c_0(x + y)] \mathrm{d}G(y) \right\} \mathrm{d}G^{(j)}(x) \\
=\ & c_K - \sum_{i=1}^{K-1} (c_K - c_i) \Bigg\{ G(Z_{i+1}) - G(Z_i) \\
&+ \int_0^{Z_1} [G(Z_{i+1} - x) - G(Z_i - x)]\, \mathrm{d}M_G(x) \Bigg\} \\
&+ \sum_{j=0}^{\infty} \int_0^{Z_1} \left[\int_x^{Z_K} \overline{G}(y - x) \mathrm{d}c_0(y) \right] \mathrm{d}G^{(j)}(x).
\end{aligned} \tag{9.24}$$

Thus, the expected cost rate is, from (9.14) and (9.24),

$$TC_3(Z_1) = \frac{\begin{aligned} c_K - \textstyle\sum_{i=1}^{K-1} (c_K - c_i) \Big\{ G(Z_{i+1}) - G(Z_i) \\ + \textstyle\int_0^{Z_1} [G(Z_{i+1} - x) - G(Z_i - x)]\, \mathrm{d}M_G(x) \Big\} \\ + \textstyle\sum_{j=0}^{\infty} \int_0^{Z_1} \left[\int_x^{Z_K} \overline{G}(y - x) \mathrm{d}c_0(y) \right] \mathrm{d}G^{(j)}(x) \end{aligned}}{1 + M_G(Z_1)}. \tag{9.25}$$

When $G(x) = 1 - \exp(-\mu x)$ and $c_0(x) = c_0 x$, (9.25) is rewritten as

$$TC_3(Z_1) = \frac{\begin{array}{c} c_K - \sum_{i=2}^{K}(c_i - c_{i-1})\left[1 - \mathrm{e}^{-\mu(Z_i - Z_1)}\right] \\ -c_0\mathrm{e}^{-\mu(Z_K - Z_1)}/\mu \end{array}}{1 + \mu Z_1} + \frac{c_0}{\mu}. \tag{9.26}$$

We find optimal Z_1^* to minimize $C_3(Z_1)$. Differentiating $C_3(Z_1)$ with respect to Z_1 and putting it equal to zero,

$$\mu Z_1 \mathrm{e}^{\mu Z_1}\left[\sum_{i=2}^{K}(c_i - c_{i-1})\mathrm{e}^{-\mu Z_i} - \frac{c_0}{\mu}\mathrm{e}^{-\mu Z_K}\right] = c_1 . \tag{9.27}$$

Denoting the left-hand side of equation (9.27) by $Q_3(Z_1)$,

$$\begin{aligned} Q_3(0) &= 0, \\ Q_3(Z_2) &= \mu Z_2(c_2 - c_1) + \mu Z_2 \mathrm{e}^{\mu Z_2}\left[\sum_{i=3}^{K}(c_i - c_{i-1})\mathrm{e}^{-\mu Z_i} - \frac{c_0}{\mu}\mathrm{e}^{-\mu Z_K}\right]. \end{aligned}$$

Furthermore, because $\exp(-\mu Z_i) > \exp(-\mu Z_K)$, if $(c_K - c_2) > c_0/\mu$, then $Q_3(Z_1)$ increases strictly with Z_1 to $Q_3(Z_2)$. Thus, if $Q_3(Z_2) > c_1$, then there exists a finite and unique Z_1^* $(0 < Z_1^* < Z_2)$ which satisfies (9.27). When $K = 2$, if $c_2 - c_1 - c_0/\mu > 0$ and

$$\mu Z_2 > \frac{c_1}{c_2 - c_1 - c_0/\mu},$$

then a finite Z_1^* exists.

9.4 MODEL 4

We consider the following maintenance policy and assumptions 1) and 3) of Model 2 and rewrite assumption 2) as follows.

2"') Degradation states of roads are denoted as discrete degradation levels Z_i $(i = 1, 2, \cdots, L-1, L, \cdots, K-1, K)$. When the cumulative damage exceeds a threshold level Z_K at time jT, the CM is done at time jT. The PM is done at time $(j+1)T$ when the cumulative damage is between $Z_1(< Z_L)$ and Z_L, and is done at time jT when the cumulative damage is between $Z_L(< Z_K)$ and Z_K.

The probability that CM is done at time $(j+2)T$ is

$$\sum_{i=1}^{L-1}\sum_{j=0}^{\infty}\int_0^{Z_1}\left[\int_{Z_i - x}^{Z_{i+1}-x}\overline{G}(Z_K - x - y)\mathrm{d}G(y)\right]\mathrm{d}G^{(j)}(x), \tag{9.28}$$

the probabilities that PM is done at time $(j+2)T$ is

$$\sum_{i=1}^{L-1}\sum_{j=0}^{\infty}\int_0^{Z_1}\left[\int_{Z_i - x}^{Z_{i+1}-x}G(Z_{i+1} - x - y)\mathrm{d}G(y)\right]\mathrm{d}G^{(j)}(x)$$

$$+\sum_{i=1}^{L-1}\sum_{j=0}^{\infty}\int_0^{Z_1}\left\{\sum_{n=i+1}^{K-1}\int_{Z_i-x}^{Z_{i+1}-x}\Big[G(Z_{n+1}-x-y)-G(Z_n-x-y)\Big]\mathrm{d}G(y)\right\}\mathrm{d}G^{(j)}(x)$$

$$+\sum_{i=L}^{K-1}\sum_{j=0}^{\infty}\int_0^{Z_1}\left[G(Z_{i+1}-x)-G(Z_i-x)\right]\mathrm{d}G^{(j)}(x), \tag{9.29}$$

where (9.12)+(9.28)+(9.29)=1.
The mean time to maintenance is

$$\sum_{j=0}^{\infty}(j+1)T\int_0^{Z_1}\overline{G}(Z_K-x)\mathrm{d}G^{(j)}(x)$$
$$+\sum_{i=1}^{L-1}\sum_{j=0}^{\infty}(j+2)T\int_0^{Z_1}\left[\int_{Z_i-x}^{Z_{i+1}-x}\overline{G}(Z_K-x-y)\mathrm{d}G(y)\right]\mathrm{d}G^{(j)}(x)$$
$$+\sum_{i=1}^{L-1}\sum_{j=0}^{\infty}(j+2)T\int_0^{Z_1}\left[\int_{Z_i-x}^{Z_{i+1}-x}G(Z_{i+1}-x-y)\mathrm{d}G(y)\right]\mathrm{d}G^{(j)}(x)$$
$$+\sum_{i=1}^{L-1}\sum_{j=0}^{\infty}(j+2)T\int_0^{Z_1}\left\{\sum_{n=i+1}^{K-1}\int_{Z_i-x}^{Z_{i+1}-x}\Big[G(Z_{n+1}-x-y)-G(Z_n-x-y)\Big]\mathrm{d}G(y)\right\}\mathrm{d}G^{(j)}(x)$$
$$+\sum_{i=L}^{K-1}\sum_{j=0}^{\infty}(j+1)T\int_0^{Z_1}\left[G(Z_{i+1}-x)-G(Z_i-x)\right]\mathrm{d}G^{(j)}(x)$$
$$=T\left[1+G(Z_L)+\int_0^{Z_1}G(Z_L-x)\mathrm{d}M_G(x)\right]. \tag{9.30}$$

The total expected cost until maintenance is

$$c_K\sum_{j=0}^{\infty}\int_0^{Z_1}\overline{G}(Z_K-x)\mathrm{d}G^{(j)}(x)$$
$$+c_K\sum_{i=1}^{L-1}\sum_{j=0}^{\infty}\int_0^{Z_1}\left[\int_{Z_i-x}^{Z_{i+1}-x}\overline{G}(Z_K-x-y)\mathrm{d}G(y)\right]\mathrm{d}G^{(j)}(x)$$
$$+\sum_{i=1}^{L-1}\sum_{j=0}^{\infty}c_i\int_0^{Z_1}\left[\int_{Z_i-x}^{Z_{i+1}-x}G(Z_{i+1}-x-y)\mathrm{d}G(y)\right]\mathrm{d}G^{(j)}(x)$$
$$+\sum_{i=1}^{L-1}\sum_{j=0}^{\infty}\int_0^{Z_1}\left\{\sum_{n=i+1}^{K-1}c_n\int_{Z_i-x}^{Z_{i+1}-x}\Big[G(Z_{n+1}-x-y)\right.$$

$$-G(Z_n - x - y)\Big]\mathrm{d}G(y)\Bigg\}\mathrm{d}G^{(j)}(x)$$

$$+\sum_{i=L}^{K-1} c_i \sum_{j=0}^{\infty}\int_0^{Z_1}[G(Z_{i+1}-x)-G(Z_i-x)]\,\mathrm{d}G^{(j)}(x)$$

$$= c_K - \sum_{i=1}^{L-1}\sum_{j=0}^{\infty}(c_K - c_i)\int_0^{Z_1}\left[\int_{Z_i-x}^{Z_{i+1}-x} G(Z_{i+1}-x-y)\mathrm{d}G(y)\right]\mathrm{d}G^{(j)}(x)$$

$$-\sum_{i=1}^{L-1}\sum_{j=0}^{\infty}\int_0^{Z_1}\Bigg\{\sum_{n=i+1}^{K-1}(c_K - c_n)\int_{Z_i-x}^{Z_{i+1}-x}\Big[G(Z_{n+1}-x-y)$$

$$-G(Z_n - x - y)\Big]\mathrm{d}G(y)\Bigg\}\mathrm{d}G^{(j)}(x)$$

$$-\sum_{i=L}^{K-1}\sum_{j=0}^{\infty}(c_K - c_i)\int_0^{Z_1}[G(Z_{i+1}-x)-G(Z_i-x)]\,\mathrm{d}G^{(j)}(x)$$

$$= c_K - \sum_{i=L+1}^{K}(c_i - c_{i-1})\Bigg\{G(Z_i) - G(Z_L)$$

$$+\int_0^{Z_1}[G(Z_i - x) - G(Z_L - x)]\,\mathrm{d}M_G(x)\Bigg\}$$

$$-\sum_{i=2}^{L}\sum_{j=i}^{K}(c_j - c_{j-1})\Bigg\{\int_{Z_{i-1}}^{Z_i} G(Z_j - x)\mathrm{d}G(x)$$

$$+\int_0^{Z_1}\left[\int_{Z_{i-1}-x}^{Z_i-x} G(Z_j - x - y)\mathrm{d}G(y)\right]\mathrm{d}M_G(x)\Bigg\}. \quad (9.31)$$

Thus, the expected cost rate is from (9.30) and (9.31),

$$TC_4(Z_1) = \frac{\begin{array}{c} c_K - \sum_{i=L+1}^{K}(c_i - c_{i-1})\Big\{G(Z_i) - G(Z_L) \\ + \int_0^{Z_1}[G(Z_i - x) - G(Z_L - x)]\,\mathrm{d}M_G(x)\Big\} \\ -\sum_{i=2}^{L}\sum_{j=i}^{K}(c_j - c_{j-1})\left\{\int_{Z_{i-1}}^{Z_i} G(Z_j - x)\mathrm{d}G(x)\right. \\ \left. + \int_0^{Z_1}\left[\int_{Z_{i-1}-x}^{Z_i-x} G(Z_j - x - y)\mathrm{d}G(y)\right]\mathrm{d}M_G(x)\right\} \end{array}}{1 + G(Z_L) + \int_0^{Z_1} G(Z_L - x)\mathrm{d}M_G(x)}. \quad (9.32)$$

Because it is difficult to discuss optimal policies analytically, it is assumed that $G(x) = 1 - \exp(-\mu x)$. Then, (9.32) is rewritten as

$$TC_4(Z_1) = \frac{\begin{array}{c} c_1 + \sum_{i=2}^{K}(c_i - c_{i-1})\mathrm{e}^{-\mu(Z_i - Z_1)} \\ + \sum_{i=2}^{L}\sum_{j=i}^{K}(c_j - c_{j-1})\mu(Z_i - Z_{i-1})\mathrm{e}^{-\mu(Z_j - Z_1)} \end{array}}{2 + \mu Z_1 - \mathrm{e}^{-\mu(Z_L - Z_1)}} \quad (9.33)$$

We seek an optimal Z_1^* which minimizes $C_4(Z_1)$. Differentiating (9.33) with respect to Z_1 and setting it equal to zero,

$$\sum_{i=2}^{L}(c_i - c_{i-1})\mathrm{e}^{-\mu(Z_i-Z_1)}\left[\frac{(1+\mu Z_1)\mu(Z_i - Z_1)}{1-\mathrm{e}^{-\mu(Z_L-Z_1)}} - 1\right]$$
$$+\left[\frac{(1+\mu Z_1)\mu(Z_L - Z_1)}{1-\mathrm{e}^{-\mu(Z_L-Z_1)}} - 1\right]\sum_{i=L+1}^{K}(c_i - c_{i-1})\mathrm{e}^{-\mu(Z_i-Z_1)} = c_1\,. \quad (9.34)$$

9.5 EXTENDED MODELS

When maintenance of mechanical systems is planned, we do not consider natural disasters, but we have to consider natural disaster damages when maintenance of social infrastructure is planned because the scale of social infrastructure is large and it is inevitably affected by natural disasters. We propose the respective extended models of Models 1, 2, 3 and 4 considering a natural disaster and its recovery.

9.5.1 MODEL 5

We make assumptions 1), 2'), and 3) of Model 1 and add assumption 4):

4) Systems are damaged by a natural disaster which occurs independently of steady deteriorations, and its disaster recovery (DR) is done. The occurrence time of a natural disaster has a distribution $F(t)$ and its average DR cost is $c_D(> c_K)$.

The probability that CM is done at time $(j+1)T$ is

$$\sum_{j=0}^{\infty}\overline{F}((j+1)T)\int_0^{Z_1}\overline{G}(Z_K - x)\mathrm{d}G^{(j)}(x)$$
$$+\sum_{j=0}^{\infty}\overline{F}((j+2)T)\int_0^{Z_1}\left[\int_{Z_1-x}^{Z_K-x}\overline{G}(Z_K - x - y)\mathrm{d}G(y)\right]\mathrm{d}G^{(j)}(x)\,, \quad (9.35)$$

the probability that PM is done at time $(j+1)T$ is

$$\sum_{j=0}^{\infty}\overline{F}((j+2)T)\int_0^{Z_1}\left[\int_{Z_1-x}^{Z_K-x}G(Z_K - x - y)\mathrm{d}G(y)\right]\mathrm{d}G^{(j)}(x)\,, \quad (9.36)$$

and the probability that DR is done at time $(j+1)T$ is

$$\sum_{j=0}^{\infty}F((j+1)T)\int_0^{Z_1}\overline{G}(Z_K - x)\mathrm{d}G^{(j)}(x)$$

$$
\begin{aligned}
&+\sum_{j=0}^{\infty} F((j+2)T) \int_0^{Z_1} \left[\int_{Z_1-x}^{Z_K-x} \overline{G}(Z_K-x-y)\mathrm{d}G(y) \right] \mathrm{d}G^{(j)}(x) \\
&+\sum_{j=0}^{\infty} F((j+2)T) \int_0^{Z_1} \left[\int_{Z_1-x}^{Z_K-x} G(Z_K-x-y)\mathrm{d}G(y) \right] \mathrm{d}G^{(j)}(x) \\
= &\sum_{j=0}^{\infty} F((j+1)T) \int_0^{Z_1} \overline{G}(Z_K-x)\mathrm{d}G^{(j)}(x) \\
&+\sum_{j=0}^{\infty} F((j+2)T) \int_0^{Z_1} \left[G(Z_K-x)-G(Z_1-x)\right] \mathrm{d}G^{(j)}(x), \qquad (9.37)
\end{aligned}
$$

where note that (9.35) + (9.36) + (9.37) = 1. The mean time to maintenance is

$$
\begin{aligned}
&\sum_{j=0}^{\infty} [(j+1)T]\overline{F}((j+1)T) \int_0^{Z_1} \overline{G}(Z_K-x)\mathrm{d}G^{(j)}(x) \\
&+\sum_{j=0}^{\infty} [(j+2)T]\overline{F}((j+2)T) \int_0^{Z_1} \left[G(Z_K-x)-G(Z_1-x)\right] \mathrm{d}G^{(j)}(x) \\
&+\sum_{j=0}^{\infty} \int_0^{(j+1)T} t\mathrm{d}F(t) \int_0^{Z_1} \overline{G}(Z_K-x)\mathrm{d}G^{(j)}(x) \\
&+\sum_{j=0}^{\infty} \int_0^{(j+2)T} t\mathrm{d}F(t) \int_0^{Z_1} \left[G(Z_K-x)-G(Z_1-x)\right] \mathrm{d}G^{(j)}(x) \\
= &\int_0^{T} \overline{F}(t)\mathrm{d}t + \sum_{j=0}^{\infty} \int_{(j+1)T}^{(j+2)T} \overline{F}(t)\mathrm{d}t \int_0^{Z_1} G(Z_K-x)\mathrm{d}G^{(j)}(x), \qquad (9.38)
\end{aligned}
$$

and the total expected cost until maintenance is

$$
\begin{aligned}
&c_K\Bigg\{ \sum_{j=0}^{\infty} \overline{F}((j+1)T) \int_0^{Z_1} \overline{G}(Z_K-x)\mathrm{d}G^{(j)}(x) \\
&+\sum_{j=0}^{\infty} \overline{F}((j+2)T) \int_0^{Z_1} \left[\int_{Z_1-x}^{Z_K-x} \overline{G}(Z_K-x-y)\mathrm{d}G(y) \right] \mathrm{d}G^{(j)}(x) \Bigg\} \\
&+c_1 \sum_{j=0}^{\infty} \overline{F}((j+2)T) \int_0^{Z_1} \left[\int_{Z_1-x}^{Z_K-x} G(Z_K-x-y)\mathrm{d}G(y) \right] \mathrm{d}G^{(j)}(x) \\
&+c_D\Bigg\{ \sum_{j=0}^{\infty} F((j+1)T) \int_0^{Z_1} \overline{G}(Z_K-x)\mathrm{d}G^{(j)}(x) \\
&+\sum_{j=0}^{\infty} F((j+2)T) \int_0^{Z_1} \left[G(Z_K-x)-G(Z_1-x)\right] \mathrm{d}G^{(j)}(x) \Bigg\}
\end{aligned}
$$

$$
\begin{aligned}
&= c_K + (c_D - c_K)\sum_{j=0}^{\infty}\Big\{F((j+1)T)\int_0^{Z_1}\overline{G}(Z_K - x)\mathrm{d}G^{(j)}(x) \\
&\quad +F((j+2)T)\int_0^{Z_1}[G(Z_K - x) - G(Z_1 - x)]\,\mathrm{d}G^{(j)}(x)\Big\} \\
&\quad -(c_K - c_1)\sum_{j=0}^{\infty}\overline{F}((j+2)T) \\
&\quad \times\int_0^{Z_1}\left[\int_{Z_1-x}^{Z_K-x} G(Z_K - x - y)\mathrm{d}G(y)\right]\mathrm{d}G^{(j)}(x). \qquad (9.39)
\end{aligned}
$$

Thus, the expected cost rate is, from (9.38) and (9.39),

$$
C_5(Z_1) = \frac{\begin{array}{l} c_K + (c_D - c_K) \\ \quad \times\sum_{j=0}^{\infty}\Big\{F((j+1)T)\int_0^{Z_1}\overline{G}(Z_K - x)\mathrm{d}G^{(j)}(x) \\ +F((j+2)T) \\ \quad \times\int_0^{Z_1}[G(Z_K - x) - G(Z_1 - x)]\,\mathrm{d}G^{(j)}(x)\Big\} \\ -(c_K - c_1)\sum_{j=0}^{\infty}\overline{F}((j+2)T) \\ \quad \times\int_0^{Z_1}\left[\int_{Z_1-x}^{Z_K-x} G(Z_K - x - y)\mathrm{d}G(y)\right]\mathrm{d}G^{(j)}(x) \end{array}}{\begin{array}{l}\int_0^T\overline{F}(t)\mathrm{d}t + \sum_{j=0}^{\infty}\int_{(j+1)T}^{(j+2)T}\overline{F}(t)\mathrm{d}t \\ \quad \times\int_0^{Z_1} G(Z_K - x)\mathrm{d}G^{(j)}(x)\end{array}}. \qquad (9.40)
$$

When $\overline{F}(t) = 1$, (9.40) agrees with (9.9).

Assuming $F(t) = 1 - \exp(-\lambda t)$ and $G(x) = 1 - \exp(-\mu x)$, (9.40) is rewritten as

$$
\frac{C_5(Z_1)}{\lambda} = \frac{\begin{array}{l} c_D - \alpha\mathrm{e}^{\alpha\mu Z_1}[(c_D - c_K)\mathrm{e}^{-\mu Z_K} \\ \quad +(c_D - c_1)\alpha\left(\mathrm{e}^{-\mu Z_1} - \mathrm{e}^{-\mu Z_K}\right) \\ \quad -(c_K - c_1)\mu\alpha(Z_K - Z_1)\mathrm{e}^{-\mu Z_K}]\end{array}}{1 - \alpha\mathrm{e}^{\alpha\mu Z_1}\left[\alpha\mathrm{e}^{-\mu Z_1} + (1-\alpha)\mathrm{e}^{-\mu Z_K}\right]}, \qquad (9.41)
$$

where $\alpha \equiv \exp(-\lambda T)$. We find optimal Z_1^* to minimize $C_5(Z_1)$. Differentiating $C_5(Z_1)$ with respect to Z_1 and putting it equal to zero,

$$
\frac{\alpha}{1-\alpha}\mathrm{e}^{-\mu(Z_K - \alpha Z_1)}\left[\mu(Z_K - Z_1)\frac{\mathrm{e}^{(1-\alpha)\mu Z_1} - \alpha}{1 - \mathrm{e}^{-\mu(Z_K - Z_1)}} - (1-\alpha)\right] = \frac{c_1}{c_K - c_1} \qquad (9.42)
$$

Letting $Q_5(Z_1)$ denote the left-hand of (9.42), and differentiating $Q_5(Z_1)$ with respect to Z_1,

$$
\begin{aligned}
&\left[-\left(\mathrm{e}^{\mu Z_1} - \alpha\mathrm{e}^{\alpha\mu Z_1}\right) + \mu(Z_K - Z_1)\left(\mathrm{e}^{\mu Z_1} - \alpha^2\mathrm{e}^{\alpha\mu Z_1}\right)\right]\left[1 - \mathrm{e}^{-\mu(Z_K - Z_1)}\right] \\
&+\mu(Z_K - Z_1)\left(\mathrm{e}^{\mu Z_1} - \alpha\mathrm{e}^{\alpha\mu Z_1}\right)\mathrm{e}^{-\mu(Z_K - Z_1)} - (1-\alpha)\alpha\mathrm{e}^{\alpha\mu Z_1}\left[1 - \mathrm{e}^{-\mu(Z_K - Z_1)}\right]^2
\end{aligned}
$$

$$> -\left(e^{\mu Z_1} - \alpha e^{\alpha\mu Z_1}\right)\left[1 - e^{-\mu(Z_K - Z_1)}\right] + \left(e^{\mu Z_1} - \alpha^2 e^{\alpha\mu Z_1}\right)\left[1 - e^{-\mu(Z_K - Z_1)}\right]^2$$
$$+ \left(e^{\mu Z_1} - \alpha e^{\alpha\mu Z_1}\right) e^{-\mu(Z_K - Z_1)}\left[1 - e^{-\mu(Z_K - Z_1)}\right]$$
$$-(1-\alpha)\alpha e^{\alpha\mu Z_1}\left[1 - e^{-\mu(Z_K - Z_1)}\right]^2 = 0\,.$$

Because $x/(1-\exp(-x))$ is strictly increasing in x from 1 to ∞, $\mu(Z_K - Z_1) > 1 - \exp[-\mu(Z_K - Z_1)]$. Thus, $Q_5(Z_1)$ is strictly increasing in Z_1, and

$$Q_5(0) = \alpha e^{-\mu Z_K}\left[\frac{\mu Z_K}{1 - e^{-\mu Z_K}} - 1\right], \quad Q_5(Z_K) = \frac{\alpha}{1-\alpha}\left[1 - e^{-(1-\alpha)\mu Z_K}\right].$$

Thus, if $Q_5(Z_K) > c_1/(c_K - c_1) > Q_5(0)$, then there exists a finite and unique $Z_1^*(0 < Z_1^* < Z_K)$ which satisfies (9.42).

9.5.2 MODEL 6

We make assumptions 1), 2”), and 3’) of Model 2 and assumption 4) of Model 5. The probability that CM is done at time $(j+1)T$ is

$$\sum_{j=0}^{\infty} \overline{F}((j+1)T)\int_0^{Z_1} \overline{G}(Z_K - x)\mathrm{d}G^{(j)}(x)\,, \tag{9.43}$$

the probability that PM is done at time $(j+1)T$ is

$$\sum_{i=1}^{K-1}\sum_{j=0}^{\infty} \overline{F}((j+1)T)\int_0^{Z_1} \left[G(Z_{i+1} - x) - G(Z_i - x)\right]\mathrm{d}G^{(j)}(x)\,, \tag{9.44}$$

and the probability that DR is done at time $(j+1)T$ is

$$\sum_{j=0}^{\infty} F((j+1)T)\int_0^{Z_1} \overline{G}(Z_1 - x)\mathrm{d}G^{(j)}(x)\,. \tag{9.45}$$

The mean time to maintenance is

$$\sum_{j=0}^{\infty}(j+1)T\overline{F}((j+1)T)\int_0^{Z_1} \overline{G}(Z_K - x)\mathrm{d}G^{(j)}(x)$$
$$+\sum_{i=1}^{K-1}\sum_{j=0}^{\infty}(j+1)T\overline{F}((j+1)T)\int_0^{Z_1} \left[G(Z_{i+1} - x) - G(Z_i - x)\right]\mathrm{d}G^{(j)}(x)$$
$$+\sum_{j=0}^{\infty}\int_0^{(j+1)T} t\mathrm{d}F(t)\int_0^{Z_1} \overline{G}(Z_1 - x)\mathrm{d}G^{(j)}(x)$$
$$= \sum_{j=0}^{\infty} G^{(j)}(Z_1)\int_{jT}^{(j+1)T} \overline{F}(t)\mathrm{d}t\,, \tag{9.46}$$

and the total expected cost until maintenance is

$$
\begin{aligned}
& c_K \sum_{j=0}^{\infty} \overline{F}((j+1)T) \int_0^{Z_1} \overline{G}(Z_K - x)\mathrm{d}G^{(j)}(x) \\
& + \sum_{i=1}^{K-1} c_i \sum_{j=0}^{\infty} \overline{F}((j+1)T) \int_0^{Z_1} \left[G(Z_{i+1} - x) - G(Z_i - x)\right] \mathrm{d}G^{(j)}(x) \\
& + c_D \sum_{j=0}^{\infty} F((j+1)T) \int_0^{Z_1} \overline{G}(Z_1 - x)\mathrm{d}G^{(j)}(x) \\
= \; & c_K - \sum_{i=1}^{K-1} (c_K - c_i) \sum_{j=0}^{\infty} \overline{F}((j+1)T) \\
& \times \int_0^{Z_1} [G(Z_{i+1} - x) - G(Z_i - x)]\mathrm{d}G^{(j)}(x) \\
& + (c_D - c_K) \sum_{j=0}^{\infty} F((j+1)T)[G^{(j)}(Z_1) - G^{(j+1)}(Z_1)] \\
= \; & c_K - \sum_{i=2}^{K} (c_i - c_{i-1}) \sum_{j=0}^{\infty} \overline{F}((j+1)T) \int_0^{Z_1} G(Z_i - x)\mathrm{d}G^{(j)}(x) \\
& + (c_K - c_1) \sum_{j=1}^{\infty} \overline{F}(jT) G^{(j)}(Z_1) \\
& + (c_D - c_K) \sum_{j=0}^{\infty} F((j+1)T)[G^{(j)}(Z_1) - G^{(j+1)}(Z_1)] \,.
\end{aligned} \tag{9.47}
$$

Thus, the expected cost rate is, from (9.46) and (9.47),

$$
C_6(Z_1) = \frac{\begin{array}{l} c_K - \sum_{i=2}^{K} (c_i - c_{i-1}) \sum_{j=0}^{\infty} \overline{F}((j+1)T) \\ \times \int_0^{Z_1} G(Z_i - x)\mathrm{d}G^{(j)}(x) \\ + (c_K - c_1) \sum_{j=0}^{\infty} \overline{F}(jT) G^{(j)}(Z_1) \\ + (c_D - c_K) \sum_{j=0}^{\infty} F((j+1)T)[G^{(j)}(Z_1) - G^{(j+1)}(Z_1)] \end{array}}{\sum_{j=0}^{\infty} G^{(j)}(Z_1) \int_{jT}^{(j+1)T} \overline{F}(t)\mathrm{d}t} \,. \tag{9.48}
$$

When $F(t) = 1 - \exp(-\lambda t)$ and $G(x) = 1 - \exp(-\mu x)$, (9.48) is

$$
\begin{aligned}
\frac{C_6(Z_1)}{\lambda} = \; & \frac{c_K - \alpha \mathrm{e}^{-(1-\alpha)\mu Z_1} \sum_{i=2}^{K} (c_i - c_{i-1}) \left[1 - \mathrm{e}^{-\mu(Z_i - Z_1)}\right]}{1 - \alpha \mathrm{e}^{-(1-\alpha)\mu Z_1}} \\
& + c_D - c_K \,,
\end{aligned} \tag{9.49}
$$

where $\alpha \equiv \exp(-\lambda T)$. We find optimal Z_1^* to minimize $C_6(Z_1)$. Differentiating

$C_6(Z_1)$ with Z_1 and putting it equal to zero,

$$\frac{\alpha}{1-\alpha}\left[1-\mathrm{e}^{-(1-\alpha)\mu Z_1}\right]\sum_{i=2}^{K}(c_i-c_{i-1})\mathrm{e}^{-\mu(Z_i-Z_1)}=c_1\,. \tag{9.50}$$

Denoting the left-hand side of equation (9.50) by $Q_6(Z_1;\alpha)$,

$$\begin{aligned} Q_6(0;\alpha) &= 0\,, \\ Q_6(Z_2;\alpha) &= \frac{\alpha}{1-\alpha}\left[1-\mathrm{e}^{-(1-\alpha)\mu Z_2}\right]\left[c_2-c_1-\sum_{i=3}^{K}(c_i-c_{i-1})\mathrm{e}^{-\mu(Z_i-Z_2)}\right]. \end{aligned}$$

Thus, if $Q_6(Z_2;\alpha) > c_1$, then there exists a finite and unique $Z_1^*(\alpha)$ $(0 < Z_1^*(\alpha) < Z_2)$ which satisfies (9.50). Furthermore, $Q_6(Z_1;\alpha)$ increases strictly with α from 0 to $\mu Z_1\sum_{j=2}^{K}(c_j-c_{j-1})\exp[-\mu(Z_j-Z_1)]$. When $\alpha=1$, (9.50) is

$$\mu Z_1\sum_{j=2}^{K}(c_j-c_{j-1})\mathrm{e}^{-\mu(Z_j-Z_1)}=c_1\,,$$

which agrees with (9.22). Thus, $Z_1^*(\alpha)$ decreases strictly with α from Z_2 to Z_1^* given in (9.22), and $Z_1^*(\alpha) > Z_1^*$ in (9.22). When $K=2$, (9.50) is equivalent to eq. (22) of [13].

9.5.3 MODEL 7

We make assumptions 1), 2"), and 3") of Model 3, and assumption 4) of Model 5. Then, the mean time to maintenance is given in (9.46) and the total expected cost until maintenance is

$$\begin{aligned} &[c_K+c_O(Z_K)]\sum_{j=0}^{\infty}\overline{F}((j+1)T)\int_0^{Z_1}\overline{G}(Z_K-x)\mathrm{d}G^{(j)}(x) \\ &+\sum_{i=1}^{K-1}\sum_{j=0}^{\infty}\overline{F}((j+1)T)\int_0^{Z_1}\left\{\int_{Z_i-x}^{Z_{i+1}-x}[c_i+c_O(x+y)]\,\mathrm{d}G(y)\right\}\mathrm{d}G^{(j)}(x) \\ &+c_D\sum_{j=0}^{\infty}F((j+1)T)\int_0^{Z_1}\overline{G}(Z_1-x)\mathrm{d}G^{(j)}(x) \\ =\;& c_K-\sum_{i=1}^{K-1}(c_K-c_i)\sum_{j=0}^{\infty}\overline{F}((j+1)T) \\ &\qquad\times\int_0^{Z_1}[G(Z_{i+1}-x)-G(Z_i-x)]\mathrm{d}G^{(j)}(x) \\ &+(c_D-c_K)\sum_{j=0}^{\infty}F((j+1)T)\left[G^{(j)}(Z_1)-G^{(j+1)}(Z_1)\right] \end{aligned}$$

$$+c_0(Z_1)\sum_{j=0}^{\infty}\overline{F}((j+1)T)\left[G^{(j)}(Z_1)-G^{(j+1)}(Z_1)\right]$$

$$+\sum_{j=0}^{\infty}\overline{F}((j+1)T)\int_0^{Z_1}\left[\int_{Z_1}^{Z_K}\overline{G}(y-x)\mathrm{d}c_0(y)\right]\mathrm{d}G^{(j)}(x). \tag{9.51}$$

Thus, the expected cost rate is, from (9.46) and (9.51),

$$C_7(Z_1)=\frac{\begin{array}{l}c_K+(c_D-c_K)\\ \quad\times\sum_{j=0}^{\infty}F((j+1)T)\left[G^{(j)}(Z_1)-G^{(j+1)}(Z_1)\right]\\ -\sum_{i=1}^{K-1}(c_K-c_i)\sum_{j=0}^{\infty}\overline{F}((j+1)T)\\ \quad\times\int_0^{Z_1}[G(Z_{i+1}-x)-G(Z_i-x)]\mathrm{d}G^{(j)}(x)\\ +c_0(Z_1)\sum_{j=0}^{\infty}\overline{F}((j+1)T)\left[G^{(j)}(Z_1)-G^{(j+1)}(Z_1)\right]\\ +\sum_{j=0}^{\infty}\overline{F}((j+1)T)\\ \quad\times\int_0^{Z_1}\left[\int_{Z_1}^{Z_K}\overline{G}(y-x)\mathrm{d}c_0(y)\right]\mathrm{d}G^{(j)}(x)\end{array}}{\sum_{j=0}^{\infty}G^{(j)}(Z_1)\int_{jT}^{(j+1)T}\overline{F}(t)\mathrm{d}t}. \tag{9.52}$$

When $\overline{F}(t)=1$, (9.52) agrees with (9.26).

When $F(t)=1-\exp(-\lambda t)$, $G(x)=1-\exp(-\mu x)$, and $c_0(x)=c_0x$, (9.52) is

$$\frac{C_7(Z_1)}{\lambda}=\frac{\begin{array}{l}c_K-\alpha\mathrm{e}^{-(1-\alpha)\mu Z_1}\sum_{i=2}^{K}(c_i-c_{i-1})\left[1-\mathrm{e}^{-\mu(Z_i-Z_1)}\right]\\ +c_0\alpha\mathrm{e}^{-(1-\alpha)\mu Z_1}\left[1-\mathrm{e}^{-\mu(Z_K-Z_1)}\right]/\mu+c_0Z_1\alpha\mathrm{e}^{-(1-\alpha)\mu Z_1}\end{array}}{1-\alpha\mathrm{e}^{-(1-\alpha)\mu Z_1}}$$
$$+c_D-c_K. \tag{9.53}$$

We find optimal Z_1^* to minimize $C_7(Z_1)$. Differentiating $C_7(Z_1)$ with Z_1 and putting it equal to zero,

$$\frac{\alpha}{1-\alpha}\left[1-\mathrm{e}^{-(1-\alpha)\mu Z_1}\right]\left[\sum_{i=2}^{K}(c_i-c_{i-1})\mathrm{e}^{-\mu(Z_i-Z_1)}-\frac{c_0}{\mu}\mathrm{e}^{-\mu(Z_K-Z_1)}\right]$$
$$+\frac{c_0}{\mu}\left[\frac{1-\alpha\mathrm{e}^{-(1-\alpha)\mu Z_1}}{1-\alpha}-\mu Z_1\right]=c_1+\frac{c_0}{\mu}. \tag{9.54}$$

When $c_0=0$, (9.54) agrees with (9.50).

9.5.4 MODEL 8

We consider assumptions 1), 2"'), and 3) of Model 4 and add assumption 4) of Model 5. The mean time to maintenance is

$$\sum_{j=0}^{\infty}(j+1)T\overline{F}((j+1)T)\int_0^{Z_1}\overline{G}(Z_K-x)\mathrm{d}G^{(j)}(x)$$

$$
\begin{aligned}
&+\sum_{j=0}^{\infty}(j+2)T\overline{F}((j+2)T) \\
&\qquad\times\sum_{i=1}^{L-1}\int_0^{Z_1}\left[\int_{Z_i-x}^{Z_{i+1}-x}\overline{G}(Z_K-x-y)\mathrm{d}G(y)\right]\mathrm{d}G^{(j)}(x) \\
&+\sum_{j=0}^{\infty}(j+2)T\overline{F}((j+2)T) \\
&\qquad\times\sum_{i=1}^{L-1}\int_0^{Z_1}\left[\int_{Z_{i-1}-x}^{Z_i-x}G(Z_{i+1}-x-y)\mathrm{d}G(y)\right]\mathrm{d}G^{(j)}(x) \\
&+\sum_{j=0}^{\infty}(j+2)T\overline{F}((j+2)T)\sum_{i=1}^{L-1}\int_0^{Z_1}\left\{\sum_{n=i+1}^{K-1}\int_{Z_i-x}^{Z_{i+1}-x}[G(Z_{n+1}-x-y)\right. \\
&\qquad\left.-G(Z_n-x-y)]\mathrm{d}G(y)\right\}\mathrm{d}G^{(j)}(x) \\
&+\sum_{j=0}^{\infty}(j+1)T\overline{F}((j+1)T)\sum_{i=L}^{K-1}\int_0^{Z_1}[G(Z_{i+1}-x)-G(Z_i-x)]\mathrm{d}G^{(j)}(x) \\
&+\sum_{j=0}^{\infty}\int_0^{(j+1)T}t\mathrm{d}F(t)\int_0^{Z_1}\overline{G}(Z_K-x)\mathrm{d}G^{(j)}(x) \\
&+\sum_{j=0}^{\infty}\int_0^{(j+2)T}t\mathrm{d}F(t)\sum_{i=1}^{L-1}\int_0^{Z_1}\left[\int_{Z_i-x}^{Z_{i+1}-x}\overline{G}(Z_K-x-y)\mathrm{d}G(y)\right]\mathrm{d}G^{(j)}(x) \\
&+\sum_{j=0}^{\infty}\int_0^{(j+2)T}t\mathrm{d}F(t)\sum_{i=1}^{L-1}\int_0^{Z_1}\left[\int_{Z_{i-1}-x}^{Z_i-x}G(Z_{i+1}-x-y)\mathrm{d}G(y)\right]\mathrm{d}G^{(j)}(x) \\
&+\sum_{j=0}^{\infty}\int_0^{(j+2)T}t\mathrm{d}F(t)\sum_{i=1}^{L-1}\int_0^{Z_1}\left\{\sum_{n=i+1}^{K-1}\int_{Z_i-x}^{Z_{i+1}-x}[G(Z_{n+1}-x-y)\right. \\
&\qquad\left.-G(Z_n-x-y)]\mathrm{d}G(y)\right\}\mathrm{d}G^{(j)}(x)
\end{aligned}
$$

$$
\begin{aligned}
&+\sum_{j=0}^{\infty}\int_0^{(j+1)T}t\mathrm{d}F(t)\sum_{i=L}^{K-1}\int_0^{Z_1}[G(Z_{i+1}-x)-G(Z_i-x)]\mathrm{d}G^{(j)}(x) \\
=&\int_0^T\overline{F}(t)\mathrm{d}t+\sum_{j=0}^{\infty}\int_{(j+1)T}^{(j+2)T}\overline{F}(t)\mathrm{d}t\int_0^{Z_1}G(Z_L-x)\mathrm{d}G^{(j)}(x)\,. \qquad (9.55)
\end{aligned}
$$

and the total expected cost until maintenance is

$$
\begin{aligned}
& c_K \sum_{j=0}^{\infty} \overline{F}((j+1)T) \int_0^{Z_1} \overline{G}(Z_K - x)\mathrm{d}G^{(j)}(x) \\
& + c_K \sum_{j=0}^{\infty} \overline{F}((j+2)T) \\
& \qquad \times \sum_{i=1}^{L-1} \int_0^{Z_1} \left[\int_{Z_i - x}^{Z_{i+1}-x} \overline{G}(Z_K - x - y)\mathrm{d}G(y) \right] \mathrm{d}G^{(j)}(x) \\
& + \sum_{j=0}^{\infty} \overline{F}((j+2)T) \\
& \qquad \times \sum_{i=1}^{L-1} c_i \int_0^{Z_1} \left[\int_{Z_{i-1} - x}^{Z_i - x} G(Z_{i+1} - x - y)\mathrm{d}G(y) \right] \mathrm{d}G^{(j)}(x) \\
& + \sum_{j=0}^{\infty} \overline{F}((j+2)T) \sum_{i=1}^{L-1} \int_0^{Z_1} \left\{ \sum_{n=i+1}^{K-1} c_n \int_{Z_i - x}^{Z_{i+1}-x} [G(Z_{n+1} - x - y) \right. \\
& \qquad \left. - G(Z_n - x - y)]\mathrm{d}G(y) \right\} \mathrm{d}G^{(j)}(x) \\
& + \sum_{j=0}^{\infty} \overline{F}((j+1)T) \sum_{i=L}^{K-1} c_i \int_0^{Z_1} [G(Z_{i+1} - x) - G(Z_i - x)]\mathrm{d}G^{(j)}(x) \\
& + c_D \left(\sum_{j=0}^{\infty} \overline{F}((j+1)T) \int_0^{Z_1} \overline{G}(Z_K - x)\mathrm{d}G^{(j)}(x) \right. \\
& + \sum_{j=0}^{\infty} \overline{F}((j+2)T) \sum_{i=1}^{K-1} \int_0^{Z_1} \left[\int_{Z_i - x}^{Z_{i+1}-x} \overline{G}(Z_K - x - y)\mathrm{d}G(y) \right] \mathrm{d}G^{(j)}(x) \\
& + \sum_{j=0}^{\infty} \overline{F}((j+2)T) \sum_{i=1}^{K-1} \int_0^{Z_1} \left[\int_{Z_{i-1} - x}^{Z_i - x} G(Z_{i+1} - x - y)\mathrm{d}G(y) \right] \mathrm{d}G^{(j)}(x) \\
& + \sum_{j=0}^{\infty} \overline{F}((j+2)T) \sum_{i=1}^{K-1} \int_0^{Z_1} \left\{ \sum_{n=i+1}^{K-1} \int_{Z_i - x}^{Z_{i+1}-x} [G(Z_{n+1} - x - y) \right. \\
& \qquad \left. - G(Z_n - x - y)]\mathrm{d}G(y) \right\} \mathrm{d}G^{(j)}(x) \\
& \left. + \sum_{j=0}^{\infty} \overline{F}((j+1)T) \sum_{i=L}^{K-1} \int_0^{Z_1} [G(Z_{i+1} - x) - G(Z_i - x)]\mathrm{d}G^{(j)}(x) \right) \\
= \; & c_K + (c_D - c_K)F(T)
\end{aligned}
$$

$$+\sum_{j=0}^{\infty}\left[(c_D-c_L)\overline{F}((j+1)T)-(c_D-c_K)\overline{F}((j+2)T)\right] \times\int_0^{Z_1} G(Z_L-x)\mathrm{d}G^{(j)}(x)$$
$$-\sum_{j=0}^{\infty}\overline{F}((j+1)T)\sum_{i=L+1}^{K}(c_i-c_{i-1})\int_0^{Z_1} G(Z_i-x)\mathrm{d}G^{(j)}(x)$$
$$-\sum_{j=0}^{\infty}\overline{F}((j+2)T)\sum_{i=2}^{L}\sum_{n=i}^{K}(c_n-c_{n-1}) \times\int_0^{Z_1}\left[\int_{Z_{i-1}-x}^{Z_i-x} G(Z_n-x-y)\mathrm{d}G(y)\right]\mathrm{d}G^{(j)}(x)\,. \quad (9.56)$$

Thus, the expected cost rate is, from (9.55) and (9.56),

$$C_8(Z_1)=\frac{\begin{array}{l}c_K+(c_D-c_K)F(T)\\ +\sum_{j=0}^{\infty}\left[(c_D-c_L)\overline{F}((j+1)T)\right.\\ \qquad\left.-(c_D-c_K)\overline{F}((j+2)T)\right]\\ \qquad\times\int_0^{Z_1}G(Z_L-x)\mathrm{d}G^{(j)}(x)\\ -\sum_{j=0}^{\infty}\overline{F}((j+1)T)\sum_{i=L+1}^{K}(c_i-c_{i-1})\\ \qquad\times\int_0^{Z_1}G(Z_i-x)\mathrm{d}G^{(j)}(x)\\ -\sum_{j=0}^{\infty}\overline{F}((j+2)T)\sum_{i=2}^{L}\sum_{n=i}^{K}(c_n-c_{n-1})\\ \qquad\times\int_0^{Z_1}\left[\int_{Z_{i-1}-x}^{Z_i-x}G(Z_n-x-y)\mathrm{d}G(y)\right]\mathrm{d}G^{(j)}(x)\end{array}}{\int_0^T\overline{F}(t)\mathrm{d}t+\sum_{j=0}^{\infty}\int_{(j+1)T}^{(j+2)T}\overline{F}(t)\mathrm{d}t\int_0^{Z_1}G(Z_L-x)\mathrm{d}G^{(j)}(x)}\,. \quad (9.57)$$

When $F(t)=1-\exp(-\lambda t)$ and $G(x)=1-\exp(-\mu x)$, (9.57) is

$$\frac{C_8(Z_1)}{\lambda}=\frac{\begin{array}{l}c_K-\alpha e^{\alpha\mu Z_1}\left\{\alpha\sum_{i=2}^{L}\sum_{j=i}^{K}(c_j-c_{j-1})\right.\\ \quad\times\left[e^{-\mu Z_{i-1}}-e^{-\mu Z_i}-\mu(Z_i-Z_{i-1})e^{-\mu Z_j}\right]\\ \quad\left.-\sum_{i=L+1}^{K}(c_i-c_{i-1})\left(e^{-\mu Z_i}-e^{-\mu Z_L}\right)\right\}\end{array}}{1-\alpha e^{\alpha\mu Z_1}\left[\alpha e^{-\mu Z_1}+(1-\alpha)e^{-\mu Z_L}\right]}+c_D-c_K\,. \quad (9.58)$$

We find optimal Z_1^* to minimize $C_8(Z_1)$. Differentiating $C_8(Z_1)$ with Z_1 and putting it equal to zero,

$$\frac{1-\alpha e^{-(1-\alpha)\mu Z_1}}{1-e^{-\mu(Z_L-Z_1)}}$$
$$\times\left\{\sum_{i=2}^{L}(c_i-c_{i-1})\left[1-e^{-\mu(Z_i-Z_1)}+\frac{\alpha}{1-\alpha}\mu(Z_i-Z_1)e^{-\mu(Z_i-Z_1)}\right]\right.$$
$$\left.+\sum_{i=L+1}^{K}(c_i-c_{i-1})\left[1-e^{-\mu(Z_L-Z_1)}+\frac{\alpha}{1-\alpha}\mu(Z_L-Z_1)e^{-\mu(Z_i-Z_1)}\right]\right\}$$

$$+\alpha e^{-(1-\alpha)\mu Z_1}\sum_{i=2}^{K}(c_i - c_{i-1})\left[1 - e^{-\mu(Z_i - Z_1)}\right] = c_K\,. \tag{9.59}$$

9.6 CONCLUSION

The maintenance of social infrastructures has some remarkable characteristics which are different from the maintenance of general mechanical systems. PM delays occur inevitably because the annual budget of local government has a maximum. Multiple degradation levels and modified costs which depend on damage levels should be introduced because social infrastructures are huge and their PM cost range changes greatly due to degradation levels. Natural disasters should be considered when the PM of social infrastructures are planned because they have huge scales and are operated for decades.

In this chapter, we have considered optimal maintenance policies considering delay, multiple degradation levels, modified costs which depend on damage levels, and disaster recovery. Eight maintenance policies have been proposed using a cumulative damage model. Model 1 has considered delayed maintenance, Model 2 has regarded multiple degradation levels, and Model 3 has introduced modified costs which depend on damage levels. Model 4 has considered delayed maintenance of multiple degradation levels. Models 5, 6, 7, and 8 are extended models of Models 1, 2, 3, and 4 in which the disaster recovery has been supposed. The expected cost rates have been obtained and optimal damage levels Z_1^* which minimize them have been discussed analytically. Although we can show that there exist optimal policies for Model 1, 2, 3, 5, 6, and 7, it is difficult to discuss optimal policies for Model 4 and 8 because they are more complicated than other models. These policies would be greatly useful when actual PMs of social infrastructures would be planned, referring to optimal policies which minimize the total cost.

REFERENCES

1. Ministry of Land. (2017). Infrastructure, transport and tourism. White Paper on Land, Infrastructure, Transport and Tourism in Japan, 2017.

2. Friesz, T., & Fernandez, J. (1979). A model of optimal transport maintenance with demand responsiveness. *Transportation Research Part B: Methodological, 13(4)*, 317-339.

3. Markow, M., & Balta, W. (1985). Optimal rehabilitation frequencies for highway pavements. *Transportation Research Record*, 1035, 31-43.

4. Tsunokawa, K., & Schofer, J. (1994). Trend curve optimal control model for highway pavement maintenance: Case study and evaluation. *Transportation Research Part A, 28(2)*, 151-166.

5. Li, Y., & Madanat, S. (2002). A steady-state solution for the optimal pavement resurfacing problem. *Transportation Research Part A, 36(6)*, 525-535.

6. Ouyang, Y., & Madanat, S. (2004). Optimal scheduling of rehabilitation activities for multiple pavement facilities: Exact and approximate solutions. *Transportation Research Part A, 38(5)*, 347-365.

7. Ouyang, Y., & Madanat, S. (2006). An analytical solution for the finite-horizon pavement resurfacing planning problem. *Transportation Research Part B, 40(9)*, 767-778.

8. Kobayashi, K., Eguchi, M., Oi, A., Aoki, K., Kaito, K., & Matsumura, Y. (2012). The Optimal Repair and Replacement Model of Pavement Structure. *Journal of Japan Society of Civil Engineers E1, 68(2)*, 54-68.

9. Kaito, K., Yasuda, K., Kobayashi, K., & Owada, K. (2005). Optimal maintenance strategies of bridge components with an average cost minimizing principles. *Proceedings of Journal of Japan Society of Civil Engineers* (801), 2005-10-21, 83-96.

10. Sakai, R., Onishi, Y., & Otsu, H. (2004). The study of the maintenance management model of the tunnel structure using stochastic process, *Proceedings of the Japan National Conference on Geotechnical Engineering*, 39th, 2-2, 1711-1712.

11. Obama, K., Kaito, K., Aoki, K., Kobayashi, K., & Fukuda, T. (2012). The optimal scrapping and maintenance model of infrastructure considering deterioration process, *Journal of Japan Society of Civil Engineers F4, 68(3)*, 141-156.

12. Tsuda, Y., Kaito, K., Aoki, K., & Kobayashi, K. (2005). Estimating Markovian transition probabilities for bridge deterioration forecasting, *Journal of Japan Society of Civil Engineers, 801/I-73, 2005.10*, 69-82.

13. Ito, K., Higuchi, Y., & Nakagawa, T. (2018). Optimal maintenance policy of coastal protection systems, *Conference Proceedings, 24th ISSAT International Conference Reliability and Quality in Design*, Toronto, Ontario, Canada, 181-184.

14. Ito, K., & Nakagawa, T. (2018). Optimal maintenance policies of social infrastructures, IEICE Technical Report 118(365), R2018-45, 13-16.

15. Kishida, T., Ito, K., Higuchi, Y., & Nakagawa, T. (2019). Optimal maintenance policy of coastal protection systems, *Conference Proceedings, 25th ISSAT International Conference Reliability and Quality in Design*, Las Vegas, Nevada, USA.

16. Kishida,T., Ito,K., Higuchi,Y. & Nakagawa,T. (2020). Optimal maintenance models of social infrastructures considering natural disasters. *Reliability and Statistical Computing: Modeling, Methods and Applications* (Springer Series in Reliability Engineering). (Pham, H. ed.) Springer Verlag, London. 245-263.

17. Nakagawa, T. (2007). *Shock and Damage Models in Reliability Theory.* Springer Verlag, London.

Section IV

Software Reliability and Testing

10 Optimal Maintenance Problem with OSS-Oriented EVM for OSS Project

Hironobu Sone
IBM Japan, Ltd., Japan

Yoshinobu Tamura
Yamaguchi University, Japan

Shigeru Yamada
Tottori University, Japan

CONTENTS

10.1 INTRODUCTION

The source code of open source software (OSS) is freely available for use, reuse, fixing, and re-distribution by users. OSS programs are used in various situations, because OSS is useful for many users for cost reduction, standardization, and quick delivery. Many OSS programs are known for their high performance and reliability, even though many OSS programs are free of charge. Furthermore, many IT companies often develop OSS programs for commercial use. In particular, OSS programs are developed using the bazaar method [1] under the situation of free and open source code. Then, OSS programs are promoted by an unspecified number of users and developers. The bug tracking system is also one of the systems used to develop OSS. A lot of fault information such as the fix status, details, and fix priorities are registered on the bug tracking system.

Recently, EVM (Earned Value Management) [2] was applied to actual software projects in various IT companies. From the characteristics of OSS development, many OSS programs are developed and maintained by several developers with many OSS users. Methods of OSS reliability assessment have been proposed [3, 4, 5]. However, it is difficult to apply EVM directly in terms of the characteristics of the OSS project. There is research on deriving EVM measurements by using the development efforts of OSS projects [8]. However, reference [8] has a problem because it was difficult to derive all the EVM values. In particular, it is important to appropriately control the quality according to the progress status of the OSS project. Also, the appropriate control of management effort for OSS will indirectly link to the quality, reliability, and cost. Moreover, it is useful for OSS project managers to decide the version upgrade if they can estimate the optimal maintenance time.

In this chapter, we examine the method of deriving EVM values and propose a method that can derive all EVM values as an OSS-oriented EVM, considering the complexity peculiar to OSS special projects. In addition, we also consider a project progress prediction method that takes into account the complexity peculiar to OSS. OSS has a support period for each version, and the end of support is called End of Life (EOL). It is dangerous in terms of vulnerability to continue using the specified version of OSS considering the EOL. Therefore, we should upgrade the version periodically. However, the maintenance cost increases with the version upgrade frequently. Therefore, it is necessary to update the OSS with low cost. Then, we find the optimum maintenance time by minimizing the total expected software maintenance effort.

10.2 RELATED RESEARCH

There are several studies on development effort in OSS [3, 4, 5]. Robles et al. [3] presented a novel approach to estimate the effort of large-scale OSS by considering data from source code management repositories. In addition, for estimating the effort, they use survey data answered by over 100 developers. Mishra et al. [4] have proposed a metric for computing the effort and contribution of a patch reviewer based on modified file size, patch size and program complexity variables. Sone et al. [5] have proposed a method of EVM assessment for OSS project. In particular, it is difficult to predict the development effort in OSS development and apply it to EVM because of the complexity of OSS development. Regarding the application of EVM in OSS development, not all EVM indicators could be derived in the EVM approach proposed by Sone et al [5].

In this chapter, we use the software reliability growth model (SRGM) with Wiener process for considering the development effort and the optimum maintenance problem. The SRGM is mainly used for software reliability assessment. The optimal maintenance problem is an application of the optimal software release problem [6, 7] proposed by Yamada et al. The optimum software release problem means deriving a time to stop software testing by minimizing the expected total software cost [6, 7]. On the other hand, the optimum maintenance problem means deriving the optimum maintenance time by minimizing the total expected software maintenance effort in OSS development.

In the past, Tamura et al. have proposed several solutions to the optimal maintenance problem in OSS development [8]. In these research papers, various prediction models have been applied to the development effort.

In this chapter, we use an earned value management (EVM) [2] in software development. EVM is one of the project management techniques for measuring project performance and progress. EVM has been developed for the success of US national projects. EVM can observe the current schedule forecast and cost of the project. EVM basically measures the project performance and progress using three indicators: Planned Value (PV), Earned Value (EV), and Actual Cost (AC). Also, we can quantitatively grasp the current status of the project by comparing three indicators as shown in Fig. 10.1. In addition, we can derive the schedule forecast, cost forecast, productivity, etc. as shown Table 10.1, by using the three indicators, PV, EV and AC. The proposed EVM for the OSS in the past had the problem that the PV could not be derived due to the characteristics of the data used. In this chapter, we solve the problem in terms of the used data and deriving methodology.

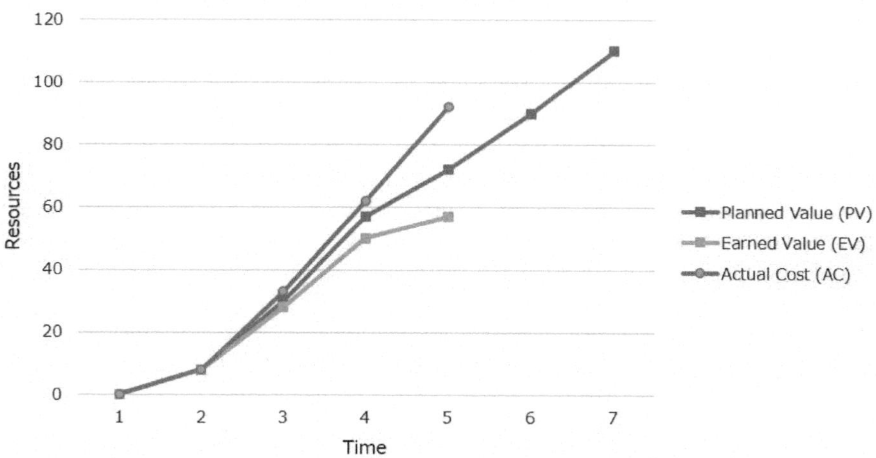

Figure 10.1 The example of EVM.

Table 10.1
Several examples of the indicators used in EVM.

EVM Elements	Explanation
Planned Value (PV)	PV is a supposed work value at any given point in the project schedule.
Earned Value (EV)	EV is a value of work progress at a given point in time.
Actual Cost (AC)	AC is an amount of resources that have been expended to date.
Budget at Completion (BAC)	BAC represents the total PV for the project.
Cost Variance (CV)	CV shows whether a project is under or over budget. CV=EV-AC
Cost Performance Index (CPI)	CPI evaluates how efficiently the project is using its resources. CPI=EV/AC
Schedure Variance (SV)	SV determines whether a project is ahead of or behind schedule. SV=EV-PV
Schedule Performance Index (SPI)	SPI evaluates how efficiently the project team is using its time. SPI=EV/PV
Estimate at Completion (EAC)	EAC shows the final cost of the project in case of continuing current performance trend. EAC=BAC/CPI
Estimate to Complete (ETC)	ETC shows what the remaining work will cost. ETC=(BAC-EV)/CPI

10.3 EFFORT ESTIMATION MODEL BASED ON STOCHASTIC DIFFERENTIAL EQUATION

Considering the characteristic of the operation phase of OSS projects, the time-dependent expenditure phenomenon of maintenance effort keeps an irregular state in the operation phase, because there is variability among the levels of developers' skill. Then, the time-dependent effort expenditure phenomenon of the maintenance phase becomes unstable.

The operation phases of many OSS projects are influenced from external factors by triggers such as the difference of skill, time lag of development and maintenance activities. Considering the above points, we apply stochastic differential equation modeling to managing the OSS project. Then, let $\Omega(t)$ be the cumulative maintenance effort, such as finding software faults and improving functionality up to operational time t $(t \geq 0)$ in the OSS project. Suppose that $\Omega(t)$ takes on continuous real values. Since the estimated maintenance efforts are observed during the operational phase of the OSS project, $\Omega(t)$ gradually increases as the operational procedures go on. Based on the software reliability growth modeling approach [9, 10, 11, 12], the following linear differential equation in terms of maintenance effort can be formulated:

$$\frac{d\Omega(t)}{dt} = \beta(t)\{\alpha - \Omega(t)\}, \tag{10.1}$$

where $\beta(t)$ is the increase rate of maintenance effort at operational time t and a non-negative function, and α means the estimated maintenance effort expenditures required until the end of operation.

Therefore, we extend Eq. (10.1) to the following stochastic differential equation with Brownian motion [13]:

$$\frac{d\Omega(t)}{dt} = \{\beta(t) + \sigma\nu(t)\}\{\alpha - \Omega(t)\}, \tag{10.2}$$

where σ is a positive constant representing a magnitude of the irregular fluctuation, and $\nu(t)$ is standardized Gaussian white noise. By using Itô's formula [14], we can obtain the solution of Eq. (10.2) under the initial condition $\Omega(0) = 0$ as follows:

$$\Omega(t) = \alpha\left[1 - \exp\left\{-\int_0^t \beta(s)ds - \sigma\omega(t)\right\}\right], \tag{10.3}$$

where $\omega(t)$ is a one-dimensional Wiener process which is formally defined as an integration of the white noise $\nu(t)$ with respect to time t. Moreover, we define the increase rate of maintenance effort in case of $\beta(t)$ as [15]:

$$\int_0^t \beta(s)ds \doteq \frac{\frac{dF_*(t)}{dt}}{\alpha - F_*(t)}. \tag{10.4}$$

In this chapter, we assume the following equations based on software reliability models $F_*(t)$ as the cumulative maintenance effort expenditures function of the proposed model:

$$F_e(t) \equiv \alpha\left(1 - e^{-\beta t}\right), \tag{10.5}$$

$$F_s(t) \equiv \alpha\left\{1 - (1 + \beta t)e^{-\beta t}\right\}, \tag{10.6}$$

where $\Omega_e(t)$ means the cumulative maintenance effort expenditures for the exponential software reliability growth model with $F_e(t)$. Similarly, $\Omega_s(t)$ is the cumulative maintenance effort expenditure for the delayed S-shaped software reliability growth model with $F_s(t)$. The reason why this research uses two models is that the exponential model and delayed S-shaped model are one of the famous software reliability growth models [16].

Therefore, the cumulative maintenance effort, Ω_* up to time t is obtained as follows:

$$\Omega_e(t) = \alpha\left[1 - \exp\{-\beta t - \sigma\omega(t)\}\right], \tag{10.7}$$

$$\Omega_s(t) = \alpha\left[1 - (1 + \beta t)\exp\{-\beta t - \sigma\omega(t)\}\right]. \tag{10.8}$$

In these models, we assume that the parameter σ depends on several noises by external factors from several triggers in open source projects. Then, the expected cumulative maintenance effort expenditures spent up to time t are respectively obtained as follows:

$$\mathrm{E}\left[\Omega_e(t)\right] = \alpha\left[1-\exp\left\{-\beta t+\frac{\sigma^2}{2}t\right\}\right], \tag{10.9}$$

$$\mathrm{E}\left[\Omega_s(t)\right] = \alpha\left[1-(1+\beta t)\exp\left\{-\beta t+\frac{\sigma^2}{2}t\right\}\right]. \tag{10.10}$$

Similarly, we consider the sample path of maintenance effort expenditures required for OSS maintenance, e.g., the needed remaining maintenance effort expenditures from time t to the end of the project, Ω_{r*} are obtained as follows:

$$\Omega_{re}(t) = \alpha\exp\left\{-\beta t-\sigma\omega(t)\right\}, \tag{10.11}$$

$$\Omega_{rs}(t) = \alpha(1+\beta t)\exp\left\{-\beta t-\sigma\omega(t)\right\}. \tag{10.12}$$

Then, the expected maintenance effort expenditures required for OSS maintenance until the end of operation time t are respectively obtained as follows:

$$\mathrm{E}\left[\Omega_{re}(t)\right] = \alpha\exp\left\{-\beta t+\frac{\sigma^2}{2}t\right\}, \tag{10.13}$$

$$\mathrm{E}\left[\Omega_{rs}(t)\right] = \alpha(1+\beta t)\exp\left\{-\beta t+\frac{\sigma^2}{2}t\right\}. \tag{10.14}$$

10.4 ASSESSMENT MEASURES FOR OSS-ORIENTED EVM

10.4.1 HOW TO USE THE OSS PROJECT DATA

In OSS-oriented EVM, the period of data used for Planned Value (PV) and Actual Cost (AC) have the different values. Both PV and AC use the data obtained from the bug tracking system and required by the fault reporters and the fault correctors. For the prediction of PV, we use Eqs. (10.7)-(10.10) and maintenance effort data up to the OSS's release. In particular, the parameter α in Eqs. (10.7)-(10.10) can be regarded as the estimated maintenance effort at the time the OSS is released. Therefore, the parameter α can be rephrased as Budget at Completion (BAC) in OSS-oriented EVM. AC uses the maintenance effort data obtained from the bug tracking system, including after the OSS release required by the fault reporter and the fault corrector. Therefore, the start time of the data used to derive PV and AC is the same.

Earned Value (EV) is the cumulative maintenance effort viewed on the same scale as the project budget (BAC). Therefore, if the OSS development effort increases but the fault is not corrected, the EV becomes small and it is regarded as an inefficient OSS project. In the derivation of the EV, the number of potential faults predicted from the fault data reported up to the time of the OSS release is used. We use Eqs. (10.7)-(10.10) to predict the number of potential faults. We derive the "fault resolving cost", the value obtained by dividing the number of potential faults from the BAC, as follows:

$$\gamma = \frac{BAC}{p}. \tag{10.15}$$

Then, γ is the fault resolving cost, and p means potential faults at OSS release. We can derive the EV in case of $F_e(t)$ and $F_s(t)$ by using the fault resolving cost γ and the cumulative number of corrected faults up to the operating time t.

$$EV_e(t) = \gamma \left[\alpha_f \left[1 - \exp\{-\beta_f t - \sigma_f \omega(t)\}\right]\right], \tag{10.16}$$

$$EV_s(t) = \gamma \left[\alpha_f \left[1 - (1 + \beta_f t) \exp\{-\beta_f t - \sigma_f \omega(t)\}\right]\right]. \tag{10.17}$$

Then, α_f, β_f, σ_f are parameters used to predict the cumulative number of correction faults at time t. Therefore, the expected EV required for OSS maintenance until the end of operation time t are respectively obtained as follows:

$$\mathrm{E}[EV_e(t)] = \gamma \left[\alpha_f \left[1 - \exp\left\{-\beta_f t + \frac{\sigma_f^2}{2} t\right\}\right]\right], \tag{10.18}$$

$$\mathrm{E}[EV_s(t)] = \gamma \left[\alpha_f \left[1 - (1 + \beta_f t) \exp\left\{-\beta_f t + \frac{\sigma_f^2}{2} t\right\}\right]\right]. \tag{10.19}$$

Then, the corrected cumulative number of faults is counted when the fault status is "Closed"in the bug tracking system.

10.4.2 HOW TO DERIVE OSS-ORIENTED EVM VALUE

Generally, the EVM is commonly applied to software development projects. However, it is difficult to directly apply the EVM to the actual OSS project, because the development cycle of the OSS project is different from the traditional software development paradigm. As the characteristics of the OSS, the OSS development project is managed by using the bug tracking system. This chapter shows the method of earned value analysis for OSS projects by using the data sets obtained from the bug tracking system.

Considering the deriving of the AC, PV and EV proposed in this chapter, we assume the following terms shown Table 10.2 as the OSS-oriented EVM for OSS development:

Then, the expected Cost Variance (CV) for OSS maintenance up to operational time t in case of $\Omega_e(t)$ and $\Omega_s(t)$ can be formulated as:

$$\mathrm{E}[CV_e(t)] = \mathrm{E}[EV_e(t)] - \mathrm{E}[\Omega_{re}^a(t)], \tag{10.20}$$

$$\mathrm{E}[CV_s(t)] = \mathrm{E}[EV_s(t)] - \mathrm{E}[\Omega_{rs}^a(t)], \tag{10.21}$$

where $\mathrm{E}[EV_e(t)]$ and $\mathrm{E}[EV_s(t)]$ are the expected maintenance effort expenditures considering the EV until the end of operation. Also, $\mathrm{E}[\Omega_{re}^a(t)]$ and

Table 10.2
Explanation for OSS-oriented EVM.

OSS-oriented EVM elements	Explanatory
Planned Value (PV)	Cumulative maintenance effort as planned value up to operational time t considering the fault reporter and fault corrector
Earned Value (EV)	Cumulative maintenance effort up to operational time t viewed on the same scale as BAC
Actual Cost (AC)	Cumulative maintenance effort up to operational time t considering the fault reporter and fault corrector
Budget at Completion (BAC)	Total budget in the end point as the specified goal of OSS project
Cost Variance (CV)	$\mathrm{E}[CV_e(t)]$ and $\mathrm{E}[CV_s(t)]$ obtained from EV-AC
Cost Performance Index (CPI)	$\mathrm{E}[CPI_e(t)]$ and $\mathrm{E}[CPI_s(t)]$ obtained from EV/AC
Schedule Variance (SV)	SV obtained from EV-PV (Explanation of formula omitted in this chapter)
Schedule Performance Index (SPI)	SPI obtained from EV/PV (Explanation of formula omitted in this chapter)
Estimate at Completion (EAC)	$\mathrm{E}[EAC_e(t)]$ and $\mathrm{E}[EAC_s(t)]$ obtained from BAC/CPI
Estimate to Complete (ETC)	$\mathrm{E}[ETC_e(t)]$ and $\mathrm{E}[ETC_s(t)]$ obtained from (BAC-EV)/CPI

$\mathrm{E}[\Omega^a_{rs}(t)]$ are the maintenance effort expenditures considering AC until the end of operation. Especially, a in case of $\mathrm{E}[\Omega^a_{re}(t)]$ and $\mathrm{E}[\Omega^a_{rs}(t)]$ comes from 'actual'. Similarly, the sample path of CV for OSS project maintenance up to operational time t in case of $\Omega_e(t)$ and $\Omega_s(t)$ are given by

$$CV_e(t) = EV_e(t) - \Omega^a_{re}(t), \tag{10.22}$$

$$CV_s(t) = EV_s(t) - \Omega^a_{rs}(t). \tag{10.23}$$

The zero point of CV $\mathrm{E}[CV_e(t)]$ and $\mathrm{E}[CV_e(t)]$ mean the starting point of surplus effort. Therefore, the OSS project managers will be able to judge the necessity of maintenance effort and stability of the OSS from the starting point of surplus effort. Moreover, we can obtain the Cost Performance Index (CPI) by using the following equations:

$$\mathrm{E}[CPI_e(t)] = \frac{\mathrm{E}[EV_e(t)]}{\mathrm{E}[\Omega^a_{re}(t)]}, \tag{10.24}$$

$$\mathrm{E}[CPI_s(t)] = \frac{\mathrm{E}[EV_s(t)]}{\mathrm{E}[\Omega^a_{rs}(t)]}. \tag{10.25}$$

The value of the CPI is derived by $\frac{EV}{AC}$. Similarly, the sample path of the CPI in case of Ω_e and Ω_s are given by

$$CPI_e(t) = \frac{EV_e(t)}{\Omega^a_{re}(t)}, \tag{10.26}$$

$$CPI_s(t) = \frac{EV_s(t)}{\Omega^a_{rs}(t)}. \tag{10.27}$$

Furthermore, we can obtain the Estimate at Completion (EAC) by using the following equations:

$$\mathrm{E}[EAC_e(t)] = \mathrm{E}[\Omega^a_{re}(t)] + \frac{(BAC - \mathrm{E}[EV_e(t)])}{\mathrm{E}[CPI_e(t)]}, \tag{10.28}$$

$$\mathrm{E}[EAC_s(t)] = \mathrm{E}[\Omega^a_{rs}(t)] + \frac{(BAC - \mathrm{E}[EV_s(t)])}{\mathrm{E}[CPI_s(t)]}. \tag{10.29}$$

The value of the EAC is derived by AC+ETC. Similarly, the sample path of the EAC in case of Ω_e and Ω_s are given by

$$EAC_e(t) = \Omega^u_{re}(t) + \frac{(BAC - EV_e(t))}{CPI_e(t)}, \tag{10.30}$$

$$EAC_s(t) = \Omega^a_{rs}(t) + \frac{(BAC - EV_s(t))}{CPI_s(t)}. \tag{10.31}$$

Finally, we can obtain the Estimate to Complete (ETC) by using the following equations:

$$\mathrm{E}[ETC_e(t)] = \frac{(BAC - \mathrm{E}[EV_e(t)])}{\mathrm{E}[CPI_e(t)]}, \tag{10.32}$$

$$\mathrm{E}[ETC_s(t)] = \frac{(BAC - \mathrm{E}[EV_s(t)])}{\mathrm{E}[CPI_s(t)]}. \tag{10.33}$$

The value of the ETC is derived by $\frac{BAC-EV}{CPI}$. Similarly, the sample path of the ETC in case of Ω_e and Ω_s are given by

$$ETC_e(t) = \frac{(BAC - EV_e(t))}{CPI_e(t)}, \tag{10.34}$$

$$ETC_s(t) = \frac{(BAC - EV_s(t))}{CPI_s(t)}. \tag{10.35}$$

In particular, Eqs. (10.24) - (10.35) are based on the EVM derivation method [2]. Also, the CPI is very important for OSS project managers to assess the stability of the OSS project.

10.5 OPTIMUM MAINTENANCE TIME BASED ON WIENER PROCESS MODELS

This section discusses the optimal maintenance time problem by minimizing the maintenance effort expenditures for the operation of OSS. Then, using the derivation method of the optimal release problem [6, 7], one of the software reliability evaluation methods, we define the following effort rate parameters:

e_1 : the maintenance effort per effort needed to operate OSS,

e_2 : the operation effort per unit time during the operation,

e_3 : the maintenance effort per effort after the upgrade task such as major version upgrade.

Then, the expected maintenance effort expenditures in the operation of OSS can be formulated as:

$$E_1(t) = e_1\mathrm{E}[\Omega_*^a(t)] + e_2 t. \tag{10.36}$$

Also, the expected software maintenance effort expenditures after the maintenance of OSS is represented as follows:

$$E_2(t) = e_3\mathrm{E}[ETC_*(t)]. \tag{10.37}$$

In particular, we consider ETC obtained from OSS-oriented EVM as the software maintenance effort after the maintenance of OSS.

Consequently, from Eqs. (10.36) and (10.37), the total expected software maintenance effort expenditures during the specified period such as the specified version is given by

$$E(t) = E_1(t) + E_2(t). \tag{10.38}$$

The optimum maintenance time t^* is obtained by minimizing $E(t)$ in Eq. (10.38).

10.6 APPLICATION OF PROPOSED METHOD TO ACTUAL DATA

10.6.1 USED DATA SET

In this chapter, we use open source project data to derive the OSS-oriented EVM and the optimum maintenance time. For applying the proposed model to actual project data, we use the data of OpenStack [17] obtained from Bugzilla. OpenStack is OSS for cloud computing. This project uses Bugzilla as an open source bug tracking system. The data about reported faults is freely available from the bug tracking system. In particular, the effort and fault data obtained from Bugzilla are version 16 (Pike). For estimating the PV and AC, in this chapter, the cumulative number of reported faults is 655 and 2249. In particular, we use project data for about 8 months before OpenStack was released to predict the PV. For prediction of the AC, we also use project data for about 16 months after OpenStack released. Also, data are weekly unit data.

Table 10.3

Parameter estimates of maintenance effort in case of OpenStack.

		Planned Value		Actual Cost	
		exponential	delayed S-shaped	exponential	delayed S-shaped
parameter	α	2.315×10^6	2.004×10^6	1.436×10^7	4.070×10^6
	β	2.879×10^{-3}	1.640×10^{-2}	1.012×10^{-3}	1.179×10^{-2}
	σ	2.315×10^{-3}	1.617×10^{-3}	4.864×10^{-4}	1.343×10^{-3}
AIC		724.297	687.640	2112.663	2035.109

Table 10.4

Parameter estimates of number of fault in case of OpenStack.

		Planned Value Estimated number of potential faults at OSS release		Estimated number of potential faults 16 months after OSS release	
		exponential	delayed S-shaped	exponential	delayed S-shaped
parameter	α	1.961×10^3	4.083×10^3	1.923×10^4	5.683×10^3
	β	1.160×10^{-2}	1.963×10^{-2}	9.978×10^{-4}	1.219×10^{-2}
	σ	1.110×10^{-2}	4.121×10^{-3}	2.202×10^{-3}	8.230×10^{-3}
AIC		319.766	306.047	1064.888	1063.093

10.6.2 NUMERICAL EXAMPLES FOR OPTIMUM MAINTENANCE TIME

Table 10.3 shows the results of parameter estimation of maintenance effort, and AIC (Akaike's Information Criterion). In terms of AIC, the delayed S-shaped model fits better than the exponential model. Also, the parameter α in the PV data can be rephrased as BAC.

In addition, we used Eqs. (10.7)-(10.10) to derive parameters for the cumulative number of faults. Table 10.4 shows the results of parameter estimation of number of faults, and AIC. In terms of AIC, the delayed S-shaped model fits better than the exponential model. Also, the parameter α in the PV data can be rephrased as potential faults at OSS release. In other words, from Eq. (10.15), we can calculate the fault resolving cost $\gamma \fallingdotseq 490$ (man $\cdot$ days).

Figures 10.2 and 10.3 show the estimated cumulative maintenance effort at operation time t in case of $\Omega_e(t)$ and $\Omega_s(t)$. There was a big difference between the PV value estimated at the time of OpenStack release and the AC value obtained 16 months after the release. We speculate that this is because more people are involved in this project after the release of OpenStack.

For deriving the EV, we need to estimate the number of faults shown in Eqs. (10.15) and (10.19). Figure 10.4 shows the estimated cumulative number of faults in case of a delayed S-shaped model in case of Eqs. (10.8) and (10.10). There is a big difference between the estimated cumulative number of faults estimated at the time of OpenStack release and the number of resolved faults

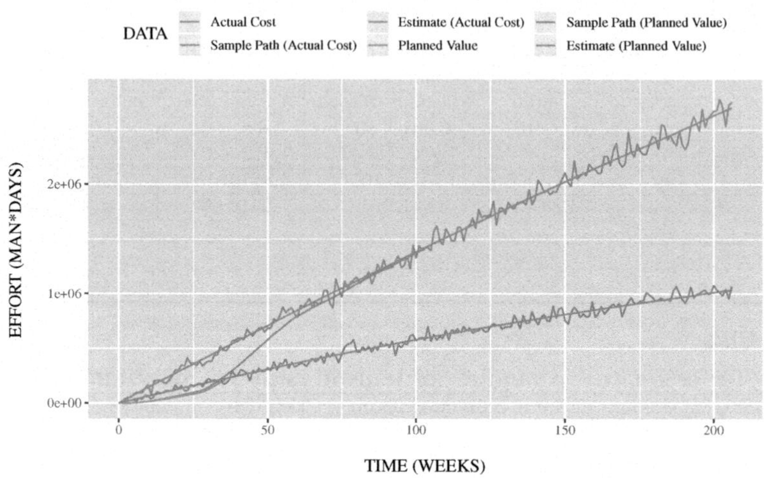

Figure 10.2 Cumulative maintenance effort of OpenStack project using exponential model in Eqs. (10.7) and (10.9).

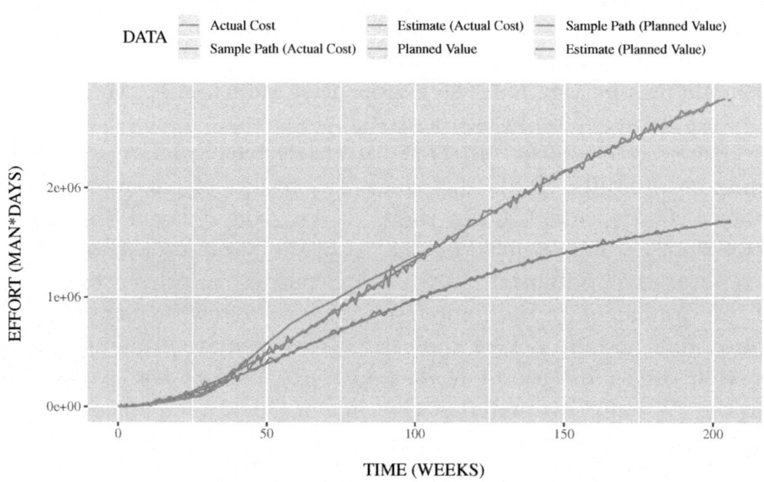

Figure 10.3 Cumulative maintenance effort of OpenStack project using S-shaped model in Eqs. (10.8) and (10.10).

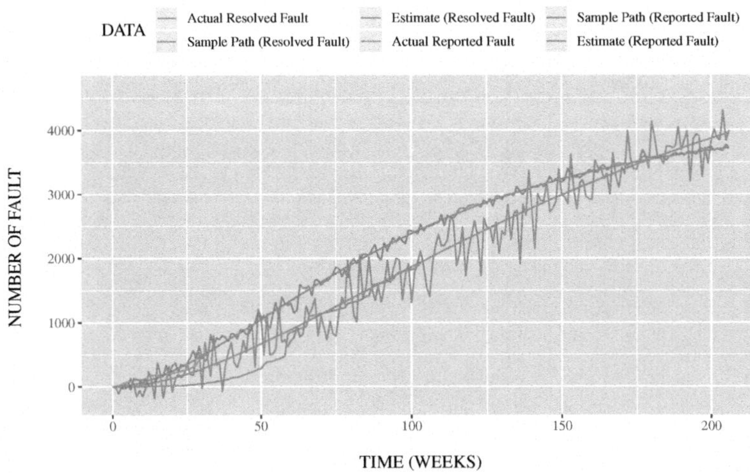

Figure 10.4 The estimated cumulative number of fault of OpenStack project using the delayed S-shaped model in Eqs. (10.8) and (10.10).

estimated 16 months after the release. We find that the OpenStack project takes a lot of time from the fault reporting to the resolution.

Figure 10.5 shows the result of deriving the EV using Eqs. (10.15) and (10.19). From Eq. (10.19), in case of more fault corrections than the number of potential faults at OSS release, the EV will be larger than the BAC. Therefore, the same result was obtained in the OpenStack project used this time.

Figures 10.6 and 10.7 show the estimated CPI and ETC value for Eqs. (10.25), (10.27), (10.33) and (10.35).

OpenStack is an inefficient project because the CPI value tends to be flat and the value is less than 1. We speculate that the reason why the OpenStack project is inefficient is that maintenance efficiency has decreased because many users started fault reporting and fault correction after the release of OpenStack. On the other hand, the reason why the ETC value is negative is that more faults than the number of potential fault reports predicted at the time of OpenStack release were reported and corrected. Therefore, at the time when ETC $= 0$, the goal at the time of OpenStack release has been achieved.

Figure 10.8 shows the estimated total software effort for the delayed S-shaped model, $\mathrm{E}[E_s(t)]$ and $E_s(t)$, respectively. In case of a delayed S-shaped model, we find that the optimum maintenance time is derived as $t^* = 8.985$ years (468.5 weeks).

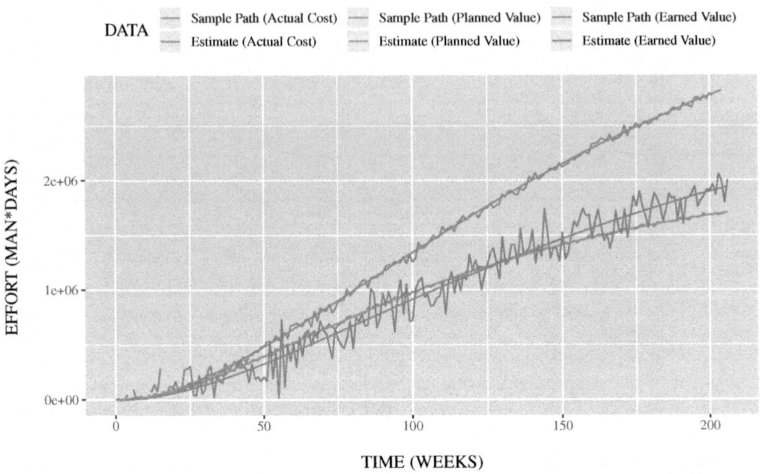

Figure 10.5 The estimated AC, PV, EV in OpenStack project using S-shaped model.

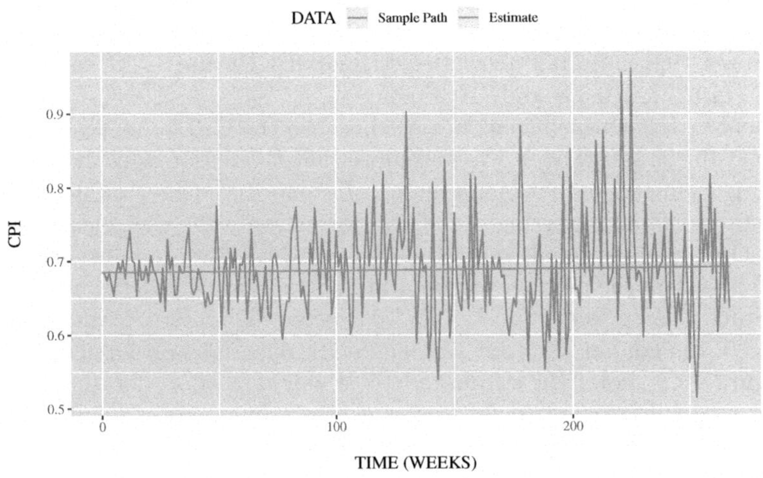

Figure 10.6 The estimated CPI in case of Eqs. (10.25) and (10.27).

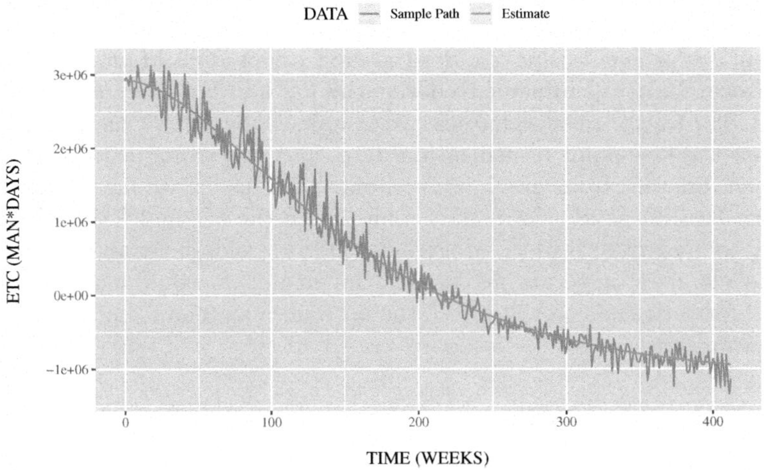

Figure 10.7 The estimated ETC in case of Eqs. (10.33) and (10.35).

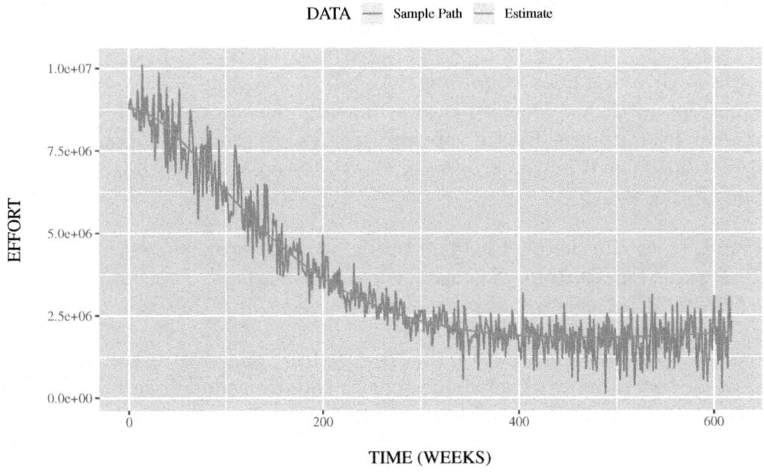

Figure 10.8 The estimated total software effort in case of $\mathrm{E}[E_s(t)]$ and $E_s(t)$.

10.7 CONCLUSION

In this chapter, we considered the derivation method of OSS-oriented EVM and verified it using actual project data. We have also examined the optimal maintenance problem using the derived OSS-oriented EVM value. In the previous research, it was difficult to derive the PV and EV when applying EVM to the OSS project. In this chapter, we examined the derivation method of the PV from the viewpoint of utilization data, and also examined the EV by an approach different from previous research. As a result, we have proposed an OSS-oriented EVM as a form that conforms to the original EVM. Therefore, we can compare the project progress of OSS development and normal software development in terms of EVM. In the future, we would like to consider a method for verifying the validity of the optimal maintenance problem derived in this chapter.

ACKNOWLEDGMENTS

This work was supported in part by the JSPS KAKENHI Grant No. 20K11799 in Japan.

REFERENCES

1. Raymond, S. E. (1999). *The Cathedral and the Bazaar: Musings on Linux and Open Source by an Accidental Revolutionary*, O'Reilly and Associates, Sebastopol, California.

2. Fleming, Q. E., & Koppelman, J. M. (2010). *Koppelman, Earned Value Project Management (4th Ed.)*, PMI, Newton Square, U.S.A.

3. Robles, G., Gonzälez-Barahona, M. J., Cervigön, C., Capiluppi, A., & Izquierdo-Cortäzar, D. (2014). Estimating development effort in Free/OSS projects by mining software repositories: a case study of OpenStack, *Proceedings of the 11th Working Conference on Mining Software Repositories*, Hyderabad. India. 222-231.

4. Mishra, R., & Sureka, A. (2014). Mining peer code review system for computing effort and contribution metrics for patch reviewers, *Proceedings of the 2014 IEEE 4th Workshop on Mining Unstructured Data*, Victoria. Canada. 11-15.

5. Sone, H., Tamura, Y., & Yamada, S. (2019). Statistical maintenance time estimation based on stochastic differential equation models in OSS development project, *Computer Reviews Journal, PURKH, 5,* pp. 126-140.

6. Yamada, S., & Osaki, S. (1985). Cost-reliability optimal software release policies for software systems, *IEEE Transactions on Reliability, R-34(5),* 422-424.

7. Yamada, S, & Osaki, S. (1987). Optimal software release policies with simultaneous cost and reliability requirements, *European Journal of Operational Research, 31(1),* 46-51.

8. Sone, H., Tamura, Y., & Yamada, S. (2018). Optimal maintenance problem with earned value requirement for OSS project, *Proceedings of the Fourteenth International Conference on Industrial Management, Hangzhou, China,* 253-258.

9. Yamada, S. (2014). *Software Reliability Modeling: Fundamentals and Applications*, Springer-Verlag, Tokyo/Heidelberg.

10. Lyu, M. R. (1996). *Handbook of Software Reliability Engineering.* IEEE Computer Society Press, Los Alamitos, CA, U.S.A.

11. Musa, J. D., Iannino, A., & Okumoto, K. (1987). *Software Reliability: Measurement, Prediction, Application.* McGraw-Hill, New York.

12. Kapur, P. K., Pham, H., Gupta, A., & Jha, P. C. (2011). *Software Reliability Assessment with OR Applications*, Springer-Verlag, London.

13. Wong, E. (1971). *Stochastic Processes in Information and Systems. McGraw-Hill*, New York.

14. Arnold, L. (1971). *Stochastic Differential Equations-Theory and Applications.* John Wiley & Sons, New York.

15. Yamada, S., Kimura, M., Tanaka, H., & Osaki, S. (1994). Software reliability measurement and assessment with stochastic differential equations, *IEICE Transactions on Fundamentals, E77-A(1),* 109-116.

16. Ranjan, K., Subhash, K., Sanjay, K. T. (2019). A study of software reliability on big data open source software, *International Journal of System Assurance Engineering and Management, 10(2),* 242-250.

17. The OpenStack Foundation, The OpenStack project, `http://www.openstack.org/`

11 Reliability Assessment Model Based on Wiener Process Considering Network Environment for Edge Computing

Yoshinobu Tamura
Yamaguchi University, Japan

Hironobu Sone
IBM Japan, Ltd., Japan

Shigeru Yamada
Tottori University, Japan

CONTENTS

11.1 INTRODUCTION

At present, several internet services are changing from cloud computing to edge computing. In particular, many open source software (OSS) components are included in the area of edge computing. In general, it is well known that many open source components are embedded in many commercial software programs. We focus on the open source software in edge computing. Network connection is a very important factor in considering edge computing. Many methods of software reliability assessment have been developed by several researchers [1, 2]. Also, several papers in terms of open source software reliability have been proposed by several researchers [3, 4].

Considering software reliability, it is important for OSS managers to appropriately control the development effort of open source software. The optimum control of management effort for OSS development will indirectly relate to the quality, reliability, and cost.

Various research papers on edge computing have been proposed several researchers [5, 6, 7, 8]. However, the dynamic reliability assessment method for edge computing has not been presented in the past. We focus on the software effort instead of the software faults. This chapter assumes that the software fault introduction in the development depends on the software effort expense.

This chapter proposes a method of OSS project management that considers irregular fluctuation in effort performance resulting from the characteristics of the network environment of OSS operation in the edge computing. In particular, an OSS project assessment method based on the Wiener process model considering the network environment in terms of effort expense is proposed in order to comprehend the cyclic situation of internet network.

11.2 WIENER PROCESS MODELING BASED ON PERIODIC WEIGHT FUNCTIONS

We apply a stochastic differential equation model with periodic weighted functions to control the development effort in the operational phase of OSS projects. Then, let $O(t)$ be the cumulative development effort expenditures up to operational time t $(t \geq 0)$ in the operation of an OSS project. Suppose that $O(t)$ takes on continuous real values. Since the estimated amount of development effort is observed during the operational phase of the OSS project, $O(t)$ gradually increases as the operational procedures go on. Then, we assume that the fault detection phenomenon is approximately equal to the effort expenditures one, because the fault detection phenomenon will depend on the effort expenditures one. Based on software reliability growth modeling [1], the following linear differential equation in terms of development effort management can be formulated:

$$\frac{dO(t)}{dt} = \beta(t)\{\alpha - O(t)\}, \tag{11.1}$$

where $\beta(t)$ is the increase rate of development effort at OSS operational time t and a non-negative function, α, means the estimated amount of development effort required until the end of operation.

Therefore, we extend Eq. (11.1) to the following stochastic differential equation with two Brownian motions [9, 10]:

$$\frac{dO(t)}{dt} = \{\beta(t) + \sigma_1 \tau_1(t) + \sigma_2 \tau_2(t)\}\{\alpha - O(t)\}, \tag{11.2}$$

where σ_1 and σ_2 are the positive constants representing a magnitude of the irregular fluctuation, and $\tau_1(t)$ and $\tau_2(t)$ are standardized Gaussian white noises, respectively [9]. We assume the following situation:

- σ_1 and $\tau_1(t)$ are the first noise in terms of the software management effort affected from the connection/processing delay for edge/IoT devices.
- σ_2 and $\tau_2(t)$ are the second noise with the cycle delay in terms of software management effort considering the connection/processing delay from the edge server.

Then, we extend to the following stochastic differential equation of an Itô type [9]:

$$\begin{aligned} dO(t) &= \left\{\beta(t) - \frac{1}{2}(\sigma_1^2 + \sigma_2^2)\right\}\{\alpha - O(t)\}dt \\ &+ s(t)\cdot\sigma_1\{\alpha - O(t)\}do_1(t) + c(t)\cdot\sigma_2\{\alpha - O(t)\}do_2(t), \quad (11.3) \end{aligned}$$

where $o_1(t)$ and $o_2(t)$ are two-dimensional Wiener processes, and are formally defined as an integration of the white noises $\tau_1(t)$ and $\tau_2(t)$ with respect to time. We assume that $[o_1(t), o_2(t)]$ are mutually independent [11, 12, 13, 14, 15, 16].

In this chapter, we assume that the increase rate of maintenance effort expense $O(t)$ can be approximately covered by the mean value function of the traditional inflection S-shaped software reliability growth model [1], because the inflection S-shaped software reliability growth model can widely cover from the exponential curve to the S-shaped one:

$$O(t) \doteq \frac{\beta}{1 + \gamma \cdot \exp(-\beta t)}, \quad (11.4)$$

where β is the changing rate of the maintenance effort expense, and γ is defined as $\frac{1-p}{p}$. The parameter p is the changing rate of the environmental factor under edge computing.

The network information is transmitted by using the communications system. Then, the actual communications system is based on the electrical signals. Therefore, it is important to focus on the electrical signals in the network environment of edge computing. In the past, we have proposed weighted functions by using exponential ones. We will be able to propose the practical assessment method for edge computing by focusing the electrical signals. In particular, the trigonometric functions are frequency used in the area of electrical signals, because the electrical signals have the characteristics of periodic functions.

Then, $s(t)$ and $c(t)$ are the weight functions at OSS operational time t. Also, we assume that $s(t)$ and $c(t)$ have a cycle delay. In particular, it is important for the communications network to use trigonometric functions, because trigonometric functions are used many science and engineering [17, 18, 19]. Therefore, the weight functions for Wiener processes are represented as follows in this chapter:

$$s(t) = \sin\left(\frac{2\pi t}{\xi T}\right), \quad (11.5)$$

$$c(t) \quad = \quad \cos\left(\frac{2\pi t}{\xi T}\right), \tag{11.6}$$

where ξ is the changing rate of the weighted Wiener processes, respectively. In Eqs. (11.5) and (11.6), $\frac{1}{T}$ means the fundamental frequency. We will be able to comprehend the characteristic change of version upgrade cycle by the trigonometric functions.

By using Itô's formula [9, 10], the solution of the former equation can be obtained as follows:

$$O(t) = \alpha\left[1 - \frac{1+\gamma}{1+\gamma\cdot\exp(-\beta t)}\exp\left\{-\beta t - s(t)\sigma_1 o_1(t) - c(t)\sigma_1 o_2(t)\right\}\right] \tag{11.7}$$

Similarly, the estimated amount of development effort required until the end of operation based on Wiener process models can give as follows:

$$O_r(t) = \alpha\left[\frac{1+\gamma}{1+\gamma\cdot\exp(-\beta t)}\exp\left\{-\beta t - s(t)\sigma_1 o_1(t) - c(t)\sigma_1 o_2(t)\right\}\right]. \tag{11.8}$$

11.3 PARAMETER ESTIMATION

Generally, the method of maximum-likelihood is well known as the method of parameter estimation [13, 14, 15, 16]. The proposed model parameters α, β, γ, σ_1, and σ_2 can be estimated by using the method of maximum-likelihood. The joint probability distribution function of the stochastic process $O(t)$ is given by

$$\begin{aligned} & P(t_1, n_1; t_2, n_2; \cdots; t_K, n_K) \\ = \; & \Pr[O(t_1) \le n_1, O(t_2) \le n_2, \cdots, O(t_K) \le n_K | O(0) = 0]. \end{aligned} \tag{11.9}$$

The probability density of Eq. (11.10) is denoted as

$$p(t_1, n_1; t_2, n_2; \cdots; t_K, n_K) = \frac{\partial^K P(t_1, n_1; t_2, n_2; \cdots; t_K, n_K)}{\partial_{n_1}\partial_{n_2}\cdots\partial_{n_K}}. \tag{11.10}$$

Since $O(t)$ are considered continuous values, the likelihood function, l_f, for the observed effort data $(t_k, n_k)(k = 1, 2, \cdots, K)$ is constructed as

$$l_f = p(t_1, n_1; t_2, n_2; \cdots; t_K, n_K). \tag{11.11}$$

For convenience in mathematical manipulations, the following logarithmic likelihood function l_{lf} is used:

$$l_{lf} = \log l_f. \tag{11.12}$$

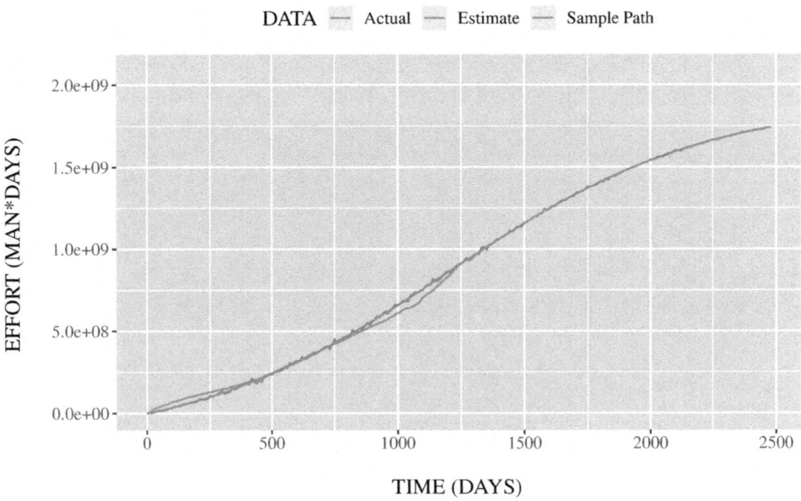

Figure 11.1 The estimated cumulative development effort expenditures in case of $O(t)$.

The maximum-likelihood estimates α^*, β^*, γ^*, σ_1^*, and σ_2^* are the values that maximize l_{lf} in Eq. (11.12). These can be obtained as the solutions of the following:

$$\frac{\partial l_{lf}}{\partial \alpha} = \frac{\partial l_{lf}}{\partial \beta} = \frac{\partial l_{lf}}{\partial \gamma} = \frac{\partial l_{lf}}{\partial \sigma_1} = \frac{\partial l_{lf}}{\partial \sigma_2} = 0. \tag{11.13}$$

Moreover, ξ included in $s(t)$ and $c(t)$ is independent on Wiener processes $o_1(t)$ and $o_2(t)$, respectively. Based on the following fitness function, ξ is determined:

$$\min_{\xi} M_i(\xi),$$
$$M_i(\xi) = \sum_{i=0}^{n} \{O(i, \xi) - n_i\}^2, \tag{11.14}$$

where $O(i)$ is the effort expense at operation time i in the proposed model, and n_i the actual software effort. Also, ξ means the changing rate parameter of Wiener processes σ_1 and σ_2 [20].

11.4 NUMERICAL EXAMPLES

We focus on the *OpenStack* Project [21]. Fig. 11.1 shows the estimated cumulative development effort expenditures in case of $O(t)$. From Fig. 11.1, we find that the noise of the Wiener process becomes gradually small as the operating time procedures go on. Similarly, Fig. 11.2 shows the estimated amount of

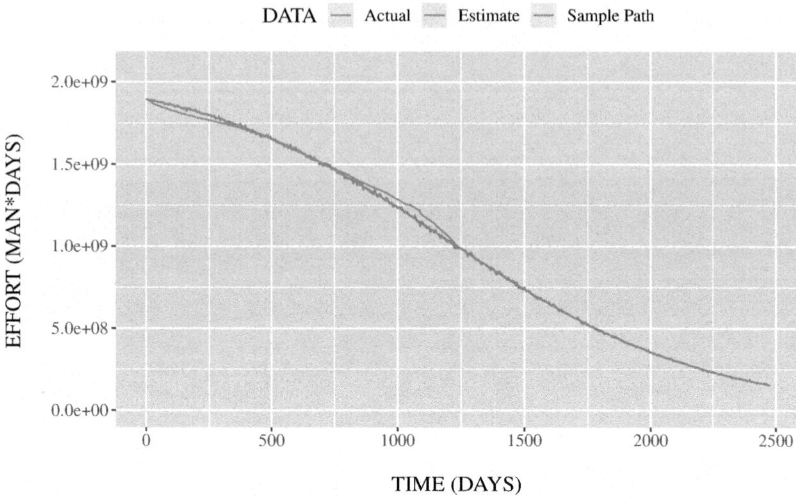

Figure 11.2 The estimated amount of development effort required until the end of operation in case of $O_r(t)$.

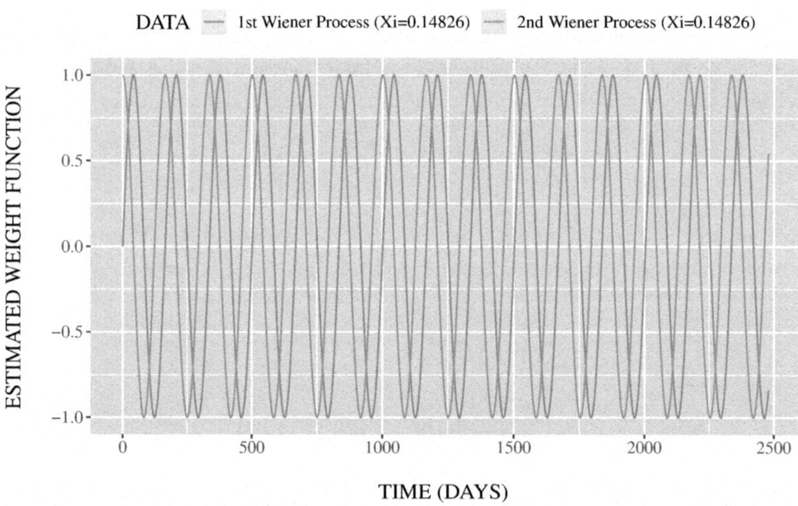

Figure 11.3 The estimated weight functions for the Wiener process at OSS operational time t.

development effort required until the end of operation in case of $O_r(t)$. From Fig. 11.2, we find that the $O_r(t)$ fits better for the actual data sets. Moreover, Fig. 11.3 shows the estimated weight functions for Wiener processes at OSS operational time t in cases of $s(t)$ and $c(t)$. From Fig. 11.3, we find that the

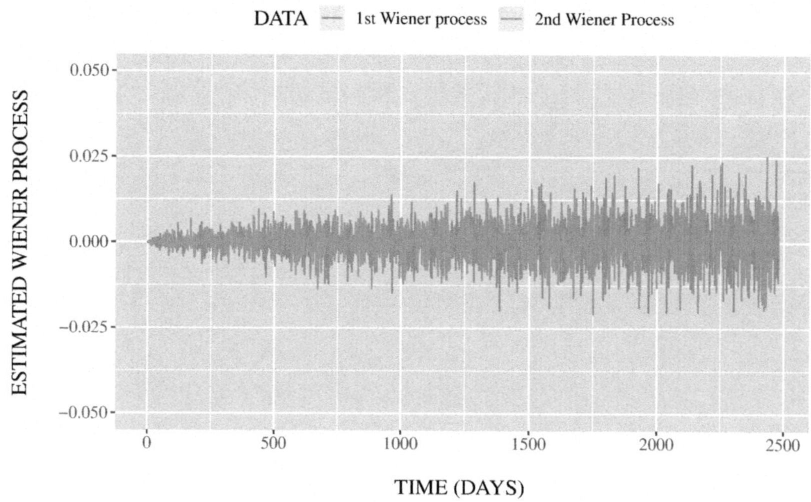

Figure 11.4 The estimated sample path in cases of $O(t)$.

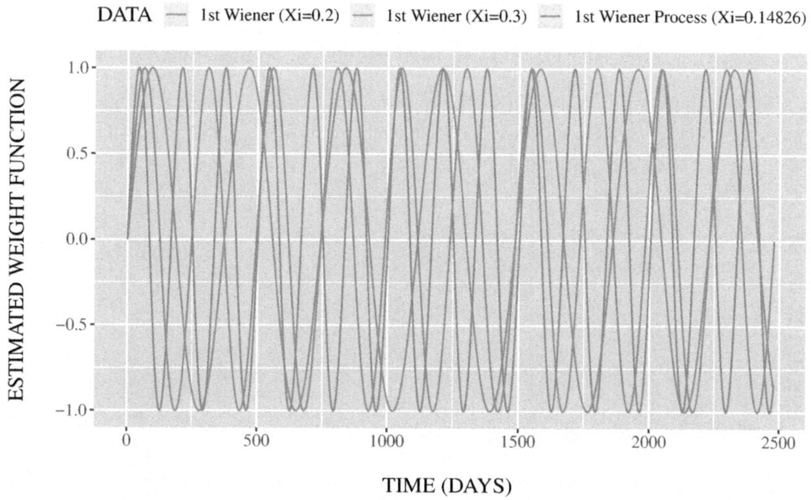

Figure 11.5 The sensitivity analysis of the weight function for the parameter ξ in cases of the 1st Wiener process.

estimated weight functions for Wiener processes at OSS operational time t in case of $O(t)$ circulate around 250 days. Furthermore, we show the estimated sample paths of the following noise terms in cases of $O(t)$ in Fig. 11.3.

$$O_s(t) = s(t)\sigma_1 o_1(t), \tag{11.15}$$

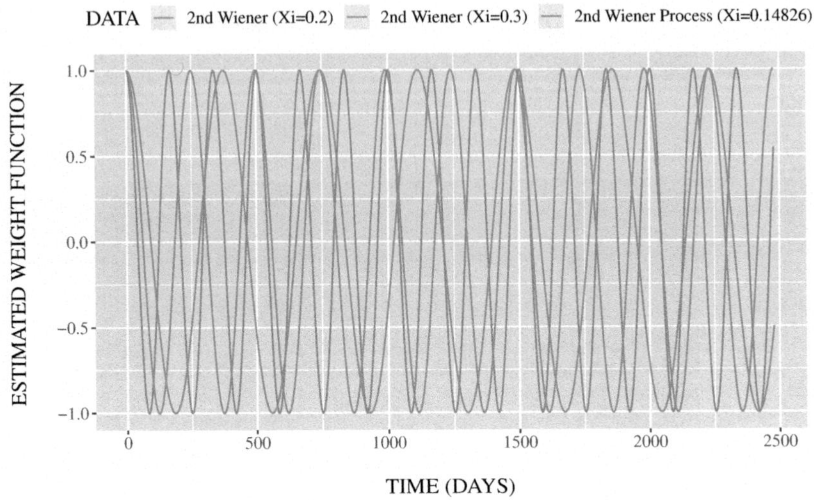

Figure 11.6 The sensitivity analysis of the weight function for the parameter ξ in cases of the 2nd Wiener process.

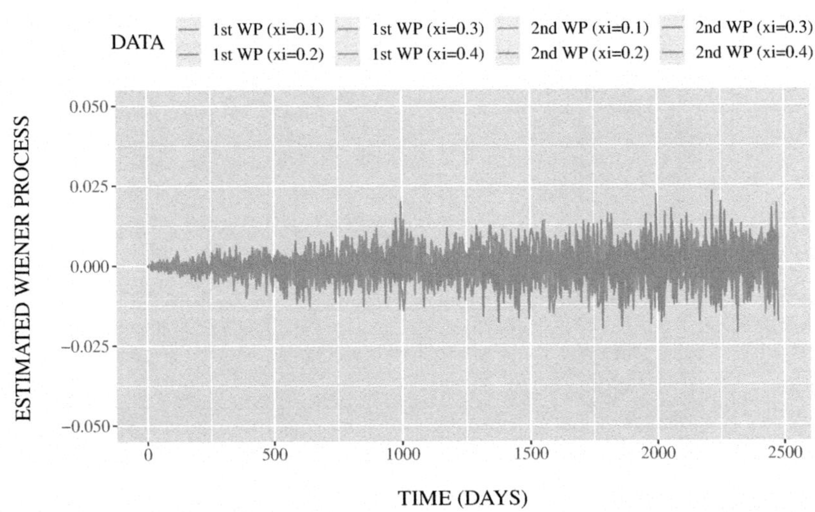

Figure 11.7 The sensitivity analysis of the estimated sample path in Eqs. (11.15) and (11.16) for the parameter ξ.

$$O_c(t) \quad = \quad c(t)\sigma_2 o_2(t). \tag{11.16}$$

From Fig. 11.3, we can confirm that the estimated sample paths are changing as the operation procedures go on.

Moreover, we show the sensitivity analysis of the weight function for the parameter ξ in Figs. 11.5 and 11.6. From Figs. 11.5 and 11.6, we find that the weight functions change with the value of parameter ξ. Furthermore, Fig. 11.7 show the sensitivity analysis of the estimated sample path in Eqs. (11.15) and (11.16) for the parameter ξ. In particular, we consider that the proposed model can adequately assess the development effort based on the electrical signals.

11.5 CONCLUDING REMARKS

The appropriate effort control considering the network environment for OSS maintenance will indirectly depend on the reliability and cost reduction of OSS. This chapter has proposed a method of OSS project assessment of edge computing considering the irregular fluctuations with Wiener continuous noise from the non-continuous characteristics of OSS development and management considering edge computing. It is difficult for the edge OSS project managers to control the progress of the OSS operation considering the network environment. This chapter's discussion used the actual development effort data as follows:

- OSS reliability assessment with cyclic time-variation
- Wiener processes considering the network environment
- Cyclic noise changing with the time-dependent
- Considering the electrical signals with the characteristics of periodic function

The proposed method will be useful for maintenance managers of edge computing as the assessment method of the effort expense progress for OSS projects in the operation phase of edge computing.

ACKNOWLEDGMENTS

This work was supported in part by the JSPS KAKENHI Grant No. 20K11799 in Japan.

REFERENCES

1. Yamada, S. (2014). *Software Reliability Modeling: Fundamentals and Applications*, Springer–Verlag, Tokyo/Heidelberg.

2. Kapur, P.K., Pham, H., Gupta, A., & Jha, P. C. (2011). *Software Reliability Assessment with OR Applications*, Springer–Verlag, London.

3. Yamada, S, & Tamura, Y. (2016). *OSS Reliability Measurement and Assessment*, Springer International Publishing, Switzerland.

4. Norris, J. (2004). Mission-critical development with open source software, *IEEE Software Magazine 21(1)*, 42-49.

5. Ozcan, M. O., Odaci, F., & Ari, I. (2019). Remote debugging for containerized applications in edge computing environments, *Proceedings of the 2019 IEEE International Conference on Edge Computing (EDGE)*, Milan, Italy, doi: 10.1109/EDGE.2019.00021, 30-32.

6. Ngoko, Y., & Cerin, C. (2017). An edge computing platform for the detection of acoustic events, *Proceedings of the 2017 IEEE International Conference on Edge Computing (EDGE)*, Honolulu, HI, USA, doi: 10.1109/IEEE.EDGE.2017.44, 240-243.

7. Caprolu, M., Di Pietro, R., Lombardi, F., & Raponi, S. (2019). Edge computing perspectives: architectures, technologies, and open security issues, *Proceedings of the 2019 IEEE International Conference on Edge Computing (EDGE)*, Milan, Italy, doi: 10.1109/EDGE.2019.00035, 116-123.

8. Dolui, K., & Datta, S. K. (2017). Comparison of edge computing implementations: Fog computing, cloudlet and mobile edge computing, *Proceedings of the 2017 Global Internet of Things Summit (GIoTS)*, Geneva, Switzerland, doi: 10.1109/GIOTS.2017.8016213, 1-6.

9. Arnold, L. (1974). *Stochastic Differential Equations–Theory and Applications.* John Wiley & Sons, New York.

10. Yamada, S., Kimura, M., Tanaka, H., & Osaki, S. (1994). Software reliability measurement and assessment with stochastic differential equations, *IEICE Transactions on Fundamentals, E77–A(1)*, 109–116.

11. Tamura, Y., Sone, H., & Yamada, S. (2019). Productivity assessment based on jump diffusion model considering the effort management for OSS project, *International Journal of Reliability, Quality and Safety Engineering (World Scientific), 26(5)*, 1950022-1–1950022-22.

12. Tamura, Y., Sone, H., Sugisaki, K., & Yamada, S. (2019). A method of parameter estimation in flexible jump diffusion process models for open source maintenance effort management, *Proceedings of the IEEE International Conference on Industrial Engineering and Engineering Management*, Macau, China, CD-ROM (Reliability and Maintenance Engineering 2).

13. Tamura, Y., & Yamada, S. (2017). Dependability analysis tool based on multi-dimensional stochastic noisy model for cloud computing with big data, *International Journal of Mathematical, Engineering and Management Sciences, 2(4)*, 273-287.

14. Tamura, Y., & Yamada, S. (2017). Open source software cost analysis with fault severity levels based on stochastic differential equation models, *Journal of Life Cycle Reliability and Safety Engineering, 6(1)*, DOI 10.1007/s41872-017-0009-5, Springer, 31-35.

15. Tamura, Y., & Yamada, S. (2017). Dependability analysis tool considering the optimal data partitioning in a mobile cloud, *in Reliability Modeling with Computer and Maintenance Applications*, World Scientific, 45-60.

16. Tamura, Y., & Yamada, S. (2018). Multi-dimensional software tool for OSS project management considering cloud with big data, *International Journal of Reliability, Quality and Safety Engineering, 25(3),* World Scientific, 1850014-1–1850014-16.

17. Malešević, B., Rašajski, M., & Lutovac, T. (2017). Refinements and generalizations of some inequalities of Shafer-Fink's type for the inverse sine function, *Journal of Inequalities and Applications, 275,* https://doi.org/10.1186/s13660-017-1554-1.

18. Makragic, M. (2017). A method for proving some inequalities on mixed hyperbolic-trigonometric polynomial functions, *Journal of Mathematical inequalities 11(3),* 817?829.

19. Rahmatollahi, G., & Abreu, G. (2012). Closed-form hop-count distributions in random networks with arbitrary routing, *IEEE Transactions on Communications, 60(2),* doi: 10.1109/TCOMM.2012.010512.110125, 429-444.

20. Tamura, Y., & Yamada, S. (2019). Maintenance effort management based on double jump diffusion model for OSS project, *Annals of Operations Research,* DOI: 10.1007/s10479-019-03170-w, Springer US, Online First, 1-16.

21. The OpenStack project, Build the future of Open Infrastructure, https://www.openstack.org/

12 Approximated Estimation of Software Target Failure Measures Conforming IEC 61508

Shinji Inoue
Kansai University

Takaji Fujiwara
SRATECH Laboratory, Inc.

Shigeru Yamada
Tottori University

CONTENTS

12.1 INTRODUCTION

Recently, the notion of functional safety is widely applied in several industrial fields due to the advancement of information processing technologies. Functional safety is maintaining a certain level of safety by functional aspects. Especially, an electrical/electronic/programmable electronic (abbreviated as E/E/PE) safety-related system is utilized for realizing functional safety of whole systems. For example, the pre-crash system and the air-bag normal expansion system of an automobile are operated by E/E/PE safety-related systems.

Along with the noticeable spread of using E/E/PE safety-related systems in several industrial areas, the International Electrotechnical Commission (abbreviated as IEC) issued the first edition of the international basic standard for the functional safety of E/E/PE safety-related systems, IEC 61508, in 2000.

Further, the second edition was issued in 2010 [5]. Recently, conformity and authentication of IEC 61508 is widely required in several industries and the other safety standards for safety-related areas have also been issued, e.g., ISO 26262, which is the international standard for functional safety of E/E/PE safety-related systems in production of automobiles. It is worth mentioning that IEC 61508 provides us with a basic standard with several fundamental techniques on safety-related issues, such as risk assessment, safety integrity level, overall safety lifecycle, and functional safety assessment, for ensuring the functional safety realized by E/E/PE safety-related systems. Especially, IEC 61508 provides us with the basic standard for analyzing the safety integrity level (abbreviated as SIL) quantitatively based on the notion of the random failure-occurrence mechanism for hardware-related harmful events in E/E/PE safety-related systems. That is, it is required to assess the safety of the hardware in E/E/PE safety-related systems based on specific probability-based measures for harmful events. Generally, probability-based safety measures are called target failure measures, and are derived by stochastic analysis methods. In recent years, several methods for calculating the target failure measures and deciding SIL for the hardware of E/E/PE safety-related systems have been proposed for enabling us to conduct quantitative safety assessment by considering several types of specific internal structures of E/E/PE safety-related systems. For example, Misumi and Sato [7] proposed methods for calculating target failure measures by considering harmful risk event occurrence mechanisms for E/E/PE safety-related systems. Kato and Sato [6] newly defined safety demand modes for E/E/PE safety-related hardware and proposed calculating methods of target failure measures for new safety demand modes. Ghadhab et al. [2] discussed the usage of a dynamic fault tree for safety analysis of vehicle guidance systems based on ISO 26262. As we discussed above, there is lots of attention on quantitative safety assessment of E/E/PE safety-related hardware.

On the other hand, it is rare for us to get a chance to read literature on quantitative safety assessment for E/E/PE safety-related software because the software failure is treated as a systematic failure in IEC 61508. This essentially means that IEC 61508 does not require any probabilistic safety analysis for E/E/PE safety-related software and considers that software failures occur inevitably due to the existence of software bugs, which are introduced during the software development process. In fact, IEC 61508 requires specific software development technologies or methods for achieving a certain level of SIL. However, for embedded software like E/E/PE safety-related software systems, it is better to consider the uncertainty of the time-varying dangerous software failure occurrence intervals because the execution frequency of program paths or software modules depend on the input to the E/E/PE safety-related software. In other words, it is difficult to obtain information on when the specific program path or software module having software bugs is executed. In that case, quantitative safety assessment considering time-varying random software failure occurrences must be effective. Recently, Fujiwara et al. [1] discussed the need for quantitative SIL assessment for E/E/PE safety-related software and proposed a basic framework for SIL-based software safety assessment. Gu [3] discussed a supporting approach for estimating software SIL for an embedded system by considering the failure-occurrence behavior between hardware and software. However, problems in SIL-based software safety assessment still remain. That is, we cannot determine how much time is needed to test the software to ensure it conforms to a certain SIL.

This chapter discusses model-based methods for conducting quantitative software safety assessment with the time-varying uncertainty of dangerous software failure occurrences of E/E/PE safety-related system by applying a software reliability modeling approach and assessment techniques. Specifically, we discuss approximated estimation methods for conducting SIL-based safety assessment for E/E/PE safety-related software by applying the notion of the software hazard rate of existing software reliability modeling technologies and assessment methods. Especially, we discuss approximated estimation methods for calculating target failure measures, which are needed for deciding the achieved SIL. Further, we show numerical examples for explaining how to use our approaches and for giving some considerations in safety assessment based on our approaches by applying software failure occurrence time data.

12.2 SIL AND TARGET FAILURE MEASURES

We give a brief introduction of the SIL and target failure measures by showing a basic configuration of the whole system including the E/E/PE safety-related system. Figure 12.1 shows the general basic configuration of the whole system with the safety-related systems. Generally, the intended function is operated by equipment, which is called equipment under control (abbreviated as EUC) and the basic control system (abbreviated as BCS). As shown in Fig. 12.1, we can see that the safety-related system is an incrementally assembled system for the main system consisting of EUC and BCS. And the safety-related system takes the safety role of the whole system finally when the main system fails the safety function, and then the safety demand is issued from the main system. Therefore, the safety assessment for the E/E/PE safety-related system is needed to consider not only the state of the E/E/PE safety-related system, but also the frequency of the safety demands to the E/E/PE safety-related system.

As we mentioned above, the main feature of IEC 61508 is to require conducting a SIL-based safety assessment for E/E/PE safety-related systems. The SIL is the graded measure divided into four safety integrity levels along with the notion of ALARP (as low as reasonably practicable) principals because hazardous risk mitigation is needed to take measures within the reasonably practicable region. Table 12.1 shows the SIL defined by IEC 61508. In Table 12.1, the SIL is measured on the basis of the target failure measures, such as the average probability of failure on demand of the safety function (abbreviated as PFD) and the average frequency of a dangerous failure of the safety

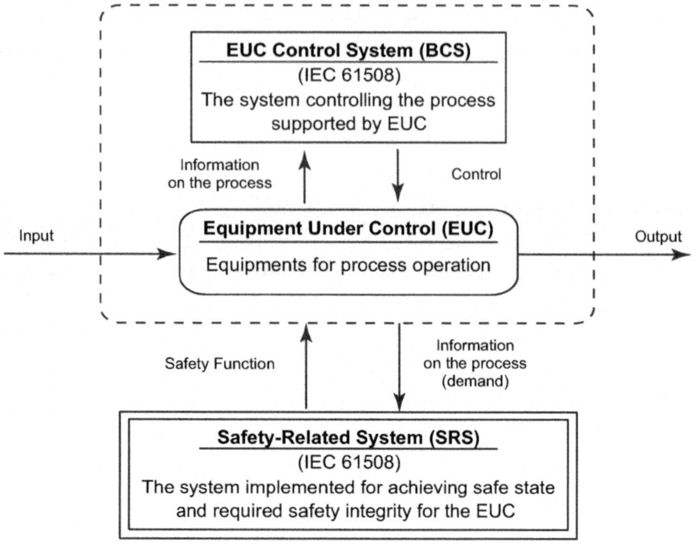

Figure 12.1 Configuration of general whole system with safety-related system.

Table 12.1
Safety Integrity Level of Functional Safety

SIL	Low Demand Mode	High Demand or Continuous Mode
4	$10^{-5} \leq \text{PFD} < 10^{-4}$	$10^{-9} \leq \text{PFH (1/hour)} < 10^{-8}$
3	$10^{-4} \leq \text{PFD} < 10^{-3}$	$10^{-8} \leq \text{PFH (1/hour)} < 10^{-7}$
2	$10^{-3} \leq \text{PFD} < 10^{-2}$	$10^{-7} \leq \text{PFH (1/hour)} < 10^{-6}$
1	$10^{-2} \leq \text{PFD} < 10^{-1}$	$10^{-6} \leq \text{PFH (1/hour)} < 10^{-5}$

function (abbreviated as PFH), depending on the operation modes. The operation modes are divided into the low demand mode and the high demand or continuous mode, which represent the possible safety demand frequency requested by the main system in operation. If the E/E/PE safety-related system is operated not greater than once a year due to safety demands, it is required to assess the SIL based on the PFD. On the other hand, the PFH is applied in the safety assessment if the E/E/PE safety-related system is demanded greater than once a year.

From the definitions of PFD mentioned above, the PFD is given by the down-time duration of the total operation time of the E/E/PE safety-related system. That is, the PFD can be essentially treated as the unavailability of the E/E/PE safety-related system. Considering the relationship between reliability and availability [1], the PFD can be given as

$$\text{PFD} = 1 - \text{Software Reliability}, \tag{12.1}$$

for the sake of simplicity. Actually, a highly reliable system has high availability if the constant down-time for debugging activities of software faults can be considered. The PFH can be essentially considered as the dangerous software failure intensity of the E/E/PE safety-related system. If the down-time for debugging activities is very short compared to the total operation time duration, the equation for approximately estimating the PFH is obtained as

$$\text{PFH} = \frac{1}{\text{MTBF}}, \tag{12.2}$$

in which the MTBF represents the mean time between software failures, which is one of the representative software reliability assessment measures. It is worth mentioning that the unit of PFH is 1/hour.

12.3 SOFTWARE HAZARD RATE MODELING

Software hazard rate models [9, 10, 11, 12] describe the behavior of the software hazard rate depending on the number of software faults in the software or depending on the number of software failure occurrences. We briefly discuss software hazard rate modeling, which is utilized for developing our mathematical models for estimating the target failure measures. Let $h_k(x)(k = 1, 2, \cdots)$ denote the software hazard rate function for the k-th software failure occurrence, and is generally formulated as

$$h_k(x) = \frac{c_k(x)}{1 - C_k(x)} = \frac{c_k(x)}{S_k(x)}, \tag{12.3}$$

where $C_k(x)$ represents the cumulative distribution function for the k-th software failure occurrence time interval, which is denoted by $X_k(k = 1, 2, \cdots)$, $c_k(x)$ the probability density function of the random variable X_k, and $S_k(x)$ the software reliability function. Therefore, $S_k(x) \equiv 1 - C_k(x)$, $C_k(x)$, $c_k(x)$, and $S_k(x)$ are given by

$$C_k(x) = 1 - \exp\Big[-\int_0^x h_k(\theta)\mathrm{d}\theta\Big], \tag{12.4}$$

$$c_k(x) = h_k(x)\exp\Big[-\int_0^x h_k(\theta)\mathrm{d}\theta\Big], \tag{12.5}$$

$$S_k(x) = \exp\Big[-\int_0^x h_k(\theta)\mathrm{d}\theta\Big], \tag{12.6}$$

respectively, by solving the differential equations in Eq. (12.3) for $S_k(x)$. It is worth mentioning that $S_k(x)$ is the time-dependent behavior of software reliability for the k-th from the $(k-1)$-th software failure occurrences. Therefore, $S_k(0) = 1$ and $S_k(\infty) = 0$, respectively. The MTBF in Eq. (12.2) can be given as the expectation of X_k within the software hazard rate modeling. Accordingly, it is formulated as

$$\mathrm{MTBF}_k \equiv \mathrm{E}[X_k] = \int_0^{\infty} S_k(\theta)\mathrm{d}\theta. \tag{12.7}$$

As for the software hazard rate, there exists a lot of so-called software hazard rate models, such as exponential, Weibull, and Pareto software hazard rate models. Regarding the exponential software hazard rate models, Jelinski and Moranda [4] proposed the model under the basic assumption that the software hazard rate is proportional to the remaining number of faults in a software system. Moranda [8] proposed a geometrically decreasing software hazard rate model by assuming that the hazard rate decreases geometrically with the number of software debugging events. These two models are the earliest ones, however, they are widely applied in practical software reliability assessment due to their applicability and the mathematical structures. For more details, the Moranda software hazard rate model is developed under the following assumptions:

(A1) The hazard rate for each software failure occurrence is constant during the software failure occurrence time interval.

(A2) The hazard rate for each software failure occurrence decreases geometrically with respect to the number of software failure occurrences observed.

(A3) Each software fault is independent and equally likely to cause a software failure.

From the assumptions above, the Moranda model is given as

$$h_k(x) \equiv h_k = B\phi^{k-1} \tag{12.8}$$
$$(B > 0, 0 < \phi < 1; k = 1, 2, \cdots),$$

where B is the initial software hazard rate for the first software failure occurrence and ϕ is the coefficient representing the decreasing rate of B. It is worth mentioning that the software hazard rate function in Eq. (12.8) does not depend on the elapsed time x.

12.4 FORMULATIONS OF TARGET FAILURE MEASURES

SIL assessment models are proposed by formulating the target failure measures, such as the PFD and PFH, based on the notion of software reliability modeling and assessment technologies focusing on the software hazard rate. For developing mathematical models, which are for estimating the software target failure measures, we incorporate the dangerous software failure occurrence rate, DFR ($0 < \text{DFR} \leq 1$), because the software hazard rate models describe the software failure occurrence phenomenon including both the safe and dangerous software failure occurrences. Incorporating the DFR and following the basic notion in Eq. (12.1), we propose the following equation for approximately estimating the PFD:

$$\text{PFD}_k \equiv \text{DFR}\cdot(1 - S_k^O(x)) \qquad (0 < \text{DFR} \leq 1), \tag{12.9}$$

where, $S_k^O(x)$ is the software reliability at the operation time x elapsed from the $(k-1)$-st software failure occurrence in the operation. $S_k^O(x)$ is obtained from the software hazard rate function for the k-th software failure occurrence in the operation, h_k^O. Regarding h_k^O, it is better to consider the difference in the software executing environment between testing and operation. Accordingly, we also incorporate the environmental coefficient representing the difference between the software hazard rates of testing and operation. That is, h_k^O is given by

$$h_k^O(x) = \text{EC} \cdot h_k^T(x) \qquad (\text{EC} > 0), \tag{12.10}$$

where $h_k^T(x)$ is the software hazard rate for the k-th software failure occurrence in the testing. If $h_k^T(x)$ is assumed to obey the Moranda model in Eq. (12.8), the corresponding probability density, cumulative distribution, reliability functions and MTBF in the operations can be derived as

$$c_k^O(x) = (\text{EC} \times B\phi^{k-1})\exp[-(\text{EC} \times Dc^{k-1})x], \tag{12.11}$$

$$C_k^O(x) = 1 - \exp[-(\text{EC} \times B\phi^{k-1})x] = C_k^T(\text{EC} \times x), \tag{12.12}$$

$$S_k^O(x) = 1 - C_k^O(x) = 1 - C_k^T(\text{EC} \times x), \tag{12.13}$$

$$\text{MTBF}_k^O = \frac{1}{\text{EC} \times B\phi^{k-1}} = \frac{1}{\text{EC}}\text{MTBF}_k^T, \tag{12.14}$$

respectively, by following Eqs. (12.4)–(12.7). From Eqs. (12.4)–(12.7), the PFD for the k-th software failure occurrence, PFD_k, is given by

$$\begin{aligned}\text{PFD}_k &= \text{DFR}\left\{1 - (1 - C_k^T(\text{EC} \times x))\right\} \\ &= \text{DFR} \cdot C_k^T(\text{EC} \times x),\end{aligned} \tag{12.15}$$

in which the Moranda model is applied to the software hazard rate. Eq. (12.15) represents the PFD at designed operation time duration x from the k-th software failure occurrence observed in the testing. Therefore, given certain designed time duration in operation, x, Eq. (12.15) is a function with respect to the number of software failure occurrences k observed in the testing. As for the PFH, which is the target failure measure for the high demand or continuous operation mode, we can obtain

$$\mathrm{PFH}_k \equiv \frac{\mathrm{DFR}}{\mathrm{MTBF}_k^O} = \frac{\mathrm{DFR} \cdot \mathrm{EC}}{\mathrm{MTBF}_k^T}, \tag{12.16}$$

by the basic notion in Eq. (12.2). In common with the PFD_k in Eq. (12.15), PFH_k in Eq. (12.16) is a function with respect to the number of software failure occurrences k observed in the testing. For more information, assuming that the software fault causing a software failure is immediately and perfectly debugged after observing the software failure, the software reliability of the E/E/PE safety-related software increases with the number of software failures that occurred. Therefore, we should note that the PFD and PFH of the E/E/PE safety-related software decreases along with the number of software failures that occurred.

12.5 NUMERICAL EXAMPLES

We show numerical examples of our approaches for approximately estimating the target failure measures, such as the PFD in Eq. (12.15) and PFH in Eq. (12.16) respectively, in case the Moranda model in Eq. (12.8) is applied for describing the software hazard rate. It should be noted that the purpose of this is just to show how to apply our approaches to software reliability data. As the first step, we estimate the values of parameters B and ϕ of the Moranda model in Eq. (12.8) by using the method of maximum likelihood. Generally, assuming that software failure occurrence time interval data have been observed in testing as $x_k (k = 1, 2, \cdots, n)$, the likelihood function, L, is obtained as

$$L = \prod_{k=1}^{n} c_k^T(x_k) = \prod_{k=1}^{n} h_k^T \exp[-h_k^T x_k], \tag{12.17}$$

where $c_k^T(x_k)$ is the probability density function for the time interval between the $(k-1)$-st and k-th software failure occurrences in the testing. Then, the log-likelihood function, in which the Moranda model is applied, is obtained as

$$\begin{aligned} \log L &= \sum_{k=1}^{n} \log(B\phi^{k-1}) - \sum_{k=1}^{n} (B\phi^{k-1}) x_k \\ &= n \log B + \frac{1}{2} n(n-1) \log \phi - B \sum_{k=1}^{n} \phi^{k-1} x_k. \end{aligned} \tag{12.18}$$

The model parameters B and ϕ can be obtained by numerically solving the simultaneous log-likelihood equations with respect to B and ϕ, which are derived from Eq. (12.18), respectively.

For the purpose of just showing our numerical examples, we apply the following software failure occurrence time data [9]: $x_k(\text{days})(k = 1, 2, \cdots, 31)$,

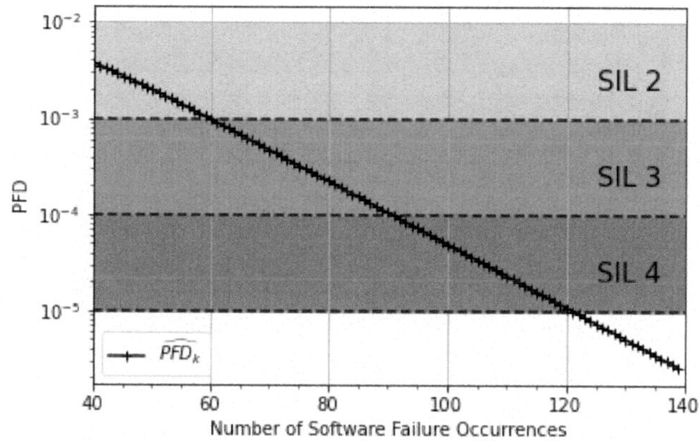

Figure 12.2 Estimated behavior of the time-averaged probability of failure on demand, $\widehat{\text{PFD}}_k$.

where x_k is the time-interval between the $(k-1)$-st and k-th software failure occurrence. It should be noted that the data applied in this chapter has not been collected from E/E/PE safety-related software practically. The parameter in Eq. (12.8) can be estimated by following the estimation method mentioned before based on the software failure occurrence time interval data shown above. As the results, $\widehat{B}$ and $\widehat{\phi}$, which are the estimates of B and ϕ, can be obtained as 0.2438 and 0.9271, respectively.

As the next step, the target failure measures, PFD and PFH, can be estimated by the estimated software hazard rate model. Just for this chapter, we set EC $= 0.1$ and DFR $= 0.01$ before estimating the PFD and PFH, respectively, for just showing numerical examples. In this case, one testing day is equivalent to 10 days in operation. However, the values of EC and DFR in Eqs. (12.15) and (12.16) are needed to set experimentally in practice. Further, we assume that the designed operation time duration is two years. From these parameter settings, the PFD can be evaluated based on the following equation:

$$\text{PFD}_k = \text{DFR} \cdot C_k^T(EC \times 730) = 0.01 \cdot C_k^T(0.1 \times 730), \tag{12.19}$$

from Eq. (12.15). And the PFH can be also estimated as

$$\text{PFH}_k = \frac{\text{DFR} \cdot \text{EC}}{\text{MTBF}_k^T} = \frac{0.01 \cdot 0.1}{\text{MTBF}_k^T}. \tag{12.20}$$

Figures 12.2 and 12.3 show the estimated behavior of PFD_k and PFH_k, which depend on the number of software failure occurrences denoted as k,

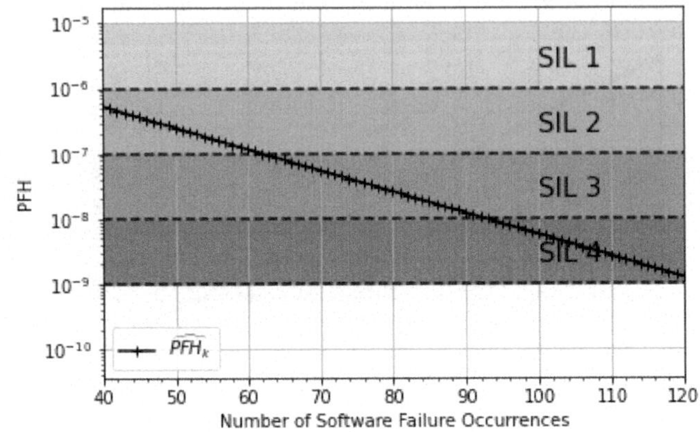

Figure 12.3 Estimated behavior of the time-averaged frequency of dangerous failure per hour, PFH_k.

Table 12.2
Estimated values of PFD$_k(k = 67, 68, \cdots, 71)$

# Software Failures	FFD
67	1.1137×10^{-3}
68	1.0593×10^{-3}
69	9.8610×10^{-4}
70	9.1766×10^{-4}
71	8.5374×10^{-4}

respectively. We note that the vertical axes in these figures follow the logarithmic axes. In Figures 12.2 and 12.3, we can see that the estimated behavior of PFD and PFH are decreasing on k. These figures enable us to see how many software faults should be detected in the software testing for achieving a certain level of SIL. For example, Table 12.2 shows the estimated values of PFDs at from the 67th to 71st software failure occurrences, respectively. From Figure 12.2 and Table 12.2, the software development manager can obtain the information that software testing is needed to conduct up to at least observing the 69th software failure occurrence and debugging the software faults causing the 69th software failure occurrence for achieving SIL 3 under the condition that the software is operated on the low demand operation mode. Further, Figure 12.3 and Table 12.3 list the estimated values of PFHs at the points from the 72th to 76th software failure occurrences, respectively. From Figure

Table 12.3
Estimated values of PFH$_k(k = 72, 73, \cdots, 76)$

# Software Failures	FFH
72	1.1134×10^{-6}
73	1.0508×10^{-6}
74	9.7426×10^{-7}
75	9.0238×10^{-7}
76	8.3746×10^{-7}

12.3 and Table 12.3, we can say that the software testing should be conducted up to observing the 74th software failure occurrence and removing the fault for ensuring SIL 3, where the system is assumed to be operated under the high demand operation or continuous operation mode.

12.6 CONCLUDING REMARKS

Calculating methods for the target failure measures, such as PFD and PFH, for the software of E/E/PE safety-related systems have been discussed in this chapter. The main features of our approaches are to develop the basic mathematical framework for calculating the target failure measures for E/E/PE safety-related software based on the notion of software reliability assessment technologies. Our calculating frameworks enable us to estimate the target failure measures by applying software reliability modeling and software reliability assessment technologies. Our basic concept is regarded as approximation methods for calculating PFD and PFH. However, our approaches yield methods for conducting quantitative SIL assessment for E/E/PE safety-related software based on the concept of software hazard (or failure) rate. Actually, our approaches enable us to obtain the information on the values of the target failure measures in software safety assessment and the information on the testing duration needed for achieving certain SIL. The methods for setting the values of DFR and EC and more plausible models for software SIL assessment must be researched as the future studies. Further, the usefulness of our approaches is also investigated by using software failure and dangerous failure data collected from E/E/PE safety-related software.

REFERENCES

1. Fujiwara, T., Kimura, M., Satoh, Y. & Yamada, S. (2011). A method of calculating safety integrity level for IEC 61508 conformity software. *Proc. 17th IEEE Pacific Rim Intern. Symp. Dependable Computing*, 269–301.

2. Ghadhab, M., Junges, S., Katoen, J.P., Kuntz, M., & Volk, M. (2019). Safety analysis for vehicle guidance systems with dynamic fault trees. *Reliability Engineering and System Safety*, 186, 37–50.

3. Gu, T. (2011). A novel approach supporting evaluation of software safety integrity level on embedded systems. *Proc. 5th Intern. Conf. New Trends in Information Science and Service Science*, 140–145.

4. Jelinski, Z. & Moranda, P.B. (1972). Software reliability research in *Statistical Computer Performance Evaluation*, Freiberger, W. (ed.), Academic Press, 465–484.

5. IEC 61508. (2010). *Functional safety of electrical / electronic / programmable electronic safety-related systems*, Edition 2.0.

6. Kato, E. & Satoh, Y. (2000). Safety integrity level model for IEC 61508 — Examination of modes of operation –. *IEICE Trans. Fundamentals of Electronics, Communications and Computer Sciences*, E83-A(5), 863–865.

7. Misumi, Y. & Sato, Y. (1999). Estimation of average hazardous-event-frequency for allocation of safety-integrity levels. *Reliability Engineering and System Safety*, 66(2), 135–144.

8. Moranda, P.D. (1975). Predictions of software reliability during debugging. *1975 Proceedings of the Annual Reliability and Maintainability Symposium*, 327–332.

9. Pham, H. (2000). *Software Reliability*. Springer-Verlag, Singapore.

10. Pham, H. (2006). *System Software Reliability*. Springer-Verlag, London.

11. Yamada, S. (2011). *Elements of software reliability: Modeling approach* in Japanese, Kyoritsu-Shuppan, Tokyo.

12. Yamada, S. (2014). *Software Reliability Modeling — Fundamentals and Applications —*. Springer, Tokyo.

Section V

Maintenance Optimization and Applications

13 Phase-Type Expansion of Markov Regenerative Processes and Its Application to Reliability Problems

Hiroyuki Okamura
Hiroshima University

Junjun Zheng
Ritsumeikan University

Tadashi Dohi
Hiroshima University

CONTENTS

13.1 INTRODUCTION

Markov modeling is one of the most important techniques to evaluate reliability and performance of systems. In a family of stochastic processes possessing some degree of Markov property, a Markov regenerative process (MRGP) is one of the widest classes of stochastic processes that are mathematically tractable.

MRGP consists of several discrete states and a time sequence of state transitions, and is an extension of both the continuous-time Markov chain (CTMC) and the renewal process. Since MRGP allows state transitions by general distributions, it is often used to analyze Petri nets. The Markov regenerative stochastic Petri net (MRSPNs), which is governed by MRGP, is a very powerful paradigm for performance and reliability [9, 4, 11].

In general, the analysis of MRGP is divided into stationary and transient analyses. Compared to the stationary analysis, the transient analysis is relatively difficult. Methods for the transient analysis of MRGP are discretization, supplementary variables and phase-type (PH) expansion.

Discretization is a method to solve the Markov renewal-type equation which includes a convolution integration. In the discretization method, we use numerical quadrature for the convolution integration, and solve the discretized linear equation. This method provides a nearly exact solution by using a large number of discretization points for any general distribution in MRGP. However, as the number of discretization points increase, time and memory requirements increase at least with the square of the number of discretization points. Also a large memory is needed to store the intermediate results which are proportional to the cube of the number of model states.

The supplementary variable approach uses the age of general distribution as a part of the state description. Thus a discrete state stochastic process is converted into one with joint continuous and discrete state space. The governing system then becomes a partial differential equation. Thus by solving the system partial differential equation, the transient probabilities as well as transient rewards are computed with less memory usage than the discretization method. This method is effective for some classes of MRGP. For example, the MRGP which is restricted to deterministic regeneration timing is efficiently solved by the supplementary variable approach [9, 8]. The supplementary variable approach is implemented in DSPNexpress [12] and TimeNET [7]. On the other hand, when we focus on the stochastic regeneration timing, the supplementary variable approach needs a truncation to obtain the function which represents the age of general distribution. It has a weakness for the general distribution having infinite support such as Weibull and log-normal distributions.

This chapter focuses on the PH expansion of MRGPs. Similar to the supplementary variable approach, the age of general distribution is expressed as the stage which consists of convolutions and mixtures of exponential distributions. That is, the general distribution is approximated as a PH distribution. The advantage of the PH expansion is to reduce the original MRGP to a time-homogeneous Markov chain, so that we could solve both stationary and transient probabilities by commonly-used techniques for CTMC. Also, this approach is promising to develop the fully-automated MRGP/MRSPN analysis tool. However, the phase expansion has not been utilized much in the analysis of MRGPs.

The performance of PH expansion depends on the PH fitting which is done to determine the parameters of PH distribution fitting to the target general distributions. The commonly used approach of PH fitting is moment matching. The usual methods focus only on the first few moments [15, 10, 13, 3]. However, even if the number of phases increases, the moment matching does not guarantee that the behavior of the probability density function (p.d.f.) and cumulative distribution function (c.d.f.) approaches the target ones. This is a weakness of moment-matching-based PH fitting.

This chapter presents an approach to use PH expansion efficiently in MRGP analysis. In particular, we present two points in the PH expansion of MRGP: (i) the systematic way to convert from MRGP to the PH-expanded MRGP, namely, CTMC and (ii) the highly-accurate PH approximation via EM (expectation-maximization) algorithm. For the systematic conversion, we first consider the structured MRGP which consists of partitioned state spaces. Moreover, the structured MRGP is divided into the MRGPs described by single general distribution and multiple general distributions. This analysis is easily performed from MRSPN modeling. Furthermore, we provide a Kronecker representation of CTMC as the PH-expanded MRGP. For the second point, we utilize the state-of-art algorithm [14] which can handle a few hundred phases to approximate a general distribution from the p.d.f. information. Since the accuracy of PH approximation depends on the number of phases, this method is expected to be applied to any kind of model-based analysis.

This chapter is organized as follows. Section 13.2 describes the definition of MRGP and the structured MRGP which is treated in this chapter. In 13.3, we introduce how to get PH parameters from the p.d.f. information via ML estimation. Furthermore, we present a Kronecker representation of approximate MRGP with PH expansion. Section 13.4 presents an illustrative example of PH expansion of MRGP. For a typical reliability model, we show how to obtain the structured MRGP from MRSPN, and demonstrate the PH expansion with Weibull distributions. In 13.5, we conclude this chapter with some remarks.

13.2 MARKOV REGENERATIVE PROCESS

Consider a stochastic process $\{S(t);\ t \geq 0\}$ with discrete state space. If $S(t)$ has time points at which the process probabilistically restarts itself, the process is called regenerative. Specifically, when state transitions at the regeneration points are governed by a discrete-time Markov chain (DTMC), the process $S(t)$ is an MRGP. Define a regeneration time sequence $T_1 < T_2 < \cdots$ and their time intervals $\Delta T_i = T_i - T_{i-1}$, $i = 1, 2, \ldots$. Then the time interval behaves as a Markov renewal sequence [5].

Suppose that the time sequence is time-homogeneous, i.e.,

$$\begin{aligned} &\Pr\{S(T_n) = j, \Delta T_n < t \mid S(T_{n-1}) = i\} \\ &\quad = \Pr\{S(T_1) = j, \Delta T_1 < t \mid S(T_0) = i\} \equiv K_{i,j}(t). \end{aligned} \tag{13.1}$$

The state probability of MRGP is given by

$$\begin{aligned} V_{i,j}(t) &= \Pr\{S(t) = j \mid S(0) = i\} \\ &= \Pr\{S(t) = j, \Delta T_1 \le t \mid S(0) = i\} + \Pr\{S(t) = j, \Delta T_1 > t \mid S(0) = i\} \\ &= \sum_l \int_0^t \Pr\{S(t-u) = j \mid S(0) = l\} dK_{i,l}(u) \\ &\quad + \Pr\{S(t) = j, \Delta T_1 > t \mid S(0) = i\}. \end{aligned} \tag{13.2}$$

In general, $\boldsymbol{K}(t)$, $\boldsymbol{V}(t)$ and $\boldsymbol{E}(t)$ are matrices whose elements are given by $K_{i,j}(t)$, $V_{i,j}(t)$ and $\Pr\{S(t) = j, \Delta T_1 > t \mid S(0) = i\}$, respectively. Then we have the Markov renewal equation for MRGP [4, 6];

$$\boldsymbol{V}(t) = \boldsymbol{E}(t) + \int_0^t d\boldsymbol{K}(u)\boldsymbol{V}(t-u), \tag{13.3}$$

where $\boldsymbol{E}(t)$ and $\boldsymbol{K}(t)$ are called local and global kernels.

13.2.1 STRUCTURED MRGP

To apply PH expansion to MRGP, we consider the following MRGP whose state space is finite and each state is classified to subspaces:

- The state space is separated into subspaces $\mathcal{S}^E$ and $\mathcal{S}_1^G, \dots, \mathcal{S}_K^G$.
- The subspace $\mathcal{S}^E$ (EXP state) consists of the states which have only EXP transitions. The CTMC generator among the states belonging to $\mathcal{S}^E$ is $\boldsymbol{D}_0$. Also the CTMC transition rates for $\mathcal{S}_1^G, \dots, \mathcal{S}_K^G$ are given by $\boldsymbol{D}_1, \dots, \boldsymbol{D}_K$, respectively.
- The subspace $\mathcal{S}_i^G$ (GEN states) consists of the states which have both non-regenerative and regenerative transitions.
    - Define the CTMC generator of non-regenerative transitions which do not transfer across different subspace $\boldsymbol{Q}_i$. On the other hand, the transition rate matrix for the non-regenerative transition to different subspace is given by $\boldsymbol{Q}_{i,j}$.
    - The regenerative transition has one general distribution; $G_i(t)$.
    - The regenerative transition has a probability matrix to determine the states just after the regeneration, which are defined as $\boldsymbol{P}_{i,0}, \dots, \boldsymbol{P}_{i,K}$. The subscript corresponds to the state of subspaces $\mathcal{S}^E$ (state 0), $\mathcal{S}_1^G, \dots, \mathcal{S}_K^G$.

The above structured MRGP representation is dominated by general distributions and matrices $\boldsymbol{D}$, $\boldsymbol{Q}$ and $\boldsymbol{P}$. Such representation can easily be generated from MRSPN analysis. According to the above descriptions, we have the

Markov renewal equations:

$$\boldsymbol{V}_{0,0}(t) = e^{\boldsymbol{D}_0 t} + \int_0^t e^{\boldsymbol{D}_0 u} \sum_{\nu=1}^{K} \boldsymbol{D}_\nu \boldsymbol{V}_{\nu,0}(t-u)du, \tag{13.4}$$

$$\boldsymbol{V}_{0,j}(t) = \int_0^t e^{\boldsymbol{D}_0 u} \sum_{\nu=1}^{K} \boldsymbol{D}_\nu \boldsymbol{V}_{\nu,j}(t-u)du, \quad j = 1, \ldots, K, \tag{13.5}$$

$$\boldsymbol{V}_{i,i}(t) = e^{\boldsymbol{Q}_i t}\overline{G}_i(t) + \int_0^t e^{\boldsymbol{Q}_i u} \sum_{\nu=0}^{K} (\boldsymbol{P}_{i,\nu} + \boldsymbol{Q}_{i,\nu})\boldsymbol{V}_{\nu,i}(t-u)dG_i(u), \tag{13.6}$$
$$i = 1, \ldots, K,$$

$$\boldsymbol{V}_{i,j}(t) = \int_0^t e^{\boldsymbol{Q}_i u} \sum_{\nu=0}^{K} (\boldsymbol{P}_{i,\nu} + \boldsymbol{Q}_{i,\nu})\boldsymbol{V}_{\nu,j}(t-u)dG_i(u), \tag{13.7}$$
$$i = 1, \ldots, K, \quad j = 0, \ldots, K, \quad i \neq j,$$

where $\overline{G}_i(t) = 1 - G_i(t)$ and $\boldsymbol{V}_{i,j}(t)$ is a submatrix of $\boldsymbol{V}(t)$.

13.2.2 STATIONARY ANALYSIS FOR STRUCTURED MRGP

The commonly used technique to compute stationary probability in MRGPs is an embedded Markov chain (EMC) approach. The EMC approach consists of two steps; the steady-state probability on regeneration points and the computation of cumulative probabilities between two successive regeneration points.

The first step utilizes the following probability matrix:

$$\tilde{\boldsymbol{Q}}_i = \int_0^\infty e^{\boldsymbol{Q}_i u} dG_i(u), \quad \bar{\boldsymbol{Q}}_i = \int_0^\infty e^{\boldsymbol{Q}_i u}\overline{G}_i(u)du. \tag{13.8}$$

This is the probability transition matrix from the states at entering $\mathcal{S}_i^G$ to the states at the instant of regeneration by the distribution $G_i(t)$. Therefore the transition probability matrix $\boldsymbol{P}^{EMC}$ at entry points to the subspaces $\mathcal{S}_1^G, \ldots, \mathcal{S}_K^G$ is partitioned by the following (i,j)-block matrices:

$$\boldsymbol{P}_{i,j}^{EMC} = \tilde{\boldsymbol{Q}}_i(\boldsymbol{P}_{i,j} + \boldsymbol{P}_{i,0}(-\boldsymbol{D}_0)^{-1}\boldsymbol{D}_j) + \bar{\boldsymbol{Q}}_i(\boldsymbol{Q}_{i,j} + \boldsymbol{Q}_{i,0}(-\boldsymbol{D}_0)^{-1}\boldsymbol{D}_j). \tag{13.9}$$

Since $\mathcal{S}_0^G$ does not include the general distribution, the above formula consists of a sum of the probabilities of the transition from the subspace i to the subspace j and the transition from j to j via the subspace $\mathcal{S}_0^G$. Using the transition probability matrix P^{EMC}, the steady-state probability $\boldsymbol{\pi}^{EMC} = \boldsymbol{\pi}^{EMC}\boldsymbol{P}^{EMC}$ is computed by numerical techniques.

In the second step, we compute cumulative probabilities, i.e., expected sojourn time of each state between two successive entry points. Using the steady-state probability vector $\boldsymbol{\pi}^{EMC}$, the expected sojourn times are derived as follows:

$$\boldsymbol{\xi}^E = \sum_{i=1}^{K} \boldsymbol{\pi}_i^{EMC}(\tilde{\boldsymbol{Q}}_i \boldsymbol{P}_{i,0} + \bar{\boldsymbol{Q}}_i \boldsymbol{Q}_{i,0})(-\boldsymbol{D}_0)^{-1}, \tag{13.10}$$

$$\boldsymbol{\xi}_i^G = \boldsymbol{\pi}_i^{EMC} \int_0^\infty \int_0^u e^{\boldsymbol{Q}_i t} dt dG_i^{(l)}(u), \tag{13.11}$$

where $\boldsymbol{\pi}_i^{EMC}$, $i = 1, \ldots, K$, is the partitioned vector of $\boldsymbol{\pi}^{EMC}$ by the subspaces $\mathcal{S}_1^G, \ldots, \mathcal{S}_K^G$. Finally, the Markov renewal reward theory [5] gives the steady-state probability $\boldsymbol{\pi}$ of MRGP as a fraction of the expected sojourn time for each state over the total time

$$\boldsymbol{\pi}^E = \frac{\boldsymbol{\xi}^E}{\boldsymbol{\xi}^E \mathbf{1} + \sum_{i=1}^K \boldsymbol{\xi}_i^G \mathbf{1}}, \tag{13.12}$$

$$\boldsymbol{\pi}_i^G = \frac{\boldsymbol{\xi}_i^G}{\boldsymbol{\xi}^E \mathbf{1} + \sum_{i=1}^K \boldsymbol{\xi}_i^G \mathbf{1}}, \tag{13.13}$$

where $\mathbf{1}$ is a column vector whose elements are 1.

13.3 PH EXPANSION OF MRGP

13.3.1 PH APPROXIMATION

The first step of PH expansion is to approximate general distributions with PH distributions. The commonly used techniques of PH approximation are moment match and maximum likelihood (ML) estimation.

The moment match is the simplest PH approximation method. The parameters of PH distribution are estimated so that the first few moments of the PH distribution match the theoretical ones. For some specified sub-class of PH distributions, the moment match gives closed forms of estimated parameters [15, 3]. The drawback of moment match is that the behavior of p.d.f.'s cannot be caught up.

The ML estimation is widely used for the parameter estimation of general distributions, and has useful properties like asymptotic consistency and normality. Some authors have discussed the ML estimation of PH distribution [2, 1]. The ML estimates of PH distribution are derived so that the mode matches the theoretical one. Therefore the estimated p.d.f.'s are more similar to theoretical p.d.f.'s than the moment match. In addition, even if the number of phases is small, it empirically provides the same first moment as the theoretical first moment [18, 14]. In particular, as the number of increases, the accuracy of fitting is improved.

This chapter presents the PH fitting from the p.d.f. information to apply the PH expansion to MRGP/MRSPN analysis. The idea behind our approach is to use the minimization of Kullback-Leibler (KL) divergence. This problem can be reduced to ML-based PH fitting with a weighted sample via the numerical integration technique.

Given an arbitrary probability density function $f(t)$, the KL divergence $KL(f, g)$ between $f(t)$ and any probability density function $g(t)$ is defined by

$$\begin{aligned} KL(f,g) &= \int_0^\infty f(t) \log \frac{f(t)}{g(t)} dt \\ &= \int_0^\infty f(t) \log f(t) dt - \int_0^\infty f(t) \log g(t) dt. \end{aligned} \tag{13.14}$$

The problem is to find $g(t)$ maximizing $\int_0^\infty f(t) \log g(t) dt$. Applying a suitable numerical integration technique, we have

$$\int_0^\infty f(t) \log g(t) dt \approx \sum_{i=1}^{K} w_i \log g(t_i), \tag{13.15}$$

where w_i, including $f(t_i)$, is a weight. The discretized points and their associated weights are determined by the numerical quadrature. Eq. (13.15) indicates that ML estimation is one approximation form of the KL divergence. Thus we can obtain an approximate PH distribution by using the weighted samples $(t_1, w_1), \ldots, (t_K, w_K)$ [14].

This paper utilizes the method to generate the weighted samples based on the double exponential (DE) formula [17]. This approach provides a more accurate approximation to many types of integral functions, compared to the trapezoidal rule, Simpson's rule, etc. The DE formula changes the original integration to an infinite integration of the function which decays according to the double exponential function. Here we use the following function

$$\phi(x) = \exp\left(\frac{\pi}{2}(x - e^{-x})\right). \tag{13.16}$$

By substituting the above function to $\int_0^\infty f(t) \log g(t) dt$, the integration is transformed to

$$\int_0^\infty f(t) \log g(t) dt = \int_{-\infty}^\infty f(\phi(x)) \log g(\phi(x)) \phi'(x) dx, \tag{13.17}$$

where $\phi'(x)$ is the first derivative of $\phi(x)$. Applying the trapezoidal rule to the above integration, we have

$$\int_{-\infty}^\infty f(\phi(x)) \log g(\phi(x)) \phi'(x) dx \approx \sum_{i=K^-}^{K^+} h\phi'(ih) f(\phi(ih)) \log g(\phi(ih)), \tag{13.18}$$

where h is a step size and $K^+(= -K^-)$ is a upper (lower) limit of discretization points. In fact, the accuracy of integration can be controlled by the parameters h and K^+. That is, given h and K^+, we generate the weighted samples $(t_1, w_1), \ldots, (t_K, w_K)$ as follows

$$t_{i-K^-+1} = \phi(ih), \tag{13.19}$$

$$w_{i-K^-+1} = h\phi'(ih)f(\phi(ih)), \quad i = K^-, \ldots, 0, \ldots, K^+, \tag{13.20}$$

where $K = K^+ - K^- + 1$.

Finally, we apply the PH fitting algorithm for weighted samples to obtain PH parameters presented in [14]. The algorithm in [14] is based on the EM algorithm for PH distribution which is originally discussed in [1]. The computation speed is further enhanced by using the sparsity of the generator $\boldsymbol{T}$. Thus the algorithm in [14] can handle CF1 with a few hundred phases to provide a highly-accurate PH approximation.

13.3.2 PH EXPANSION

This chapter presents a Kronecker representation based on the approximate MRGP with PH expansion. To obtain the Kronecker representation, we focus on the fact that regeneration point processes are regarded as PH renewal processes in the PH expansion. Assume that the general distribution $G_i(t)$ approximates a PH distribution with an initial probability vector $\boldsymbol{\alpha}_i$, an infinitesimal generator $\boldsymbol{T}_i$ and an exit rate vector $\boldsymbol{\xi}_i = -\boldsymbol{T}_i\boldsymbol{1}$.

Let $X(t)$ be the number of regenerations experienced before time t. Define $\boldsymbol{D}_0^{PH}$ and $\boldsymbol{D}_1^{PH}$ as an infinitesimal generator and rate matrix for the occurrence of regenerative points in the approximate MRGP. Taking superposition of the non-regeneration point process and the PH renewal processes, we have

$$\boldsymbol{D}_0^{PH} = \begin{pmatrix} \boldsymbol{D}_0 & \boldsymbol{D}_1 \otimes \boldsymbol{\alpha}_1 & \cdots & \boldsymbol{D}_K \otimes \boldsymbol{\alpha}_K \\ & \boldsymbol{Q}_1 \oplus \boldsymbol{T}_1 & & \\ & & \ddots & \\ & & & \boldsymbol{Q}_k \oplus \boldsymbol{T}_K \end{pmatrix}, \tag{13.21}$$

$$\boldsymbol{D}_1^{PH} = \begin{pmatrix} \boldsymbol{O} & \boldsymbol{O} & \cdots & \boldsymbol{O} \\ \boldsymbol{P}_{1,0} \otimes \xi_1 & \boldsymbol{P}_{1,1} \otimes \boldsymbol{\Lambda}_{1,1} & \cdots & \boldsymbol{P}_{1,K} \otimes \boldsymbol{\Lambda}_{1,K} \\ \vdots & \vdots & \ddots & \vdots \\ \boldsymbol{P}_{K,0} \otimes \xi_K & \boldsymbol{P}_{K,1} \otimes \boldsymbol{\Lambda}_{K,1} & \cdots & \boldsymbol{P}_{K,K} \otimes \boldsymbol{\Lambda}_{K,K} \end{pmatrix}, \tag{13.22}$$

$$\boldsymbol{D}_2^{PH} = \begin{pmatrix} \boldsymbol{O} & \boldsymbol{O} & \cdots & \boldsymbol{O} \\ \boldsymbol{Q}_{1,0} \otimes \boldsymbol{1} & \boldsymbol{Q}_{1,1} \otimes \bar{\boldsymbol{\Lambda}}_{1,1} & \cdots & \boldsymbol{Q}_{1,K} \otimes \bar{\boldsymbol{\Lambda}}_{1,K} \\ \vdots & \vdots & \ddots & \vdots \\ \boldsymbol{Q}_{K,0} \otimes \boldsymbol{1} & \boldsymbol{Q}_{K,1} \otimes \bar{\boldsymbol{\Lambda}}_{K,1} & \cdots & \boldsymbol{Q}_{K,K} \otimes \bar{\boldsymbol{\Lambda}}_{K,K} \end{pmatrix}, \tag{13.23}$$

where $\otimes$ and $\oplus$ are the Kronecker product and sum, $\boldsymbol{\Lambda}_{i,j} = \boldsymbol{\xi}_i\boldsymbol{\alpha}_j$ and $\bar{\boldsymbol{\Lambda}}_{i,j} = \boldsymbol{1}\boldsymbol{\alpha}_j$. Thus the infinitesimal generator of MRGP with PH expansion is given by $\boldsymbol{D}_0^{PH} + \boldsymbol{D}_1^{PH} + \boldsymbol{D}_2^{PH}$. The transient probability of $X(t)$ can be derived by the commonly-used CTMC analysis such as the uniformization. To relax the number of states, we can also apply the Krylov-based CTMC analysis [16]. Also the steady-state analysis can be performed on the PH-expanded MRGP to get stationary measures approximately.

13.4 ILLUSTRATIVE EXAMPLES

13.4.1 MRSPN TO MRGP

In this section, we present an illustrative example of MRGP modeling in the reliability domain. Suppose that the system consists of n units. All the units fail according to the identical exponential distribution with rate λ. If k or more units fail, the system also fails, i.e., k-out-of-n system. The failed unit will be repaired by one repairman. That is, the failed unit is buffered and the repairman fixes the buffered failed units one by one. The time to repair one unit follows a general distribution having the c.d.f. $G(t)$.

To obtain the MRGP representation, we first consider the MRSPN for this system. Petri net (PN) is a directed bipartite graph with two types of nodes; place and transitions. Places and transitions in PNs are represented by circles and rectangles, respectively. Directed arcs connect places to transitions, and transitions to places. The place that connects to a transition is called an input place of the transition. On the other hand, the place that be connected from a transition is called an output place of the transition. Tokens (represented by dots) are located at places in PNs. When a transition fires, it removes a token from each input place of the transition, and puts a token to each output place of the transition. The firing of a transition occurs only when there is at least one token for each input place of the transition. Then the transition is said to be enabled. A marking of a PN is given by a vector that represents the number of tokens for all the places. In the PN modeling, markings provide the state of a target system. A stochastic PN (SPN) is defined as the PN which has random firing times. In a deterministic PN, transitions immediately fire when the transitions are enabled. SPNs allow random delay times of firing from the time when transitions are enabled. In particular, random firing times are allowed to obey exponential distributions. The transition with such a random firing time is called an EXP transition. On the other hand, MRSPNs (Markov regenerative stochastic Petri nets) are a versatile tool for model-based performance evaluation more than SPNs [4]. A MRSPN is defined as a super-class of SPNs, and it has at most one GEN transition, which has the random firing time following a general probability distribution.

Figure 13.1 shows the MRSPN where circles, a white rectangle and a black rectangle indicate place, EXP transition and GEN transition, respectively. The place may have tokens. EXP and GEN transitions move tokens from input and output. Then the system state is represented by a vector of the number of tokens in two places. The number of tokens in the place `Pnormal` indicates the number of working units and the number of tokens in the place `Pfail` means

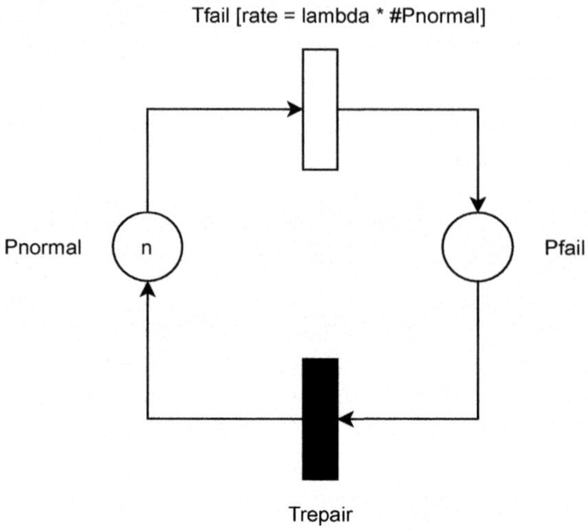

Figure 13.1 MRSPN model for k-out-of-n system.

the number of failed units. The transitions `Tfail` and `Trepair` correspond to the events of unit failure and repair. In this model, since EXP and GEN transition are non-regenerative and regenerative transition in MRGP, the EXP state becomes $\mathcal{S}_E = \{(n, 0)\}$, where $(n, 0)$ means the state in which `Pnormal` and `Pfail` have n and 0 tokens. Similarly, the GEN state is $\mathcal{S}_1^G = \{(n-1, 1), \ldots, (1, n-1), (0, n)\}$. Therefore, we have

$$\boldsymbol{D}_0 = \begin{pmatrix} -n\lambda \end{pmatrix}, \quad \boldsymbol{D}_1 = \begin{pmatrix} n\lambda & 0 & \cdots & 0 \end{pmatrix} \tag{13.24}$$

$$\boldsymbol{Q}_1 = \begin{pmatrix} -(n-1)\lambda & (n-1)\lambda & & & \\ & -(n-2)\lambda & (n-2)\lambda & & \\ & & \ddots & & \\ & & & -\lambda & \lambda \\ & & & & 0 \end{pmatrix}, \tag{13.25}$$

$$\boldsymbol{P}_{1,1} = \begin{pmatrix} 0 & & & \\ 1 & 0 & & \\ & \ddots & \ddots & \\ & & 1 & 0 \end{pmatrix}, \quad \boldsymbol{P}_{1,0} = \begin{pmatrix} 1 \\ 0 \\ \vdots \\ 0 \end{pmatrix} \tag{13.26}$$

Based on these matrices, we can obtain some performance indices from the formula systematically.

Next we consider the replacement problem for this k-out-of-n system. Figure 13.2 illustrates the MRSPN model for the k-out-of-n system with replacement. When the number of failed units attain a certain threshold u, the

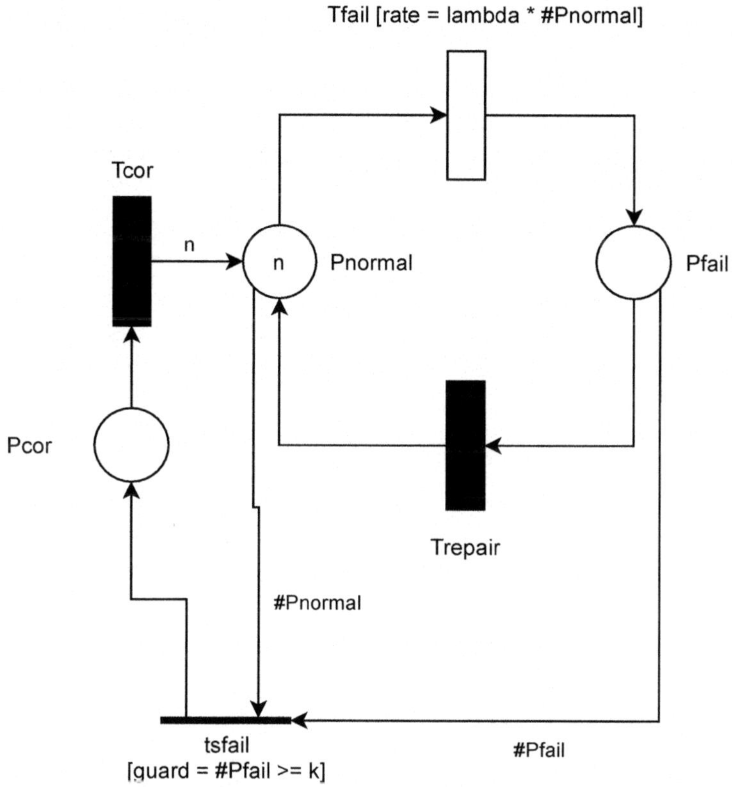

Figure 13.2 MRSPN model for k-out-of-n system with replacement.

system is replaced as good as new. To achieve such system behavior, we add an immediate transition `tsfail` with guard condition. This transition works only when the guard condition holds and the immediate transition has the priority for EXP and GEN transitions. Therefore, when the system failure occurs (`#Pfail >= u`), all the tokens move to `Pcor` and `Pcor` has one token. `Tcor` represents the event for replacement. Once the corrective maintenance (replacement) is finished, `Tcor` transition generates n tokens to `Pnormal`. In this case, there are two GEN transitions, and thus we have $\mathcal{S}_E = \{(n, 0, 0)\}$, $\mathcal{S}_1^G = \{(n-1, 1, 0), \ldots, (n-u-1, u+1, 0)\}$ and $\mathcal{S}_2^G = \{(0, 0, 1)\}$ where the vector (x, y, z) means `Pnormal`, `Pfail` and `Pcor` have x, y and z tokens, respectively. Also the matrices of MRGP are

$$\boldsymbol{D}_0 = \begin{pmatrix} -n\lambda \end{pmatrix}, \quad \boldsymbol{D}_1 = \begin{pmatrix} n\lambda & 0 & \cdots & 0 \end{pmatrix} \tag{13.27}$$

Table 13.1
General distributions and the number of weighted samples required.

	Distribution	Mean	CV2	3rd	# of samples
$G_1(t)$	Weibull	10 (hours)	0.1	1300.77	299
$G_2(t)$	Weibull	24 (hours)	0.5	40187.40	206

$$\boldsymbol{Q}_1 = \begin{pmatrix} -(n-1)\lambda & (n-1)\lambda & & \\ & \ddots & & \\ & -(n-u-2)\lambda & (u+2)\lambda \\ & & -(n-u-1)\lambda \end{pmatrix}, \tag{13.28}$$

$$\boldsymbol{Q}_{1,2} = \begin{pmatrix} 0 \\ \vdots \\ 0 \\ (n-u-1)\lambda \end{pmatrix}, \quad \boldsymbol{P}_{1,1} = \begin{pmatrix} 0 & & & \\ 1 & 0 & & \\ & \ddots & \ddots & \\ & & 1 & 0 \end{pmatrix}, \quad \boldsymbol{P}_{1,0} = \begin{pmatrix} 1 \\ 0 \\ \vdots \\ 0 \end{pmatrix} \tag{13.29}$$

$$\boldsymbol{Q}_2 = \begin{pmatrix} 0 \end{pmatrix}, \quad \boldsymbol{P}_{2,0} = \begin{pmatrix} 1 \end{pmatrix}. \tag{13.30}$$

13.4.2 PH EXPANSION

Consider the PH expansion of MRGP. In this experiment, we have two general distributions for the repair time distribution for one unit $G_1(t)$ and the replacement time distribution for the corrective maintenance $G_2(t)$. They are given by Weibull distribution presented in Table 13.1. In the table, the column 'Mean' indicates the mean time of repair and the mean time of replacement, the columns 'CV2' and '3rd' mean the squared coefficient of variation and the third moment, respectively. The column '# of samples' indicates the number of weighted samples to perform the PH approximation based on the DE formula. Also, based on the stationary analysis of MRGP, we compute the system availability when $k = 3$ and $n = 10$. Table 13.2 shows the system availability for the level of replacement varies from $u = 3$ to $u = 10$. From this table, we find the system availability is maximized when $u = 6$.

Table 13.3 shows the result of PH fitting with the algorithm of [14]. This table shows CV2 and third moments of approximate PH distributions. In addition, the table shows the number of iterations and the computation times measured in seconds. The approximate moments get close to the theoretical ones as the number of phases increase. This indicates that the number of phases is an important factor to control the accuracy of approximation. Also the computation time increases as the number of phases becomes large. That

Table 13.2
System availability.

u	Availability
3	0.9977478092568584
4	0.9999007889342432
5	0.9999812441622042
6	0.9999824591309038
7	0.9999824561298188
8	0.9999824555367032
9	0.9999824555229462
10	0.9999824555227907

Table 13.3
PH fitting results.

	# of phases	CV2	3rd	iteration	time (sec.)
$G_1(t)$	10	0.1168	1369.48	299	2.51
	50	0.1001	1301.30	80	12.03
	100	0.1000	1300.92	30	36.06
	200	0.1000	1300.77	20	137.35
$G_2(t)$	10	0.5015	40416.45	80	0.89
	50	0.5000	40185.90	10	8.50
	100	0.5000	40187.28	10	32.39
	200	0.5000	40187.31	10	129.57

is, we control both accuracy and computation speed only by adjusting the number of phases in the ML-based PH fitting.

Figures 13.3 and 13.4 illustrate the p.d.f.'s for the exact and PH approximated distributions. In Figs. 13.3 and 13.4, we find that the PH approximation with a large number of phases is quite accurate, so that the p.d.f. coincides with the exact Weibull distribution. In particular, PH approximation becomes accurate as the variation of coefficient is close to 1.

Figure 13.5 depicts the transient availabilities with PH approximation. For example, 'PH (10 phases)' indicates the approximated transient availability when both $G_1(t)$ and $G_2(t)$ are approximated by PH distributions with 10 phases. The transient availabilities are computed by CTMC analysis for the PH expanded CTMC presented in Sec. 13.3.2. From the result, the approximated transient availabilities with 50, 100 and 200 phases take almost the same values. Since $G_1(t)$ and $G_2(t)$ are well approximated by PH distributions

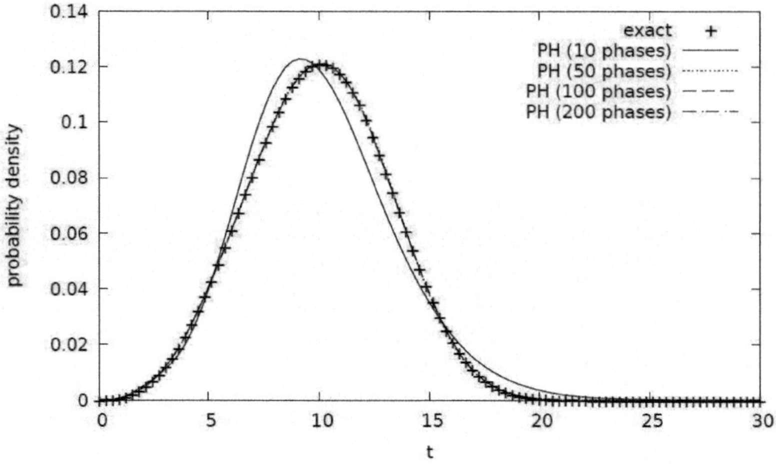

Figure 13.3 The p.d.f.'s of exact and PH approximated distributions ($G_1(t)$).

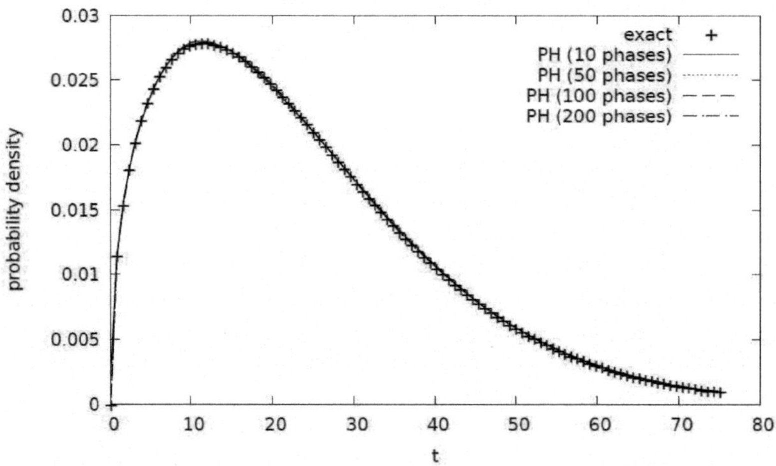

Figure 13.4 The p.d.f.'s of exact and PH approximated distributions ($G_2(t)$).

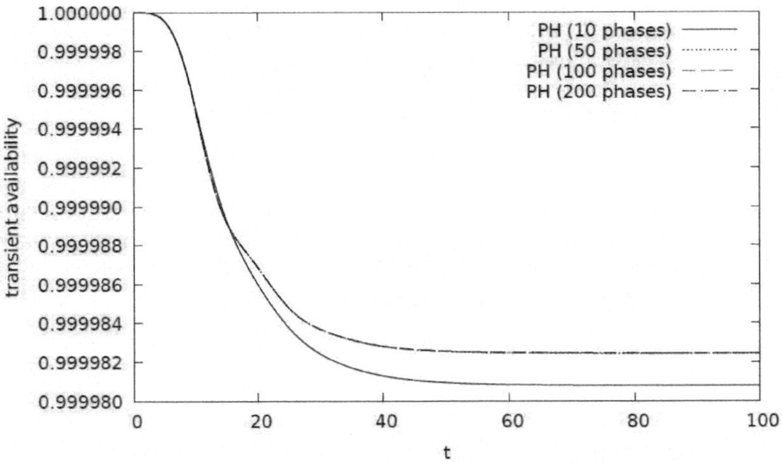

Figure 13.5 The transient system availability with PH approximation.

with 50, 100 and 200 phases, these values may be close to the exact transient availability.

13.5 CONCLUSIONS

In this chapter, we have discussed the PH expansion of MRGP. In particular, we have presented the structured MRGP consisting of transition rate matrices, and how to apply the PH expansion of the structured MRGP. Also, we have introduced the PH approximation from a given probability density function. In the example, we have considered the typical reliability model and exhibited the PH expansion of the model. As a result, the PH expansion is useful to obtain the transient results such as transient availability. On the other hand, as an issue for the PH expansion, we do not know the best way to determine the number of phases, although the number of phases strongly affects the accuracy of PH approximation. In the future, we will consider the method to determine the optimal number of phases by using some information criteria.

REFERENCES

1. Asmussen, S., Nerman, O., & Olsson, M. (1996). Fitting phase-type distributions via the EM algorithm. *Scandinavian Journal of Statistics*, 23(4), 419–441.

2. Bobbio, A., & Cumani, A. (1992). ML estimation of the parameters of a PH distribution in triangular canonical form. In Balbo, G., & Serazzi, G., editors, *Computer Performance Evaluation*, 33–46, Elsevier Science Publishers.

3. Bobbio, A., Horváth, A., & Telek, M. (2005). Matching three moments with minimal acyclic phase type distributions. *Stochastic Models*, 21(2-3), 303–326.

4. Choi, H., Kulkarni, V. G., & Trivedi, K. S. (1994). Markov regenerative stochastic Petri nets. *Performance Evaluation*, 20, 337–357.

5. Çinlar, E. (1969). Markov renewal theory. *Advances in Applied Probability*, 1, 123–187.

6. Fricks, R., Telek, M., Puliafito, A., & Trivedi, K. S. (1998). Markov renewal theory applied to performability evaluation. In Bagchi, K., & Zobrist, G., editors, *State-of-the Art in Performance Modeling and Simulation. Modeling and Simulation of Advanced Computer Systems: Applications and Systems*, 193–236, Gordon and Breach Publishers, Newark, NJ.

7. German, R., Kelling, C., Zimmermann, A., & Hommel, G. (1995). TimeNET: a tool for evaluating non-Markovian stochastic Petri nets. *Performance Evaluation*, 24, 69–87.

8. German, R., & Lindemann, C. (1994). Analysis of stochastic Petri nets by the method of supplementary variables. *Performance Evaluation*, 20, 317–335.

9. German, R., Logothetis, D., & Trivedi, K. S. (1995). Transient analysis of Markov regenerative stochastic Petri nets: a comparison of approaches. *In Proceedings of the 6th International Conference on Petri Nets and Performance Models*, 103–112.

10. Johnson, M. A., & Taaffe, M. R. (1989). Matching moments to phase distributions: Mixtures of erlang distribution of common order. *Stochastic Models*, 5, 711–743.

11. Kulkarni, V. G. (1995). *Modeling and Analysis of Stochastic Systems.* Chapman and Hall, New York.

12. Lindemann, C. (1995). DSPNexpress: a software package for the efficient solution of deterministic and stochastic Petri nets. *Performance Evaluation*, 22, 3–21.

13. Marie, R. (1980). Calculating equilibrium probabilities for $\lambda(n)/C_k/1/N$ queues. *In Proc. Int. Symp. Computer Performance Modelling, Measurement and Evaluation*, 117–125, New York, ACM Press.

14. Okamura, H., Dohi, T., & Trivedi, K. S. (2011). A refined EM algorithm for PH distributions. *Performance Evaluation*, 68(10), 938–954.

15. Osogami T., & Harchol-Balter, M. (2006). Closed form solutions for mapping general distributions to minimal PH distributions. *Performance Evaluation*, 63(6), 524–552.

16. Saad, Y. (1992). Analysis of some Krylov subspace approximations to the matrix exponential operator. *SIAM Journal on Numerical Analysis*, 209–228.

17. Takahasi, H., & Mori, M. (1974). *Double Exponential Formulas for Numerical Integration.* Publ. RIMS, Kyoto Univ., 9, 721–741.

18. Thümmler, A., Buchholz, P., & Telek, M. (2006). A novel approach for phase-type fitting with the EM algorithm. *IEEE Transactions on Dependable and Secure Computing*, 3(3), 245–258.

14 A Hybrid Model Fitting Framework Considering Accuracy and Performance

Vidhyashree Nagaraju
University of Tulsa, OK, USA

Lance Fiondella
University of Massachusetts Dartmouth, MA, USA

CONTENTS

14.1 INTRODUCTION

Software reliability growth models [31] are fit to time series data associated with failures experienced during testing in order to predict measures such as the time required to achieve a desired failure intensity or time between failures. Historically, numerical algorithms such as Newton's method were employed, which required good initial parameter estimates and therefore a high level of expertise to apply SRGM. Recent approaches to overcome the instability of classical numerical methods include techniques such as metaheuristic optimization [34] and swarm intelligence [24, 47, 21], which exhibit robust global search. However, these techniques can require significant computing resources and time to converge to a precise optimum, which is important for SRGM because some model parameters are very sensitive to accurate estimates of other parameters. Moreover, most past research applying metaheuristic optimization and swarm intelligence techniques do not explicitly consider the runtime required, yet both stability and performance are important, especially when an algorithm is to be implemented in a tool so that nonexperts can apply SRGM quickly and confidently without needing to learn the underlying mathematics. To achieve this pragmatic goal, more comprehensive methods are needed to assess the consistency of convergence as well as the runtime of these algorithms.

Existing research to fit SRGM with soft computing techniques [26, 22] as well as AI and swarm intelligence [35] abound. Additional examples of machine learning techniques include neural networks [25, 19] and support vector machines [51, 40]. Applications of metaphor-based meta-heuristics and evolutionary algorithms include genetic algorithms [34, 11, 17], genetic programming [15, 16], harmony search [5, 13], and gravitational search [12]. Applications of swarm intelligence algorithms to software reliability include particle swarm optimization [47, 24], artificial bee colony [45], ant colony optimization [54, 44], cuckoo search [1], grey wolf optimization [48], firefly [2, 14], ant lion optimization [3], and whale optimization [30].

Among existing swarm intelligence methods, particle swarm optimization [27] has been identified as a stable and efficient algorithm [9] to identify maximum likelihood estimates of SRGM [47, 7, 18]. The first application of PSO to NHPP SRGM [47, 49] was motivated by the efficiency of PSO on related software defect prediction problems. Subsequent studies applying PSO to identify the MLEs of traditional NHPP SRGM include [8, 6] as well as models with a larger number of parameters such as those incorporating a testing effort function [24]. Due to the slower runtime of PSO on more complex models, studies have also compared PSO runtime with alternative algorithms, including artificial bee colony (ABC) optimization [54] and the genetic algorithm [20]. Other studies combine SI methods into hybrid algorithms to accelerate convergence. Examples of hybrid algorithms include PSO with GA [33, 41, 28], PSO with the gravitational search algorithm [43], and PSO with ABC [29] or grey wolf optimization [4].

Numerical and statistical methods for software reliability model fitting include the expectation-maximization (EM) algorithm [39], expectation conditional maximization (ECM) algorithm [36, 53], and Newton-Raphson method [10]. When initial parameter estimates are far from the maximum, the EM and ECM algorithms are stable, but can exhibit slow convergence, whereas the Newton-Raphson method often diverges. To overcome these limitations, this paper implements experimental methods that combine the global search properties of PSO to obtain a near optimal solution, followed by a numerical or statistical method to obtain a precise optimum. Runtime is explicitly considered in order to identify the number of iterations and PSO population that converge consistently while minimizing overall runtime. Experiments are performed in the context of an NHPP SRGM [52] and a more advanced NHPP SRGM incorporating covariates [38]. Our results indicate that our hybrid approach converges more rapidly on NHPP SRGM incorporating covariates than either alternative in isolation.

The remainder of the chapter is organized as follows: Section 14.2 reviews software reliability growth models, while Section 14.3 describes parameter estimation methods. Section 14.4 presents a sequence of illustrative examples demonstrating how particle swarm optimization is combined with traditional methods in a manner to conduct rigorous tradeoff assessment between convergence and runtime. Section 14.5 offers conclusions and suggests future research.

14.2 SOFTWARE RELIABILITY GROWTH MODELS

This section describes two classes of models, namely the nonhomogeneous Poisson process and a discrete Cox proportional hazard (DCPH) NHPP SRGM model incorporating covariates to characterize testing activities performed to expose defects during testing. Each subsection provides a brief overview of the mathematical formulation, culminating in the statement of a log-likelihood function that will serve as the optimization objective to which particle swarm optimization as well as particle swarm optimization in combination with traditional methods is applied.

14.2.1 NONHOMOGENEOUS POISSON PROCESS SOFTWARE RELIABILITY GROWTH MODELS

This section describes nonhomogeneous Poisson process software reliability growth models. The NHPP [42] counts the number of events that occur by time t. In the context of SRGM, this counting process $N(t)$ corresponds to the number of unique software defects detected by time t. The expected value or average of the counting process is denoted as $m(t) := E[N(t)]$. For example, the mean value function of the Weibull SRGM [52] is

$$m(t) = a\left(1 - e^{-bt^c}\right) \tag{14.1}$$

where a represents the number of faults that would be discovered if the testing was continued indefinitely, while b and c are the scale and shape parameter respectively. Many models may be expressed in the general form $a \times F(t)$, where $F(t)$ is a cumulative distribution function (CDF).

The instantaneous failure rate of an SRGM is $\lambda(t) := \frac{\partial m(t)}{\partial t}$. Therefore, the instantaneous failure rate of the Weibull SRGM is

$$\lambda(t) = abct^{c-1}e^{-bt^{c}}. \tag{14.2}$$

Given the vector of failure times $\mathbf{T} = \langle t_1, t_2, \ldots, t_n \rangle$, fitting a model requires the maximization of the log-likelihood function.

$$LL(\theta, a) = -m(t_n) + \sum_{i=1}^{n} \log\left(\lambda(t_i)\right), \tag{14.3}$$

where t_n indicates the time at which the nth failure was observed.

14.2.2 DISCRETE COX PROPORTIONAL HAZARD NHPP SOFTWARE RELIABILITY GROWTH MODELS

This section describes a discrete Cox proportional hazard NHPP SRGM [38] and presents an example in the context of the geometric hazard function. DCPH NHPP models are composed of n discrete time intervals. Each interval i is characterized by the vector $\mathbf{x_i} = (x_1, x_2, \ldots, x_j)$, which represents the amount of each of j testing activities (covariates) performed in that interval and Y_i, the corresponding number of defects detected such that $\sum_{i=1}^{n} Y_i$ is the total number of defects discovered through the first n intervals.

The mean value function

$$H_{n;a,\theta} = a \sum_{i=1}^{n} p_{i,\mathbf{x_i}:\theta,\beta} \tag{14.4}$$

characterizes the average number of faults detected in the first n intervals, where $p_{i,\mathbf{x_i}:\theta,\beta}$ is the discrete Cox proportional hazard function given by

$$p_{i,\mathbf{x_i}:\theta,\beta} = \left(1 - (1 - h^0_{i:\theta})^{g(\mathbf{x_i},\beta)}\right) \prod_{k=1}^{i-1} (1 - h^0_{k:\theta})^{g(\mathbf{x_k},\beta)} \tag{14.5}$$

and $g(\mathbf{x_i}, \beta) = \exp(\mathbf{x_i}, \beta)$. Vector $\beta = (\beta_1, \beta_2, \ldots, \beta_j)$ denotes parameters associated with the j test activities and $h^0_{i:\theta}$ is the baseline hazard function possessing parameters θ. For example, the geometric hazard function is

$$h^0_{i;b} = b \tag{14.6}$$

where $b \in (0, 1)$.

The log-likelihood function of the DCPH NHPP SRGM is

$$\begin{aligned} LL(\theta, \beta, a) &= -a\sum_{i=1}^{n} p_{i,\mathbf{x_i}:\theta,\beta} + \sum_{i=1}^{n} y_i \ln(a) \\ &+\sum_{i=1}^{n} y_i \ln(p_{i,\mathbf{x_i}:\theta,\beta}) - \sum_{i=1}^{n} \ln(y_i!). \end{aligned} \tag{14.7}$$

14.3 PARAMETER ESTIMATION ALGORITHMS

This section discusses techniques to establish initial parameter estimates based on the expectation maximization algorithm as well as a strategy to combine this technique with particle swarm optimization. Next, methods to precisely identify the maximum likelihood estimates of model parameters, including the expectation conditional maximization algorithm and Newton's method, are discussed.

14.3.1 INITIAL PARAMETER ESTIMATES

The EM algorithm [39] provides a calculus-based method to obtain initial estimates of a model. The initial estimate of parameter a is

$$a^{(0)} = n. \tag{14.8}$$

The remaining parameters of the distribution function $F(\bullet;\theta)$ are computed as

$$\theta^{(0)} := \sum_{i=1}^{n} \frac{\partial}{\partial \theta} \log\left[f(\bullet;\theta)\right] = \mathbf{0}, \tag{14.9}$$

where $\mathbf{0}$ is the zero vector of length $|\theta|$, the number of parameters to be estimated. For covariate models, $F(\bullet;\theta)$ is replaced with $\sum_{i=1}^{n} p_{(\bullet:\theta,\beta)}$ in Equation (14.9).

Differentiating the portion of Equation (14.1) that corresponds to $F(t)$ produces $f(t)$, which is substituted into Equation (14.9) to obtain

$$b^{(0)} = \frac{n}{\sum_{i=1}^{n} t_i^c} \tag{14.10}$$

and $c^{(0)} = 1$, where c has been set to the feasible initial value one because Equation (14.9) lacks a closed form expression, but $c^{(0)} = 1$ simplifies to the special case of the exponential model. These initial estimates can be used with Newton's method, the ECM algorithm, or any other strategy such as particle swarm optimization to maximize the log-likelihood function. For example, one possible set of bounds for the initial positions of the population of particles in particle swarm optimization are uniform random numbers in the interval $\left(\frac{1}{\alpha}\theta_i^{(0)}, \alpha\theta_i^{(0)}\right)$, where $\alpha > 1$ is a scaling parameter.

In the covariate model with geometric hazard rate, parameters $b^{(0)}$ as well as $\beta^{(0)}$ lack a closed form expression and must therefore be solved numerically to identify initial values to define intervals for the PSO algorithm. Alternatively, $\beta^{(0)} = 0$, which simplifies the model to a discrete geometric model without covariates. Then, $b^{(0)} = 1 - \exp(-2/(n+1))$. PSO or other metaheuristic algorithms can also be applied directly to Equation (14.9) with a uniform random variable in the interval $(0, 1 - \exp(-2/(n+1)))$ to identify feasible initial estimates.

14.3.2 PARTICLE SWARM OPTIMIZATION (PSO)

Particle swarm optimization [27] is a computational method to iteratively search an m-dimensional space for the global optimum of an objective function $f(\mathbf{x})$ such as a log-likelihood function. Toward this end, PSO maintains a population of p candidate solutions (particles), each of which possesses a position and velocity. The velocity vector of the ith particle at discrete time step $(t+1)$ is

$$\mathbf{v}_i^{t+1} = a\mathbf{v}_i^t + c_1\nu_1(\mathbf{p}_i - \mathbf{x}_i^t) + c_2\nu_2(\mathbf{g} - \mathbf{x}_i^t) \tag{14.11}$$

where $\mathbf{v}_i^t$ is the velocity vector of the particle i at time step t, $\mathbf{p}_i$ is the best position visited by the ith particle in all time steps up to and including the present time so that $f(\mathbf{p}_i) \geq f(\mathbf{p}_i^t)$, while $\mathbf{g}$ is the global best position visited by any particle within the population. Since $\mathbf{x}_i^t$ denotes the position of the ith particle at time step t, $(\mathbf{p}_i - \mathbf{x}_i^t)$ and $(\mathbf{g} - \mathbf{x}_i^t)$ are the vectors pointing from a particle's present position to the direction of the particle and global best respectively. $(\mathbf{p}_i - \mathbf{x}_i^t)$ provides the particle with memory of the direction of the best solution relative to its present position, while $(\mathbf{g} - \mathbf{x}_i^t)$ directs each particle within the swarm toward the global best. In this manner, velocity is determined by a weighted average of the particle's present velocity as well as the direction to the particle and global best. In the classical formulation of PSO, these weights are constants a, c_1, and c_2, while ν_1 and ν_2 are uniform random numbers in the interval $(0, 1)$ generated at each time step to introduce randomness into the search process.

The position of the ith particle at time step $(t+1)$ is simply the sum of the particle's present position and velocity in time step $(t+1)$ or

$$\mathbf{x}_i^{t+1} = \mathbf{x}_i^t + \mathbf{v}_i^{t+1}. \tag{14.12}$$

The initial locations of the particles within the population are generated according to user-specified constraints and their fitness $f(\mathbf{p}_i^0)$ evaluated and assigned to the particle's best (and only) location seen so far. The best of these initial positions constitutes the initial value of the global best. The body of the PSO algorithm either executes a pre-specified number of times (T) or until no significant improvement is attained in one or more successive time steps. In the case where the number of iterations is constant, the complexity of PSO is the product of the number of iterations, population, dimension of the search space (model parameters), and number of failures or intervals, $O(Tpmn)$. Moreover, the complexity of covariate models also contains the term j for the number of covariates observed in each of the n intervals. Thus, the global search capability of particle swarm optimization must compensate for the time spent evaluating the log-likelihood functions in Equations (14.3) and (14.7).

14.3.3 EXPECTATION CONDITIONAL MAXIMIZATION (ECM) ALGORITHM

The expectation conditional maximization algorithm [36, 53] simplifies maximum likelihood estimation by dividing a single M-step of the EM algorithm [53] into $|\theta|$ conditional-maximization (CM) steps. EM algorithms apply $|\theta|$ update rules in a single M-step, whereas the CM-steps of the ECM algorithm update one parameter at a time holding the $|\theta| - 1$ other parameters constant at their present values. This reduces maximum likelihood estimation to a sequence of $|\theta|$ one-dimensional problems.

The CM-steps of the Weibull software reliability growth model are [36]

$$b^{''} = \frac{n}{\sum_{i=1}^{n} t_i^{c'} - \frac{n t_n^{c'}}{1-e^{b'' t_n^{c'}}}}, \tag{14.13}$$

$$c^{''} = \frac{-n}{\sum_{i=1}^{n}\left(\log(t_i)(1-b' t_i^{c''})\right) + \frac{n b' t_n^{c''}\log(t_n)}{1-e^{b' t_n^{c''}}}}. \tag{14.14}$$

Similarly, the CM-steps of the covariate model with geometric hazard function with j covariates is obtained by substituting the hazard function for the geometric model of Equation (14.6) into Equation (14.5) and then substituting the result into Equation (14.7), producing

$$\begin{aligned} LL(a,b,\beta) &= -a\sum_{i=1}^{n}\left(\left(1-(1-b)^{e^{\sum_{l=1}^{j}\mathbf{x_{1,i}}\beta_1}}\right)\prod_{k=1}^{i-1}(1-b)^{e^{\sum_{l=1}^{j}\mathbf{x_{1,k}}\beta_1}}\right) \\ &\quad + \log(a)\sum_{i=1}^{n} y_i - \sum_{i=1}^{n}\log(y_i!) + \sum_{i=1}^{n}(y_i \\ &\quad \log\left(\left(1-(1-b)^{e^{\sum_{l=1}^{j}\mathbf{x_{1,i}}\beta_1}}\right)\prod_{k=1}^{i-1}(1-b)^{e^{\sum_{l=1}^{j}\mathbf{x_{1,k}}\beta_1}}\right)\right). \end{aligned} \tag{14.15}$$

Differentiating Equation (14.15) with respect to a and solving,

$$a^{''} = \frac{\sum_{i=1}^{n} y_i}{\sum_{i=1}^{n}\left(\left(1-(1-b')^{e^{\sum_{l=1}^{j}\mathbf{x_{1,i}}\beta_1'}}\right)\prod_{k=1}^{i-1}(1-b')^{e^{\sum_{l=1}^{j}\mathbf{x_{1,k}}\beta_1'}}\right)}. \tag{14.16}$$

Substituting $a^{''}$ into Equation (14.15) reduces the number of parameters in the log-likelihood function. Differentiating the reduced log-likelihood function with respect to b and β produces CM-steps for these parameters, which are not presented here due to their complexity.

14.3.4 NEWTON'S METHOD (NM)

Newton's method [10] identifies the maximum likelihood estimates of a model's parameters by solving a system of non-linear equations obtained by taking partial derivatives of the log-likelihood function in Equation (14.3) with respect to individual model parameters and equating these partial derivatives to zero.

$$\frac{\partial}{\partial\theta}LL(\theta) = \mathbf{0}. \tag{14.17}$$

The maximum likelihood estimates of the Weibull software reliability growth model are

$$a^{''} = \frac{n}{1 - e^{-b^{'}t_n^{c^{'}}}}, \tag{14.18}$$

$$b^{''} = \frac{n}{a^{'}t_n^{c^{'}}e^{-b^{''}t_n^{c^{'}}} + \sum_{i=1}^{n} t_i^{c^{'}}}, \tag{14.19}$$

$$c^{''} = \frac{n}{a^{'}b^{'}t_n^{c^{''}}\log(t_n)e^{-b^{'}t_n^{c^{''}}} - \sum_{i=1}^{n}\log(t_i) + b^{'}\sum_{i=1}^{n} t_i^{c^{''}}\log(t_i)}, \tag{14.20}$$

which can be simplified by substituting Equation (14.18) into Equations (14.19) and (14.20), simultaneously solving for b and c, and substituting these results into Equation (14.18) to identify numerical values of the maximum likelihood estimates.

14.4 ILLUSTRATIONS

This section examines several alternative combinations of algorithms to fit traditional NHPP and covariate software reliability growth models. We first conduct a detailed analysis in the context of particle swarm optimization, studying the tradeoff between iterations and population size. In light of this analysis, the runtime performance and convergence of model fitting algorithms including and excluding PSO are compared. Specifically, Table 14.1 lists the combinations considered and identifies each combination with a unique numeric value referred to as a Design for the sake of clarity.

Table 14.1 indicates that all designs utilize the EM algorithm to obtain initial estimates because good initial estimates often accelerate convergence significantly. Designs I, II, and III represent cases where a single method is employed, including PSO, Newton's method, and the expectation conditional maximization algorithm, respectively. On the other hand, Designs IV and

Table 14.1
Combination of algorithms

EM	PSO	NM	ECM	Design
X	X	-	-	I
X	-	X	-	II
X	-	-	X	III
X	X	X	-	IV
X	X	-	X	V

V employ PSO followed by Newton's method or the expectation conditional maximization algorithm in order to assess if the global search properties of PSO coupled with a numerical method lowers the time to converge to a near optimal solution.

14.4.1 NONHOMOGENEOUS POISSON PROCESS SOFTWARE RELIABILITY GROWTH MODELS

14.4.1.1 PSO Tradeoff Analysis

This example assesses the tradeoff between the number of iterations (T) and number of particles (p) in the swarm. Toward this end, the first example holds the number of function evaluations constant to compare the performance of PSO, ranging from many iterations on a smaller number of particles to a few iterations with a large number of particles. The second example considers a range of function evaluations to assess how many particles and iterations are needed to achieve a near optimal estimate, which would be sufficient to identify initial estimates for faster but less stable numerical methods.

The examples are performed in the context of the Weibull software reliability growth model. Specifically, the PSO algorithm is applied to maximize the log-likelihood function given in Equation (14.3), after substituting Equations (14.1) and (14.2) for the mean value function and failure intensity respectively. The SYS1 failure times data set [32], which consists of $n = 136$ failures, is then substituted for t_i. The PSO parameters (a, c_1, and c_2) were set to 0.5 and the coefficient for initial positions (α) to 2.0 was applied to determine initial particle locations around initial estimates $a^{(0)} = 136$, $b^{(0)} = 4.04 \times 10^{-5}$, and $c^{(0)} = 1.0$ computed from Equations (14.8)-(14.10) of the EM algorithm.

Constant Number of Function Evaluations

This section explores the tradeoff between the number of iterations and population size when the number of function evaluations is held constant at $1,024$.

Figure 14.1 shows the average log-likelihood attained after a specified number of iterations for populations ranging from 1 to 256, where averages in this and subsequent experiments were computed from 100 independent runs in order to illustrate trends more clearly.

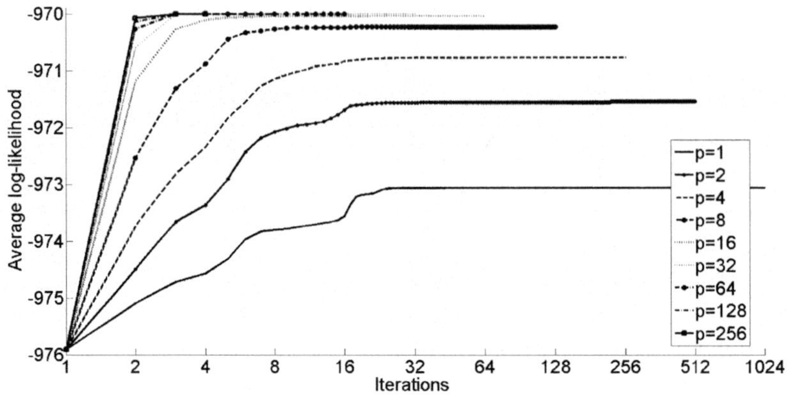

Figure 14.1 Tradeoff between iterations and population on average log-likelihood for fixed number of function evaluations (Design I)

The x-axis shows the number of iterations on an exponential scale because smaller populations ran for a greater number of iterations such that $p \times i = 1024$. For example, the lower curve shows the average for the special case possessing a population of one, in which the particle and global best are the same. Due to the degenerate size of the population, the single particle is unable to benefit from sharing across particles in the swarm and was therefore only able to reach an average log-likelihood value of -973. Similarly, the line above shows the progress made by a population of two particles over 512 iterations. Thus, the right endpoints of the curves for a population of size one and size two correspond to a total of $1,024$ function evaluations. The population of two ends on average approximately one and a half points higher, suggesting that increasing the population and decreasing the number of iterations produces a better result, despite approximately equal work overall. The curves above these two show the average performance of populations between 4 and 256, decreasing the number of iterations in order to hold the number of function evaluations constant.

Figure 14.1 also indicates that the marginal utility between a population of two and four, measured as the difference in log-likelihood at the right endpoints, is smaller than the increase when the population is doubled from one to two. Moreover, the marginal utility decreases as the population increases from 2^i to 2^{i+1}, approaching 0 as the population increases from 16 to 32. Moving left along the curves that achieve a near optimum (populations of 16 and greater), we see that it took approximately eight iterations for the population of 16 to reach an average log-likelihood of approximately -970, whereas a population of 32 took approximately 4 iterations. Thus, approximately 128 function evaluations were sufficient to reach a near optimal value, indicating that 7/8ths of the function evaluations contributed very little to convergence, but consumed nearly 90% of the computational power and processing time. Moreover, none of these combinations is sufficient to consistently achieve the maximum likelihood value of −966.0803, confirming that PSO may be a good global search method, but that classical optimization techniques must also be employed to reach the maximum.

Variable Number of Function Evaluations

To reduce the number of function evaluations, this section explores the tradeoff between population and iterations for a range of function evaluations.

Figure 14.2 shows the average log-likelihood attained for anywhere from $2^1 = 2$ to $2^{10} = 1024$ function evaluations, identified by separate curves in the legend. The x-axis indicates the population size, whereas the number of iterations is implicit. For example, the curve with 8 function evaluations possesses three points. From left to right, these include four iterations with a population of two, two iterations with a population of four, and one iteration with a population of eight. Thus, to hold the number of function evaluations constant, we halve the number of iterations when doubling the population.

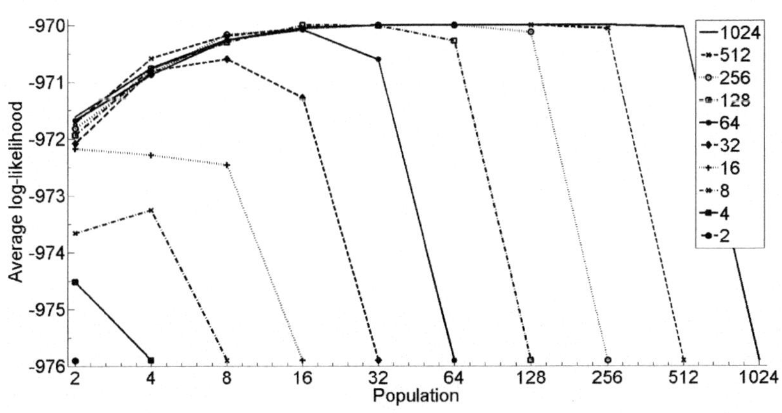

Figure 14.2 Average log-likelihood function for variable number of function evaluations (Design I)

The curve with 8 function evaluations indicates that the extremes of many iterations with a small population and a single iteration with a larger population are suboptimal. This trend is also present in the curves for 32 function evaluations and above. Moreover, we observe that one iteration performs worst for all function evaluations considered and that two iterations with a population half the size performs substantially better. However, two iterations (second point from right) on the curves with 1024, 512, and 256 function evaluations achieve nearly identical average log-likelihoods close to −970, confirming that these additional function evaluations are of negligible utility. Examining the 128 function evaluations curve from right to left, suggests that four iterations with a population of 32 two performs as well as 256 function evaluations or higher, which agrees with the observations made in Figure 14.1.

14.4.1.2 Performance assessment

This section assesses the runtime performance and convergence of alternative designs given in Table 14.1. All designs include initial estimates with the expectation maximization algorithm, which only required 0.00015 seconds on average and therefore constituted a negligible portion of the total runtime.

PSO performance

Figure 14.3 compares Designs I-III of Table 14.1, reporting the average runtime of the PSO algorithm (Design I) in seconds as a function of the number of iterations and population sizes considered previously. The solid and dashed horizontal lines in Figure 14.1 correspond to the average time required for Newton's method (Design II) and the ECM algorithm (Design III), which respectively required an average of 0.5591 and 0.9581 seconds to converge to the MLE from the initial estimates. Here, convergence occurred when the difference in the log-likelihood function evaluated with parameter values from two successive iterations was less than a small positive constant $\varepsilon \geq 10^{-10}$. Thus, while it is possible to allow PSO to run for many iterations, Figure 14.3 indicates that combinations of iterations and populations of PSO that exceed the times required of Newton's method or the ECM algorithm cannot be justified for multistage methods, since Newton's method and the ECM algorithm exhibited stable convergence on each of the hundred runs performed. In other words, excessive global search with PSO is not justified.

PSO+NM Performance

Figure 14.4 shows the average runtime of Design IV, where the PSO algorithm ran for a specified number of iterations on populations ranging from 1 to 256 and the parameters with the best log-likelihood so far are used as input to Newton's method and run to convergence to identify MLEs of model parameters. In other words, the independent variable of Figure 14.4 is the number of iterations of PSO performed and Newton's method is always run until convergence was attained. Figure 14.4 indicates that the lowest runtimes occur when no more than 32 iterations are performed and the best runtime was 1.0205 seconds when two iterations ($i = 2$) of PSO were performed on a population of $p = 64$ particles. However, this best time for Design IV is 1.83 times higher than Design II, which excludes PSO. This result suggests that application of the PSO to fit the Weibull SRGM to the SYS1 dataset is not appropriate because it is always slower than applying Newton's method

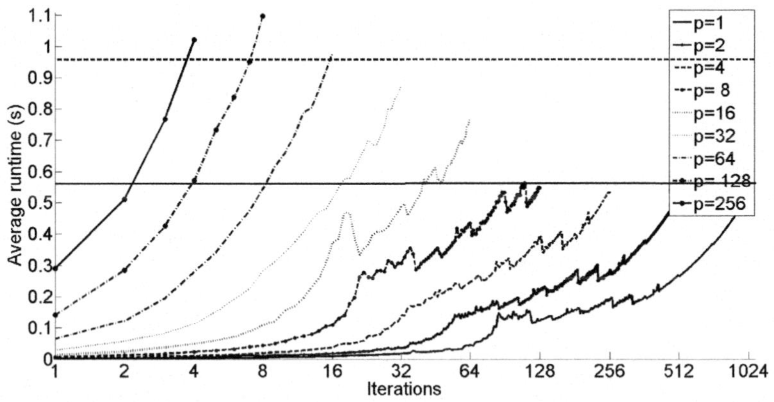

Figure 14.3 Average runtime of PSO (Design I) vs. Newton's method (Design II) and expectation conditional maximization algorithm (Design III)

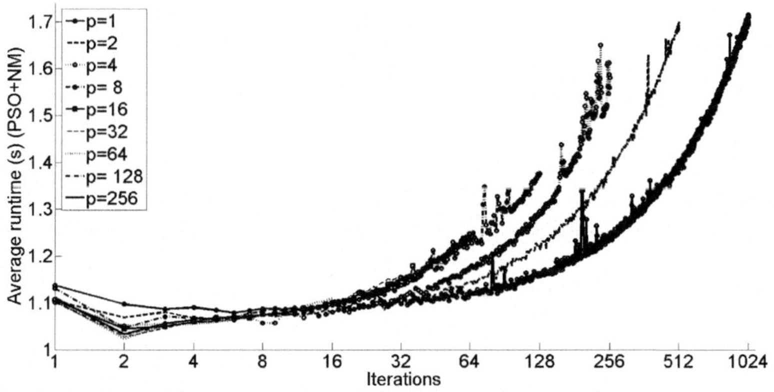

Figure 14.4 Average runtime of PSO+NM (Design IV)

with initial estimates obtained from the EM algorithm. In this case, the relatively smooth nature of the search space coupled with the low dimension of the problem are the primary reasons why PSO does not provide any tangible benefit. However, PSO coupled with Newton's method may be desirable when initial estimates obtained from the EM algorithm are not sufficient for Newton's method to converge consistently or PSO can quickly identify improved estimates that enable faster convergence. Therefore, Design IV may still be justified on more complex models and datasets.

PSO+ECM performance

Using the same methodology described above to compare Designs II and IV, Figure 14.5 shows the average runtime of Design V. The solid vertical line corresponds to the average time required by the ECM algorithm (Design III).

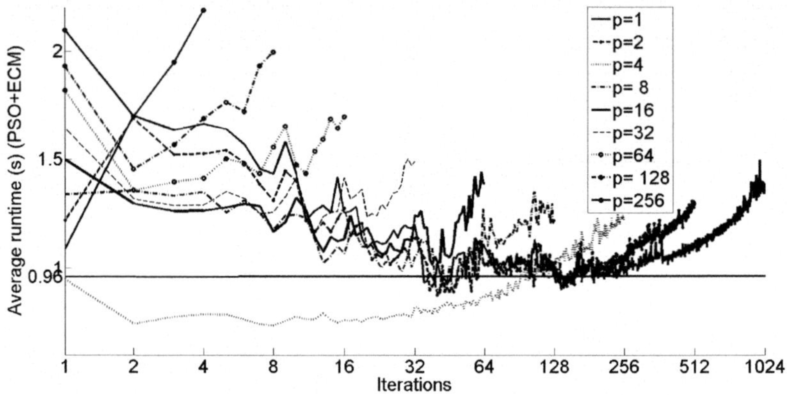

Figure 14.5 Average runtime of PSO+ECM (Design V)

Figure 14.5 illustrates that most combinations of PSO iterations and population sizes required more time than Design III, which does not employ a PSO stage. However, a population of $p = 4$ particles and eight iterations of PSO followed by the expectation maximization algorithm exhibited an average runtime of 0.7345 seconds, representing a speedup factor of 1.3 (0.9581/0.7345) over Design III. Despite the fact that Design II outperformed Design III and V, Figure 14.5 demonstrates that hybrid designs composed of PSO plus a traditional algorithm can achieve speedup, but that sensible application of PSO must take the additional runtime into consideration.

14.4.2 DISCRETE COX PROPORTIONAL HAZARD NHPP SOFTWARE RELIABILITY GROWTH MODELS

14.4.2.1 Constant and Variable Average Number of Function Evaluations

This section provides an abbreviated examination of the tradeoffs between iterations and population size in the context of the DCPH NHPP software reliability growth model. Thus, we seek to maximize Equation (14.7) given DS2 (the second data set originally reported by Shibata et al. [50]) composed of $n = 14$ weeks of observations, including three covariates, execution time (E) in hours, failure identification work (F) in person hours, and computer time failure identification (C) in hours.

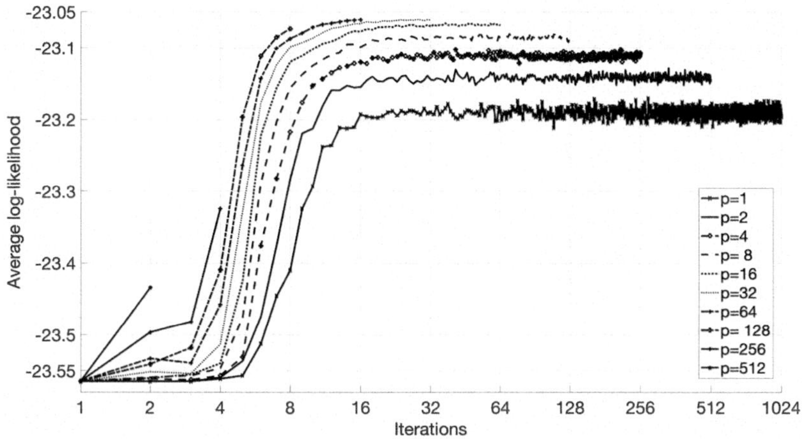

(a) Tradeoff between iterations and populations on average log-likelihood for fixed number of function evaluations

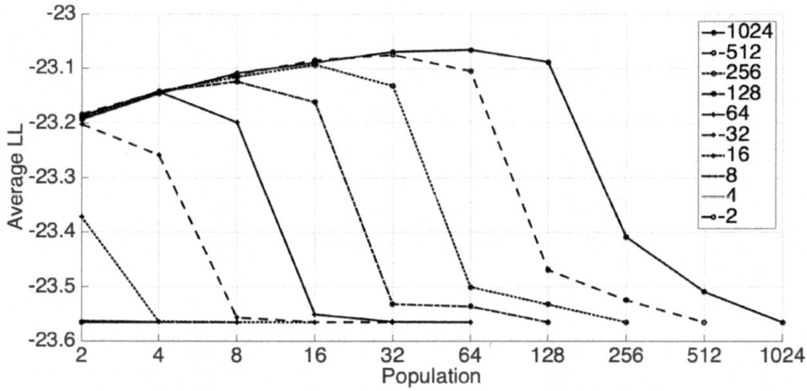

(b) Average log-likelihood function for variable number of function evaluations

Figure 14.6 PSO tradeoff analysis (Design I)

The EM initial estimates were computed by solving Equation (14.9) for the geometric hazard rate model, which took 0.014 seconds on average. The initial estimates obtained for this set of examples are $b^{(0)} = 0.058545$, $\beta_1^{(0)} = 0.144189$, $\beta_2^{(0)} = 0.039082$, and $\beta_3^{(0)} = 0.080035$. The maximum log-likelihood value of -23.006729 was also predetermined in order to assess the performance of PSO with different numbers of iterations and population sizes.

Figure 14.6(a) shows the average maximum log-likelihood attained after a specified number of iterations for populations ranging from 1 to 256 (Design I) and Figure 14.6(b) shows the average maximum log-likelihood attained

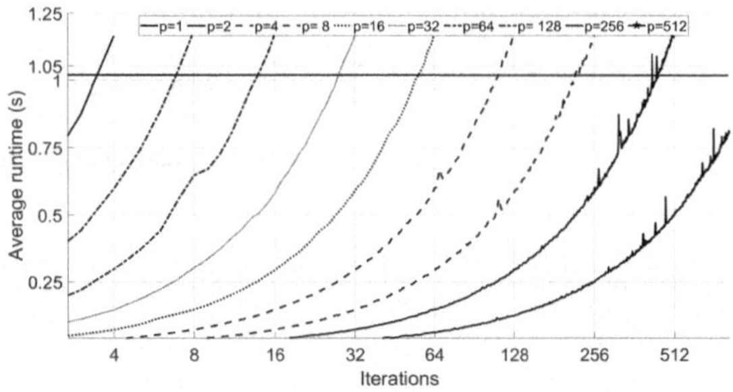

(a) Average runtime of PSO (Design I) vs. Newton's method (Design II) and expectation maximization algorithm (Design III)

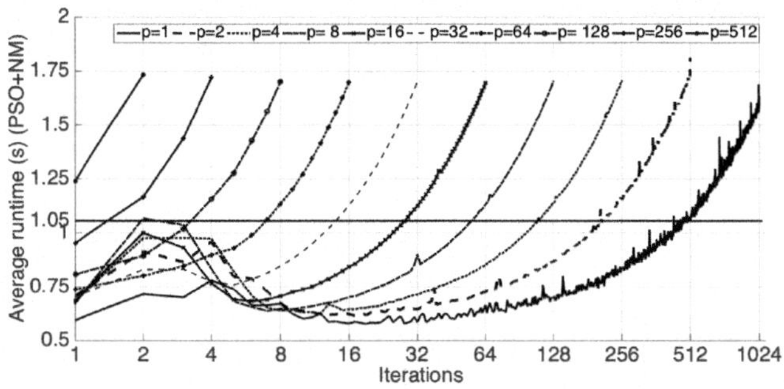

(b) Average runtime of PSO+NM (Design IV)

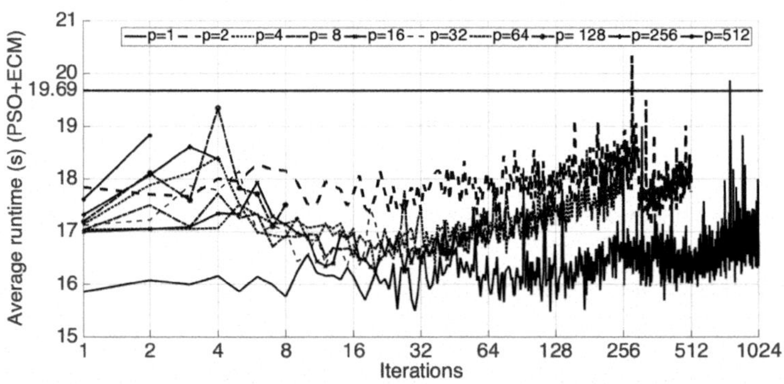

(c) Average runtime of PSO+ECM (Design V)

Figure 14.7 Performance assessment

for anywhere from $2^1 = 2$ to $2^{10} = 1024$ function evaluations, identified by the individual curves. The trends in Figures 14.6(a) and 14.6(b) are similar to those in Figures 14.1 and 14.2. Small populations with many iterations and large populations with few iterations are suboptimal, failing to reach the maximum likelihood estimate. For example, Figure 14.6(b) suggests that a good balance of as few as 64 function evaluations are sufficient to approach the maximum, implying that as little as 6.25% of 1,024 function evaluations used in the final curve may be adequate.

14.4.2.2 Performance Assessment

This section assesses the runtime performance of Design II through V on the DCPH NHPP SRGM when applied to the DS2 data set. The average runtimes of Design II (Newton's method) and Design III (ECM algorithm) were 1.0558 and 19.6913 seconds respectively.

Figure 14.7(a) compares Designs I (PSO) and Design II (Newton's method). Design III (ECM algorithm) was not competitive. Figure 14.7(b) compares Design IV (PSO+NM) with Design II, while Figure 14.7(c) compares Design IV (PSO+ECM) with Design III.

Figure 14.7(b) indicates that the runtime of Design IV (PSO+NM) with a population of $p = 512$ for $16 \leq i \leq 32$ iterations obtained good initial estimates, which reduced the overall time to converge to as little as 54.7% (0.5771/1.0558) of the time required by Newton's method when run directly on EM initial estimates.

14.5 CONCLUSION AND FUTURE WORK

This chapter implemented experimental methods that combine the global search properties of PSO to obtain a near-optimal solution, followed by a numerical or statistical method to obtain a precise optimum. Runtime was explicitly considered in order to identify the number of iterations and PSO population that converged consistently while minimizing overall runtime. Experiments were performed in the context of an NHPP SRGM and more advanced NHPP SRGM incorporating covariates. Our results indicated that hybrid methods are not suitable for models possessing a small number of parameters because the time consumed by PSO is not needed to enhance the stability of convergence, whereas hybrid methods consisting of the PSO followed by Newton's method reduced the average time to fit the discrete Cox proportional hazard NHPP software reliability growth model by nearly 50%.

Future research will assess if the hybrid techniques proposed here achieve more significant speed up on higher dimensional problems such as the discrete Cox proportional hazard NHPP software reliability growth model with a larger number of covariates.

ACKNOWLEDGEMENT

This material is based upon work supported by the National Science Foundation under Grant Number (#1749635). Any opinions, findings, and conclusions or recommendations expressed in this material are those of the authors and do not necessarily reflect the views of the National Science Foundation.

REFERENCES

1. Al-Saati, N. & Abd-AlKareem, M. (2013). The use of cuckoo search in estimating the parameters of software reliability growth models. *International Journal of Computer Science and Information Security*, 11.

2. Al-Saati, N. & Alabajee, M. (2016). On the performance of firefly algorithm in software reliability modeling. *Proc. International Journal of Recent Research and Review*, 9(4):1–9.

3. Alabajee, M. & Alreffaee, T. (2018). Exploring ant lion optimization algorithm to enhance the choice of an appropriate software reliability growth model. *International Journal of Computer Applications*, 182(4):1–8.

4. Alneamy, J. S. & Dabdoob, M. M. (2017). The use of original and hybrid grey wolf optimizer in estimating the parameters of software reliability growth models. *International Journal of Computer Applications*, 167(3):12–21.

5. Altaf, I., Majeed, I. & Iqbal, K. A. (2016). Effective and optimized software reliability prediction using harmony search algorithm. In *Proc. International Conference on Green Engineering and Technologies*, pages 1–6.

6. Altaf, I., Rashid, F., Dar, J. A. & Rafiq, M. (2015). Survey on parameter estimation in software reliability. In *2015 International Conference on Soft Computing Techniques and Implementations (ICSCTI)*, pages 22–27. IEEE.

7. Bidhan, K. & Awasthi, A. (2014). Estimation of reliability parameters of software growth models using a variation of particle swarm optimization. In *IEEE International Conference-Confluence The Next Generation Information Technology Summit (Confluence)*, pages 800–805.

8. Bidhan, K. & Awasthi, A. (2014). A review on parameter estimation techniques of software reliability growth models. *International Journal of Computer Applications Technology and Research*, 3(4):267–272.

9. Bonyadi, M. & Michalewicz, Z. (2015). Analysis of stability, local convergence, and transformation sensitivity of a variant of the particle swarm optimization algorithm. *IEEE Transactions on Evolutionary Computation*, 20(3):370–385.

10. Burden, R. & Faires, J. (2004). *Numerical Analysis*. Brooks/Cole, Belmont, CA, 8th edition.

11. Chen, M., Wu, H. & Shyur, H. (2001). Analyzing software reliability growth model with imperfect-debugging and changepoint by genetic algorithms. In *Proc. International Conference on Computers and Industrial Engineering*, pages 520–526, Montreal, Canada.

12. Choudhary, A., Baghel, A. S. & Sangwan, O. P. (2017). An efficient parameter estimation of software reliability growth models using gravitational search algorithm. *International Journal of System Assurance Engineering and Management*, 8(1):79–88.

13. Choudhary, A., Baghel, A. S. & Sangwan, O. P. (2017). Efficient parameter estimation of software reliability growth models using harmony search. *IET Software*, 11(6):286–291.

14. Choudhary, A., Baghel, A. S. & Sangwan, O. P. (2018). Parameter estimation of software reliability model using firefly optimization. In *Proc. Data Engineering and Intelligent Computing*, 407–415. Springer.

15. Costa, E. O., Vergilio S. R., Pozo, A. T. R. & Souza, G. A. (2005). Modeling software reliability growth with genetic programming. In *Proc. International Symposium on Software Reliability Engineering*, Chicago, IL.

16. Costa, E. O., Souza, G. A., Pozo, A. T. R. & Vergilio, S. R. (2007). Exploring genetic programming and boosting techniques to model software reliability. *IEEE Transactions on Reliability*, 56(3):422–434.

17. Dai, Y., Xie, M., Poh, K. & Yang, B. (2003). Optimal testing resource allocation with genetic algorithm for modular software systems. *Journal of Systems and Software*, 66(1):47–55.

18. Diwaker, C., Tomar, P., Poonia, R. & Singh, V. (2018). Prediction of software reliability using bio inspired soft computing techniques. *Journal of Medical Systems*, 42(5):93.

19. Dohi, T., Nishio, Y. & Osaki, S. (1999). Optimal software release scheduling based on artificial neural networks. *Annals of Software Engineering*, 8.

20. Govindasamy, P. & Dillibabu, R. (2018). Maximum likelihood estimation and optimisation of parameters of software reliability models using evolutionary optimisation techniques. *TAGA*, 14:2529–2542.

21. Hassanien, A. & Emary, E. (2016). *Swarm Intelligence: Principles, Advances, and Applications*. CRC Press.

22. Hudaib, A. & Moshref, M. (2018). Survey in software reliability growth models: Parameter estimation and models ranking. *International Journal of Computer Systems*, 5(5):11–25.

23. Jin, C. & Jin, S. (2016). Parameter optimization of software reliability growth model with s-shaped testing-effort function using improved swarm intelligent optimization. *Applied Soft Computing*, 40:283–291.

24. Jin, C. & Jin, S. (2016). Parameter optimization of software reliability growth model with s-shaped testing-effort function using improved swarm intelligent optimization. *Applied Soft Computing*, 40:283–291.

25. Karunanithi, N., Whitley, D. & Malaiya, Y. K. (1992). Prediction of software reliability using connectionist models. *IEEE Transactions on Software Engineering*, 18(7):563–574.

26. Kaswan, K., Choudhary, S. & Sharma, K. (2015). Software reliability modeling using soft computing techniques: Critical review. *Journal of Information Technology and Software Engineering*, 5:144.

27. Kennedy, J. & Eberhart, R. (1995). Particle swarm optimization (PSO). In *IEEE International Conference on Neural Networks, Perth, Australia*, 1942–1948.

28. Kumar, A., Tripathi, R. P., Saraswat, P. & Gupta, P. (2017). Parameter estimation of software reliability growth models using hybrid genetic algorithm. In *Proc. IEEE International Conference on Image Information Processing*, 1–6.

29. Li, Z., Yu, M., Wang, D. & Wei, H. (2019). Using hybrid algorithm to estimate and predicate based on software reliability model. *IEEE Access.*

30. Lu, K. & Ma, Z. (2018). Parameter estimation of software reliability growth models by a modified whale optimization algorithm. In *Proc. IEEE International Symposium on Distributed Computing and Applications for Business Engineering and Science*, 268–271.

31. Lyu, M, editor. (1996). *Handbook of Software Reliability Engineering.* McGraw-Hill, New York, NY.

32. Lyu, M. (2005). *Handbook of Software Reliability Engineering: Data directory.* `http://www.cse.cuhk.edu.hk/~lyu/book/reliability/`, [Online; accessed 23-May-2005].

33. Rao, K. M. & Anuradha, K. (2016). A new method to optimize the reliability of software reliability growth models using modified genetic swarm optimization. *International Journal of Computer Applications*, 145(5):1–8.

34. Minohara, T. & Tohma, Y. (1995). Parameter Estimation of Hyper-Geometric Distribution Software Reliability Growth Model by Genetic Algorithms. In *Proc. International Symposium on Software Reliability Engineering*, pages 324–329, Toulouse, France.

35. Mohanty, R., Ravi, V. & Patra, M. R. (2010). The application of intelligent and soft-computing techniques to software engineering problems: A review. *International Journal of Information and Decision Sciences*, 2(3):233–272.

36. Nagaraju, V., Fiondella, L., Zeephongsekul, P., Jayasinghe, C. & Wandji, T. (2017). Performance optimized expectation conditional maximization algorithms for nonhomogeneous Poisson process software reliability models. *IEEE Transactions on Reliability*, 66(3):722–734.

37. Nagaraju, V., Fiondella, L., Zeephongsekul, P., Jayasinghe, C. & Wandji, T. (2017) Performance optimized expectation conditional maximization algorithms for nonhomogeneous Poisson process software reliability models. *IEEE Transactions on Reliability*, 66(3):722–734.

38. Nagaraju, V., Jayasinghe, C. & Fiondella, L. (2020). Optimal test activity allocation for covariate software reliability and security models. *Journal of Systems and Software*, 110643.

39. Okamura, H., Watanabe, Y. & Dohi, T. (2003) An iterative scheme for maximum likelihood estimation in software reliability modeling. In *IEEE International Symposium on Software Reliability Engineering*, 246–256.

40. Pai, P. & Hong, W. (2006). Software reliability forecasting by support vector machines with simulated annealing algorithms. *Journal of Systems and Software*, 79:747–755.

41. Rao, M. & Anuradha, K. (2016). A hybrid method for parameter estimation of software reliability growth model using modified genetic swarm optimization with the aid of logistic exponential testing effort function. In *Proc. IEEE International Conference on Research Advances in Integrated Navigation Systems*, 1–8.

42. Ross, S. (2003). *Introduction to Probability Models.* Academic Press, New York, NY, 8th edition.

43. Sangeeta, Sharma, K. & Bala, M. (2017). Exhausting meta-heuristic nature inspired approaches for the parameter estimation analysis of software reliability growth model. *Advanced Science and Technology Letters*, 144:1–8.

44. Shanmugam, L. & Florence, L. (2012). A comparison of parameter best estimation method for software reliability models. *International Journal of Software Engineering and Applications*, 3:91–102.

45. Sharma, T., Pant, M. & Abraham, A. (2011). Dichotomous search in abc and its application in parameter estimation of software reliability growth models. In *Proc. World Congress on Nature and Biologically Inspired Computing*, 207–212.

46. Sheta, A. (2006). Reliability growth modeling for software fault detection using particle swarm optimization. In *Proc. IEEE Congress on Evolutionary Computation*, 3071–3078.

47. Sheta, A. (2006). Reliability growth modeling for software fault detection using particle swarm optimization. In *IEEE International Conference on Evolutionary Computation*, 3071–3078.

48. Sheta, A. & Abdel-Raouf, A. (2016). Estimating the parameters of software reliability growth models using the grey wolf optimization algorithm. *International Journal of Advanced Computer Science and Applications*, 7(4):499–505.

49. Sheta, A. & Al-Salt, J. (2007). Parameter estimation of software reliability growth models by particle swarm optimization. *Management*, 7:14.

50. Shibata, K., Rinsaka, K. & Dohi, T. (2006) Metrics-based software reliability models using non-homogeneous Poisson processes. In *IEEE International Symposium on Software Reliability Engineering*, pages 52–61.

51. Xing, F. & Guo, P. (2005) Support vector regression for software reliability growth modeling and prediction. In *Proc. Advances in Neural Networks*, pages 925–930.

52. Yamada, S. & Osaki, S. (1983). Reliability growth models for hardware and software systems based on nonhomogeneous Poisson processes: A survey. *Microelectronics Reliability*, 23(1):91–112.

53. Zeephongsekul, P., Jayasinghe, C., Fiondella, L. & Nagaraju, V. (2016). Maximum-likelihood estimation of parameters of NHPP software reliability models using expectation conditional maximization algorithm. *IEEE Transactions on Reliability*, 65(3):1571–1583.

54. Zheng, C., Liu, X., Huang, S. & Yao, Y. (2011). A parameter estimation method for software reliability models. *Procedia Engineering*, 15:3477–3481.

15 Alternating α-Series Process

Richard Arnold
Victoria University of Wellington, New Zealand

Stefanka Chukova
Victoria University of Wellington, New Zealand

Yu Hayakawa
Waseda University, Japan

Sarah Marshall
Auckland University of Technology, New Zealand

CONTENTS

15.1 INTRODUCTION

The reliability of a system is of interest to both its users as well as its manufacturers [19], which is why system reliability modelling has attracted the attention of many practitioners and researchers over the years. In this chapter, we consider a repairable system, which upon failure, undergoes a repair. The repair restores the system to a functioning state, taking into account possible system ageing. The lifetime of such a repairable system can be modelled as an alternating sequence of operational times and repair times. Often, for these systems, the number of failures or number of completed repairs over a given time interval is of interest. For example, in warranty cost analysis, the number of failures over the warranty period usually corresponds to the number of warranty claims, which is of major interest to manufacturers. The usual assumption regarding the repair times, while modelling these types of systems, is to consider them negligible, which most of the time is a reasonable assumption. However, this is not the case if the repair times are lengthy or the cost of down-time during the repair is substantial. Then, in order to provide a better model of the system, it is important that these repair times are not ignored and the system is modelled using an alternating stochastic process.

A variety of models have been used to study this type of alternating system. Under the usual assumption of independent and identically distributed (i.i.d.) operational times and i.i.d. repair times, an alternating renewal (AR) process can be used to model the system's lifetime [20]. Under an AR process the operational times and repair times are each modelled by a renewal process (RP). For example, [8, 9] studied models based on an AR process and used them to evaluate the warranty costs over a finite time horizon. For further details on the RP and AR process, see [23]. For systems which are impacted by ageing, operational times tend to decrease and repair times tend to increase over time. For such systems the assumption of i.i.d. operational and repair times is no longer appropriate and may lead to inaccurate estimation of the number of failures in a given time interval, or in the case of warranty analysis, of the warranty cost. For a comprehensive discussion on system ageing and its implications on reliability modelling, refer to [16].

System ageing has been incorporated into reliability models in a variety of ways, for example, via extensions of the alternating renewal process. [18] consider such an extension, called a generalised alternating renewal (GAR) process, in which the operational times are i.i.d. and follow an RP, but the repair times follow an increasing geometric process (GP). Like an RP, the GP has independent interevent times, however, these times are not identically distributed. Instead the time argument of the interevent-time distribution is transformed after each event, which enables monotonic trends to be modelled [17]. In the case of the GAR process, this enables system ageing to be modelled through stochastically increasing repair times. The GAR process has been studied over a finite time horizon and applied to warranty cost analysis [18]. The GAR process was extended by [1], by allowing the operational times to be also affected by ageing. In this extended model, called an alternating geometric (AG) process, the operational times are modelled by a decreasing GP and the repair times are modelled by an increasing GP. Reference [1] studies the AG process over a finite time horizon and applies it to warranty cost analysis, and [3] studies two counting processes associated with the AG process.

In the aforementioned work, system ageing has been accounted for via stochastically increasing (decreasing) repair (operational) times, which have been modelled using an increasing (decreasing) GP. However, there are a large number of other stochastic processes that can be used to model trends in interevent times. In [2] and [24] the authors review extensions of the GP which enable greater flexibility in modelling monotonic trends. An example of such a process is the α-series (AS) process, which was introduced by [6]. The main advantage of the AS process over the GP is that, under certain conditions, the number of events observed by a given time under a decreasing AS process has a finite expected value, unlike the decreasing GP. Further properties of the AS process have been studied by [7]. In addition, several statistical inference studies on the AS process have been published in recent years such as [11, 13, 14, 15]. In [28] a repairable multi-state system with a general α-series process and an order-replacement policy is considered. An explicit expression for the long-run expected cost function is developed to search the optimal order-replacement policy.

The main goal of this chapter is to introduce and study the alternating α-series (AAS) process. Extending the results of [1, 3], two counting processes associated with the AAS process are also discussed. The remainder of this chapter is structured as follows. In Section 15.2 the AS process is defined, and then in Section 15.3, the AAS process is introduced and two key counting processes associated with the AAS process are defined. In Section 15.4 two approaches for computing the mean and variance of the two counting processes are provided. The accuracy of these approaches is demonstrated using numerical examples in Section 15.5. In Section 15.6, the AAS process is applied to automotive warranty data and is compared with AR process and AG process models. Section 15.7 concludes the chapter.

15.2 α-SERIES PROCESS

In this section the α-series (AS) process, which was first studied by [6], is introduced. Two equivalent definitions are provided below.

Definition 1. Let $\{X_n, n = 1, 2, \ldots\}$ be a sequence of independent, non-negative random variables. If the distribution function of X_n is given by $F(n^\alpha x)$ for $n = 1, 2, \ldots$, where $\alpha \in \mathbb{R}$, then $\{X_n, n = 1, 2, \ldots\}$ is called an α-series process.

Definition 2. A stochastic process $\{X_n\}_1^\infty$ is referred to as an α-series process (AS process) with parameter α, if there exists a real number α such that $\{n^\alpha X_n\}_1^\infty$ is a renewal process (RP).

It can be easily seen that the expected value of X_n is given by $\mathrm{E}(X_n) = E(X_1)/(n^\alpha)$. An AS process is stochastically increasing if $\alpha < 0$ and stochastically decreasing if $\alpha > 0$. If $\alpha = 0$, then the AS process is an RP. If X_1 is exponentially distributed and $\alpha = 1$, then the process becomes a linear birth process [6].

15.3 ALTERNATING α-SERIES PROCESS

15.3.1 INTRODUCTION

In this section the alternating α-series (AAS) process is introduced and two counting processes associated with the AAS process are defined.

Consider a repairable item, which initially operates for a length of time X_1 and then fails. It is then repaired for a length of time Y_1. After the repair, the item is again operational for a time X_2, which is followed by a repair for a time Y_2 and so on. The process is defined by a sequence of alternating operational and repair times, and so is called an alternating process.

Definition 3. Let $\{X_n\}_1^\infty$ and $\{Y_n\}_1^\infty$ be independent sequences of random variables. If the sequence of the operational times $\{X_i\}_1^\infty$ is a stochastically decreasing AS process with parameters $\{\alpha, F_{X_1}(t)\}$, $\alpha > 0$ and the sequence of repair times $\{Y_i\}_1^\infty$ is a stochastically increasing AS process with parameters $\{\beta, F_{Y_1}(t)\}$, $\beta < 0$, then a corresponding alternating process is referred to as an alternating α-series (AAS) process with parameters $\{\alpha, F_{X_1}(t);\, \beta, F_{Y_1}(t)\}$.

The AAS process can be used to model ageing systems in which stochastically decreasing operational times and stochastically increasing repair times are observed.

15.3.2 COUNTING PROCESS 1: N(t) NUMBER OF CYCLES COMPLETED BY TIME t

Consider an AAS process with parameters $\{\alpha, F_{X_1}(t);\, \beta, F_{Y_1}(t)\}$, with $\alpha > 0$ and $\beta < 0$. Let a cycle be defined as a period of time consisting of an operational time followed by the corresponding repair time. Denote by $Z_n = X_n + Y_n$, the length of the nth cycle, i.e., the sum of the nth operational and n^{th} repair times, with the cumulative distribution function (CDF) $H_n(t)$, where

$$H_n(z) = F_{X_n} * F_{Y_n}(z), \tag{15.1}$$

and "*" denotes a convolution. Let $T_n = \sum_{i=1}^{n}(X_i + Y_i)$. Then, the CDF of T_n is given by

$$G_n(t) = P(T_n \le t) = H_1 * H_2 * \cdots * H_n(t). \tag{15.2}$$

An example of the AAS process cycles is shown in Figure 15.1. The number of AAS process cycles completed by time t is given by

$$N(t) = \sup\{n : T_n \le t\}.$$

Now, let $t > 0$ be the length of a finite period of time. Then the following, well-known, result holds [23]

$$\{N(t) \ge n\} \iff \{T_n \le t\}. \tag{15.3}$$

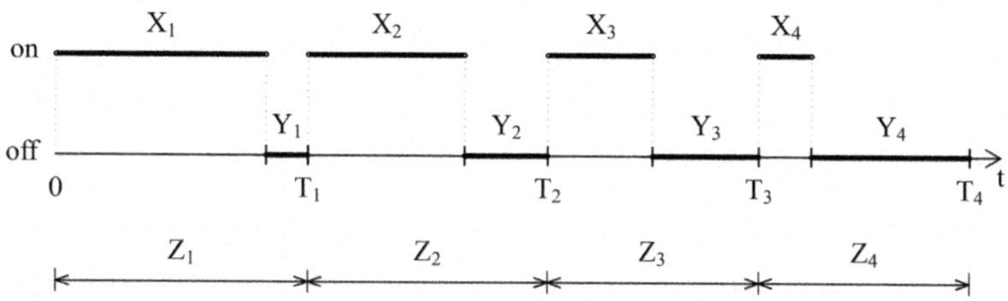

Figure 15.1 Example of AAS process cycles, showing operational (on) times and repair (off) times.

15.3.3 COUNTING PROCESS 2: M(t) NUMBER OF FAILURES UP TO TIME t

Consider an AAS process with parameters $\{\alpha, F_{X_1}(t); \beta, F_{Y_1}(t)\}$, with $\alpha > 0$ and $\beta < 0$. Now consider another counting process, say $M(t)$, related to the underlying AAS process, such that $M(t)$ represents the number of failures occurring before time t. Denote by $Z'_n = Y_n + X_{n+1}$ the length of the nth shifted cycle, i.e., the sum of the nth repair and $(n+1)^{\text{th}}$ operational times, $n = 1, 2, 3 \ldots$, with CDF $H'_n(t)$.

$$H'_n(z) = F_{Y_n} * F_{X_{n+1}}(z). \tag{15.4}$$

The time until the completion of the $(n-1)^{\text{th}}$ shifted cycle is

$$T'_n = X_1 + \sum_{i=1}^{n-1} Z'_i, \tag{15.5}$$

for $n = 1, 2, \ldots$, with the empty sum for $n = 1$ equal to 0. Denote by $G'_n(t)$ the CDF of T'_n, then

$$G'_n(t) = P(T'_n \leq t) = F_1 * H'_1 * H'_2 * \cdots * H'_{n-1}(t). \tag{15.6}$$

An example of the AAS process shifted cycles is shown in Figure 15.2. The number of failures, $M(t)$, which have occurred by time t is defined as follows

$$M(t) = \sup\{n : T'_n \leq t\}.$$

Now, let $t > 0$ be the length of a finite period of time. Then, the following, well-known result holds [23]

$$\{M(t) \geq n\} \iff \{T'_n \leq t\}. \tag{15.7}$$

15.4 MEAN AND VARIANCE OF THE COUNTING PROCESSES N(t) AND M(t)

In this section, two approaches are proposed for the computation of the mean and variance of the counting processes, which were defined in Sections 15.3.2

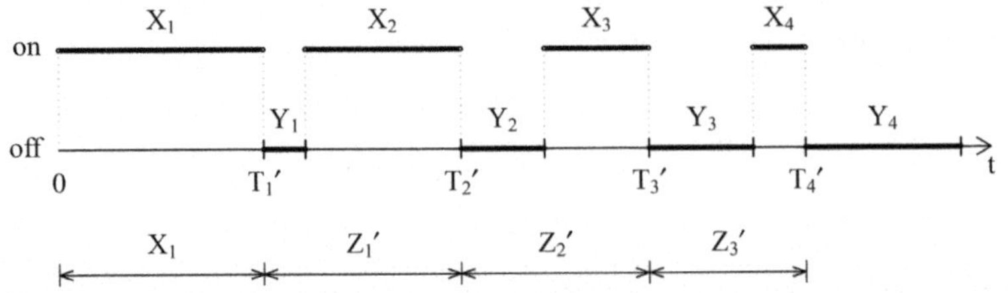

Figure 15.2 Example of AAS process shifted cycles, showing operational (on) times and repair (off) times.

and 15.3.3. These approaches extend a method used to compute the mean and variance for the GP [4] and the AG process [3]. In what follows, we provide a brief overview of these approaches. For further details, refer to [3].

15.4.1 COMPUTING E(N(t)) AND Var(N(t))

Let the mean and variance of the number of cycles, $N(t)$, be denoted by $\mathrm{E}(N(t))$ and $\mathrm{Var}(N(t))$ respectively.

Using (15.7) and the standard approach for deriving results for the renewal function [23] and the geometric function [17], the following formulae for the mean $\mathrm{E}(N(t))$ and the variance $\mathrm{Var}(N(t))$ functions are obtained:

$$\mathrm{E}(N(t)) = \sum_{k=1}^{\infty} G_k(t), \quad t \geq 0 \tag{15.8}$$

and

$$\mathrm{Var}(N(t)) = 2\sum_{k=1}^{\infty} k\, G_k(t) - \mathrm{E}(N(t))(1 + \mathrm{E}(N(t))), \quad t \geq 0. \tag{15.9}$$

Similar to [3], we consider two approaches for the computation of $\mathrm{E}(N(t))$ and $\mathrm{Var}(N(t))$.

Approach A

In this section we adapt the approach outlined in [4] in order to approximate $\mathrm{E}(N(t))$ and $\mathrm{Var}(N(t))$. Consider a uniform partition of $[0, t]$ into m equal sub-intervals, such that $\{0, \frac{t}{m}, \frac{2t}{m}, \ldots, \frac{(m-1)t}{m}, t\}$, then for $n \geq 2$ and $i = 1, 2, \ldots, m$, $G_n(t_i)$ in (15.2) can be approximated as follows

$$\tilde{G}_n(t_i) = \sum_{j=1}^{i} \frac{\tilde{G}_{n-1}(t_{i-j+1}) + \tilde{G}_{n-1}(t_{i-j})}{2} \left(H_n(t_j) - H_n(t_{j-1})\right), \tag{15.10}$$

where $t_i = \frac{it}{m}$ and $\tilde{G}_1(t_i)$ is equal to $H_1(t_i)$ for $i = 1, 2, \ldots, m$. Using (15.8) and (15.9), we can calculate approximations of $\mathrm{E}(N(t))$ and $\mathrm{Var}(N(t))$ as follows:

$$\mathrm{E}(N(t)) \approx \sum_{k=1}^{m_1} \tilde{G}_k(t), \quad t \geq 0 \tag{15.11}$$

$$\mathrm{Var}(N(t)) \approx 2\sum_{k=1}^{m_1} k\tilde{G}_k(t) - \sum_{k=1}^{m_1} \tilde{G}_k(t)\left(1 + \sum_{k=1}^{m_1} \tilde{G}_k(t)\right), \quad t \geq 0, \tag{15.12}$$

where m_1 is chosen such that the general term in (15.11) is sufficiently small, i.e., such that $\tilde{G}_{m_1}(t) < \epsilon$.

A uniform partition of $[0, t]$ has been considered here for the sake of simplicity. However, the approach can be generalised to a non-uniform partition, see [4] for further details. In the approach proposed above, the trapezoidal integration rule is used to approximate $G_n(t)$, as shown in (15.10), and in [4]. Alternative methods to approximate $G_n(t)$ could also be used, e.g., similar work related to the renewal function [25] uses the midpoint (or rectangle) rule. For more on numerical integration see [12] and [21].

Approach B

In this section an alternative approach for computing $\mathrm{E}(N(t))$ and $\mathrm{Var}(N(t))$ is considered. This approach was proposed in relation to the AG process in [3] and follows from the definition of the expected value and variance of a discrete random variable.

If $t \geq 0$, and the distribution $P(N(t) = k)$, $k = 0, 1, 2, \ldots$ is known, then the expected value of $N(t)$ is equal to

$$\mathrm{E}(N(t)) = \sum_{k=0}^{\infty} k\, P(N(t) = k) \tag{15.13}$$

and the variance of $N(t)$ is given by

$$\mathrm{Var}(N(t)) = \sum_{k=0}^{\infty} k^2\, P(N(t) = k) - (E(N(t)))^2. \tag{15.14}$$

The distribution of $N(t)$ can be obtained using the well-known result for counting processes in (15.7) as follows:

$$\begin{aligned} P(N(t) = k) &= P(N(t) \geq k) - P(N(t) \geq k + 1) \\ &= P(T_k \leq t) - P(T_{k+1} \leq t) \\ &= G_k(t) - G_{k+1}(t). \end{aligned} \tag{15.15}$$

Using an appropriate method to approximate $G_k(t)$ (such as (15.10)), and with the appropriate truncation of the infinite series in (15.13) and (15.14), the mean $\mathrm{E}(N(t))$ and variance $\mathrm{Var}(N(t))$ can be computed.

15.4.2 COMPUTING E(M(t)) AND Var(M(t))

The computation of $\mathrm{E}(M(t))$ and $\mathrm{Var}(M(t))$ follows the same ideas outlined in Section 15.4.1 for the computation of $\mathrm{E}(N(t))$ and $\mathrm{Var}(N(t))$, by replacing $G(\cdot)$ with $G'(\cdot)$ and by making some minor adjustments to (15.10), as follows. Consider a uniform partition of $[0, t]$, such that $\{0, \frac{t}{m}, \frac{2t}{m}, \ldots, \frac{(m-1)t}{m}, t\}$, then for $n \geq 2$ and $i = 1, 2, \ldots, m$, $G'_n(t)$ can be approximated by

$$\tilde{G}'_n(t_i) = \sum_{j=1}^{i} \frac{\tilde{G}'_{n-1}(t_{i-j+1}) + \tilde{G}'_{n-1}(t_{i-j})}{2} \left(H'_{n-1}(t_j) - H'_{n-1}(t_{j-1}) \right), \tag{15.16}$$

where $t_i = \frac{it}{m}$ and $\tilde{G}'_1(t_i) = F_{X_1}(t_i)$ for $i = 1, 2, \ldots, m$. Approximations for $\mathrm{E}(M(t))$ and $\mathrm{Var}(M(t))$ can be obtained by adapting (15.11) and (15.12) (Approach A) and (15.13), (15.14) and (15.15) (Approach B).

15.5 NUMERICAL RESULTS

In this section we compare the computation of the mean and variance function of the two counting processes, $N(t)$ and $M(t)$, associated with the AAS process using simulation and the two numerical approaches outlined above. Then, we explore some properties of the AAS process for a variety of parameter values.

Table 15.1 contains the values of $\mathrm{E}(N(t))$ and $\mathrm{Var}(N(t))$ and Table 15.2 contains the values of $\mathrm{E}(M(t))$ and $\mathrm{Var}(M(t))$ computed via the two numerical approaches and simulation, for an AAS process with $\mathrm{E}(X_1) = 3$, $\mathrm{Var}(X_1) = 9$, $\alpha = 1$, $\mathrm{E}(Y_1) = 0.01$, $\mathrm{Var}(Y_1) = 0.0001$, $\beta = -1$, F_{X_1} and F_{Y_1} are the exponential CDFs with the corresponding parameters. The numerical approaches and the simulation were performed using R 4.0.4 [22]. A uniform partition of $[0, t]$, such that $\{0, \frac{t}{m}, \frac{2t}{m}, \ldots, \frac{(m-1)t}{m}, t\}$ with step $\frac{t}{m} = 0.01$ was used, and the infinite series in (15.11) and (15.12) and the equivalent expressions for $M(t)$ are truncated using $\epsilon = 10^{-3}$. The R function `integrate` (part of the `stats` package [22]) is used to compute $H_n(z)$, with a tolerance value (`abs.tol`) of $\eta = 10^{-16}$. The simulation values are the average values across 10,000,000 simulation runs. As shown in Tables 15.1 and 15.2, the numerical approaches produce values that are close to the simulation. We found that the mean and variance computed using Approach B begin to deviate as t increases, compared with the values computed using Approach A and simulation. We expect that this is due to computing error accumulation, however this is an area for further investigation.

To demonstrate the variety of monotonic trends that can be modelled using an AAS process, we have computed the expected number of cycles $\mathrm{E}(N(t))$ and the expected number of failures $\mathrm{E}(M(t))$ for a range of parameter values. The results presented below were computed using both numerical approaches and simulation, however due to the similarity of the results, only those for Approach A are shown in the plots.

Table 15.1

Comparison of numerical approaches and simulation for an AASP with $E(X_1)$ = 3, $\text{Var}(X_1)$ = 9, α = 1, $E(Y_1)$ = 0.01, $\text{Var}(Y_1)$ = 0.0001, β = -1, F_{X_1}, F_{Y_1} = exponential cdf. Settings for numerical approaches: $\epsilon=10^{-3}$, t/m = 0.01, $\eta = 10^{-16}$. Settings for simulation: n = 10000000

	$E(N(t))$			$\text{Var}(N(t))$		
Time	Approach A	Approach B	Simulation	Approach A	Approach B	Simulation
0.6	0.213682	0.213682	0.213738	0.250136	0.250136	0.250140
1.0	0.382684	0.382684	0.382808	0.504318	0.504318	0.504426
2.0	0.907106	0.907106	0.907360	1.582393	1.582393	1.583126
5.0	3.670939	3.670939	3.672390	12.160205	12.160205	12.167529
10.0	13.030337	13.030337	13.031078	59.528072	59.528072	59.546670
12.0	17.890105	17.890105	17.890020	76.948257	76.948257	76.967922
15.0	25.393798	25.393794	25.394476	89.810023	89.809996	89.826206
18.0	32.578429	32.575922	32.578501	89.169183	89.187123	89.171773
20.0	37.054373	37.004514	37.056137	84.715775	85.516449	84.763496

Table 15.2

Comparison of numerical approaches and simulation for an AASP with $E(X_1)$ = 3, $\text{Var}(X_1)$ = 9, α = 1, $E(Y_1)$ = 0.01, $\text{Var}(Y_1)$ = 0.0001, β = -1, F_{X_1}, F_{Y_1} = exponential cdf. Settings for numerical approaches: $\epsilon=10^{-3}$, t/m = 0.01, $\eta = 10^{-16}$. Settings for simulation: n = 10000000

	$E(M(t))$			$\text{Var}(M(t))$		
Time	Approach A	Approach B	Simulation	Approach A	Approach B	Simulation
0.6	0.219208	0.219208	0.219247	0.261090	0.261090	0.261077
1.0	0.390343	0.390343	0.390475	0.524666	0.524666	0.524703
2.0	0.923025	0.923025	0.923286	1.651363	1.651363	1.651975
5.0	3.754913	3.754913	3.756531	12.895593	12.895593	12.904556
10.0	13.397308	13.397308	13.398132	63.089088	63.089088	63.109118
12.0	18.383714	18.383714	18.383884	81.017003	81.017003	81.040221
15.0	26.042233	26.042230	26.042833	93.502801	93.502778	93.520386
18.0	33.332423	33.330080	33.332591	91.870768	91.888722	91.874823
20.0	37.856610	37.809023	37.858466	86.768550	87.561627	86.814588

In Figure 15.3, the columns depict two different values of $\text{E}(X_1)$ and the rows depict two different values of β. Firstly, let us look at the plot on the top left of Figure 15.3. In this scenario, the expected time until the first failure, $\text{E}(X_1) = 0.3$, is similar to the first expected repair time, $\text{E}(Y_1) = 0.01$. The AS process parameter for the repair times is $\beta = -5$, which means that the repair times increase rapidly. Consequently, the first two repairs occur very quickly as both $\text{E}(X_1)$ and $\text{E}(Y_1)$ are small. However, $\text{E}(Y_n)$ soon begins to increase rapidly and becomes significantly larger than $\text{E}(X_n)$. Since the system cannot fail while being repaired, the number of cycles increases at a slower rate for $n > 2$. The AS process parameter for the operational times α has little impact on the expected number of cycles as $\mathrm{E}(Y_n)$ is larger than $\mathrm{E}(X_n)$ for $n > 2$. A similar trend in $\mathrm{E}(N(t))$ (an initial steep increase, followed by a more gradual increase) is also shown in the plot in the top right of Figure 15.3. However, in this case, due to the longer expected time until the first failure, $\mathrm{E}(X_1) = 3$, the impact is less severe. Notice that α also has an impact here, with higher values of α leading to a more rapid decrease in the operational times, and thus a higher number of cycles.

Now, let us look at the bottom row of Figure 15.3. In these two plots, $\beta = -1$, which means the increase of the repair times is slower than the increase of the repair time in the top row. Once again, when $\mathrm{E}(X_1)$ is small, α does not have much impact as $\mathrm{E}(X_n)$ are small compared with $\mathrm{E}(Y_n)$. However, since $\mathrm{E}(Y_n)$ increases more slowly than in the plots in the top row of Figure 15.3, $\mathrm{E}(N(t))$ is much larger. When $\mathrm{E}(X_1) = 3$, with $\alpha = 0.1$, the operational time decreases very slowly, leading to a low value of $\mathrm{E}(N(t))$. Similar trends were observed for $\mathrm{E}(M(t))$.

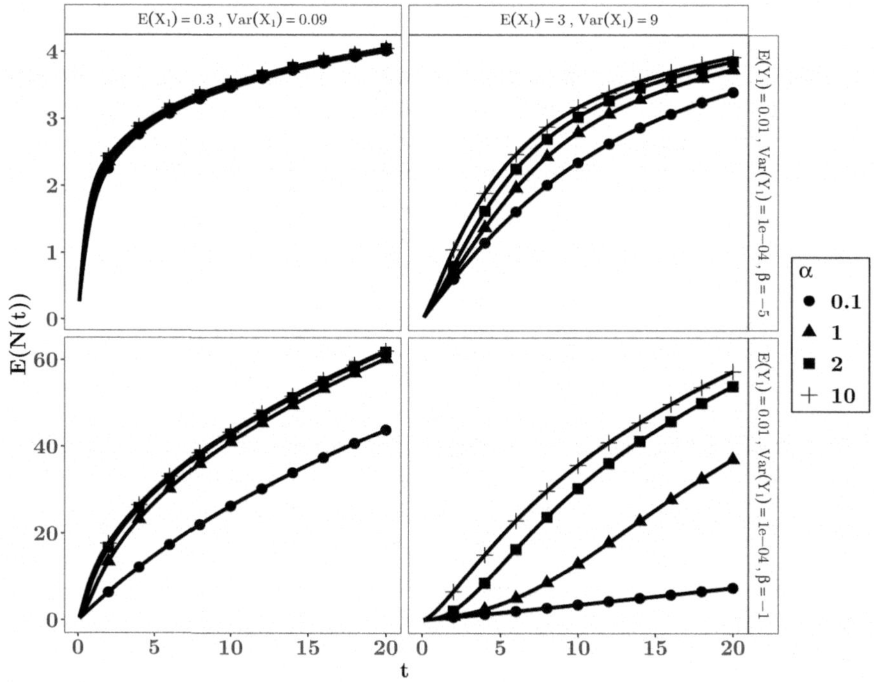

Figure 15.3 Expected number of completed AAS process cycles, $\mathrm{E}(N(t))$, for $\alpha \in \{0.1, 1, 2, 10\}$ for F_{X_1}, F_{Y_1} = exponential CDF. $\mathrm{E}(N(t))$ was computed using numerical approach A, with $\epsilon = 10^{-3}$, $t/m = 0.01$, and $\eta = 10^{-16}$.

15.6 APPLICATION OF AN AAS PROCESS TO MODELLING WARRANTY DATA

In this section we fit an AAS process to warranty claims data from an automotive manufacturer. We begin by summarising a process for fitting an AS process to data, then provide an overview of the warranty data, before finally fitting the AAS process to the data.

15.6.1 PROCEDURE FOR FITTING AN AAS PROCESS

Following [5], we complete the following steps to fit an AS process to a random sample of data $\{X_1, X_2, \ldots, X_n\}$:

Step 1: Testing for a trend in the data using the Mann Test

- $H_0 : \{X_1, X_2, \ldots, X_n\}$ comes from an RP;
- $H_1 : \{X_1, X_2, \ldots, X_n\}$ has a monotone trend

Step 2: Testing whether the data come from an AS process (graphical test)

- Visual inspection of $\ln X_n$ against $\ln k$ should reveal a linear relationship if the data are consistent with an AS process [5].

Step 3: Parameter estimation

- Estimate mean and variance of X_1, and α

Step 4: Testing whether the data come from an AS process (parametric test)

- $H_0 : \alpha = 0$
- $H_1 : \alpha \neq 0$

For further details on these steps, including the test statistics and their distributions, refer to [5]. To fit an AAS process to data, complete the steps outlined above for the operational times and then for the repair times.

15.6.2 WARRANTY DATA

The warranty database contains over 200,000 claims from vehicles manufactured between 1998 and 2001. The warranty database contains the age and mileage at the time of the claim, as well as the costs (labour, parts and other) associated with that claim. Due to commercial confidentiality, the costs in the database have been masked by the provider. For previous analysis of this dataset refer to [1, 18].

In order to fit the AAS process described in Section 15.3 to these data, we need to extract the operational times X_i and repair times Y_i, which are not explicitly included in the warranty database. So, following [1], we have estimated/modelled them as follows. We assume that the labour cost of a claim is related to the repair time and we use this relationship to estimate the repair times $Y_i, i = 1, 2, \ldots, n$. Using a linear transformation, the labour costs are converted to repair times. A minimum repair time of 1 day and a maximum repair time of 90 days were matched with the minimum and maximum labour costs across all claims in the database. For technical reasons, a small random error, generated from $U(\pm 1 \times 10^{-7})$ days, was added to each repair time to prevent ties. We assume that the age of the vehicle at the ith claim is

$$T_{i-1} + X_i,$$

where $T_{i-1} = \sum_{j=1}^{i-1}(X_j + Y_j)$. Using this relationship, the operational times $X_i, i = 1, 2, \ldots, n$ can be identified.

To demonstrate the process of fitting an alternating α-series process to data, we have selected two vehicles with at least 9 claims, Vehicles A and B, from this database. In Section 15.6.3, we fit an AAS process, as well as an alternating renewal process (ARP) and an alternating geometric process (AGP) to Vehicles A and B. To fit GPs to the operational and repair times data we follow a procedure outlined by Lam [17, §4.2, pp. 101-104]. This procedure was used to assess the fit of an AGP to the same vehicles from the warranty database in [1]. To fit an ARP to the data, the mean and variance are estimated using the sample mean and sample variance of the observed/calculated operational and repair times.

To compare the models we compute the mean squared error of

- the fitted values for each of the processes (renewal (RP), geometric (GP) and α-series (ASP)), $(\hat{X}_{kRP}, \hat{Y}_{kRP}), (\hat{X}_{kGP}, \hat{Y}_{kGP}), (\hat{X}_{kASP}, \hat{Y}_{kASP})$ against the observed operational and repair times (X_k, Y_k), $k = 1, 2, \ldots n$, and
- the cumulative fitted values $S_{X_j}^{(n)} = \sum_{i=1}^{n} \hat{X}_{ij}$ for $j = \{RP, GP, ASP\}$ and $S_{Y_j}^{(n)} = \sum_{i=1}^{n} \hat{Y}_{ij}$ for $j = \{RP, GP, ASP\}$ against the observed cumulative operational and repair times $S_X^{(n)} = \sum_{i=1}^{n} X_i$ and $S_Y^{(n)} = \sum_{i=1}^{n} Y_i$.

15.6.3 FITTING AN AAS PROCESS TO THE WARRANTY CLAIMS DATA

In order to apply an AAS process to the warranty claims data, we aim to demonstrate that the operational times form a stochastically decreasing AS process and the repair times form a stochastically increasing AS process. We complete the procedure outlined in Section 15.6.1 for the operational times and the repair times.

The hypothesis test results and parameter estimates computed as part of fitting an AS process are provided in Table 15.3. The parameter estimates for the RP and the GP are also shown in Table 15.3. The observed/calculated cumulative operational and repair times, along with the corresponding fitted values for the α-series process (ASP), RP and GP models are shown in Figure 15.4. We define the following notation, for each process $i \in \{\text{ASP, RP, GP}\}$:

- $\text{E}(X_1) = \lambda_i$: expected value of the first operational time
- $\text{E}(Y_1) = \mu_i$: expected value of the first repair time
- $\text{Var}(X_1) = \sigma^2_{X,i}$: variance of the first operational time
- $\text{Var}(X_1) = \sigma^2_{Y,i}$: variance of the first repair time

First observe the p-values associated with the Mann Test (Step 1). As shown in Table 15.3, H_0 is rejected at the 10% level for the repair times for both vehicles and for the operational times for Vehicle B. Closer inspection of Figure 15.4 suggests that, while the operational times for Vehicle A do have a decreasing trend, this trend may not be monotonic.

Now, consider Figure 15.5 (Step 2). This plot shows the relationship between the natural log of the operational and repair times, and the natural log of the failure number. If an ASP is a good fit to the data, then a linear trend should be observed, see [5]. The small sample size in these plots makes it difficult to determine linearity. Notice that the two operational times plots both show a negative correlation, while the repair times plots both show a positive correlation. This is consistent with our expectation relating to decreasing and increasing ASPs, respectively.

The parameter estimates for the ASP, RP and GP are provided in Table 15.3 (Step 3). As expected, the parameter estimates for the operational times are consistent with a decreasing GP and ASP, and for the repair times are consistent with an increasing GP and ASP.

Using the estimated values for α and β, we can test if the data are consistent with an AS process (Step 4). As shown in Table 15.3, H_0 is rejected at the 5% level for the repair times for both vehicles and the operational times for Vehicle A, and at the 10% level for the operational times for Vehicle B.

The fit of the models against the cumulative operational and repair times is shown in 15.4. As expected both the ASP and GP models are a better fit than the RP for the operational and repair times for both vehicles. The ASP model appears to be better than the GP model for the cumulative operational times for both vehicles, whereas the GP model appears to be better than the ASP model for the cumulative repair times. This is confirmed by examining the mean squared error for the cumulative times in Table 15.3. The mean squared error for the operational and repairs times indicate the superiority of the ASP and GP models over the RP model. The nature of the trend

Table 15.3

Hypothesis Tests and Parameter Estimation for the AASP, ARP and AGP for two vehicles from a warranty claims database.

	Operational Times			Repair Times		
		A	B		A	B
	n	9	11	n	9	11
Mann Test	p-value	0.348	0.043	p-value	0.029	0.087
ASP	$\hat{\lambda}_{ASP}$	291.874	274.685	$\hat{\mu}_{ASP}$	1.638	1.776
	$\widehat{\sigma^2}_{X,ASP}$	66047.829	21471.819	$\widehat{\sigma^2}_{Y,ASP}$	0.212	0.869
	$\hat{\alpha}$	1.278	0.924	$\hat{\beta}$	-0.246	-0.507
$H_0 : \alpha = 0$ $H_1 : \alpha \neq 0$	p-value	0.045	0.000	p-value	0.087	0.048
RP	$\hat{\lambda}_{RP}$	80.337	81.268	$\hat{\mu}_{RP}$	2.348	4.455
	$\widehat{\sigma^2}_{X,RP}$	9547.458	4842.573	$\widehat{\sigma^2}_{Y,RP}$	0.696	15.805
GP	$\hat{\lambda}_{GP}$	186.075	184.704	$\hat{\mu}_{GP}$	1.650	2.078
	$\widehat{\sigma^2}_{X,GP}$	25007.671	11445.363	$\widehat{\sigma^2}_{Y,GP}$	0.150	1.194
	$\hat{a}$	1.358	1.226	$\hat{b}$	0.921	0.880
MSE	X_k vs $\hat{X}_{kRP}$	8486.630	4402.339	Y_k vs $\hat{Y}_{kRP}$	0.619	14.368
	X_k vs $\hat{X}_{kGP}$	3849.489	1308.586	Y_k vs $\hat{Y}_{kGP}$	0.282	8.917
	X_k vs $\hat{X}_{kASP}$	4310.665	1973.130	Y_k vs $\hat{Y}_{kASP}$	0.411	10.626
MSE	$S_X^{(n)}$ vs $S_{X_{RP}}^{(n)}$	34863.499	33319.745	$S_Y^{(n)}$ vs $S_{Y_{RP}}^{(n)}$	2.773	63.951
	$S_X^{(n)}$ vs $S_{X_{GP}}^{(n)}$	5976.129	1491.396	$S_Y^{(n)}$ vs $S_{Y_{GP}}^{(n)}$	0.317	4.638
	$S_X^{(n)}$ vs $S_{X_{ASP}}^{(n)}$	3811.975	1315.226	$S_Y^{(n)}$ vs $S_{Y_{ASP}}^{(n)}$	0.855	8.775

observed in the operational times, particularly in the first few observations, mean that the mean squared error for the operational times is lower for the GP compared with the ASP, but for the cumulative operational times is lower for the ASP compared with the GP. However, as shown in Figure 15.4, the fit of the ASP and GP models to the cumulative operational and repair times are very similar, particularly for later failures. This suggests that in warranty analysis, both the AAS process and AGP models can be used to model the operational and repair times. However, given the time horizon here is relatively short and the sample time is small, further investigation is needed to compare the behaviour of the AAS process and AGP models over longer time periods.

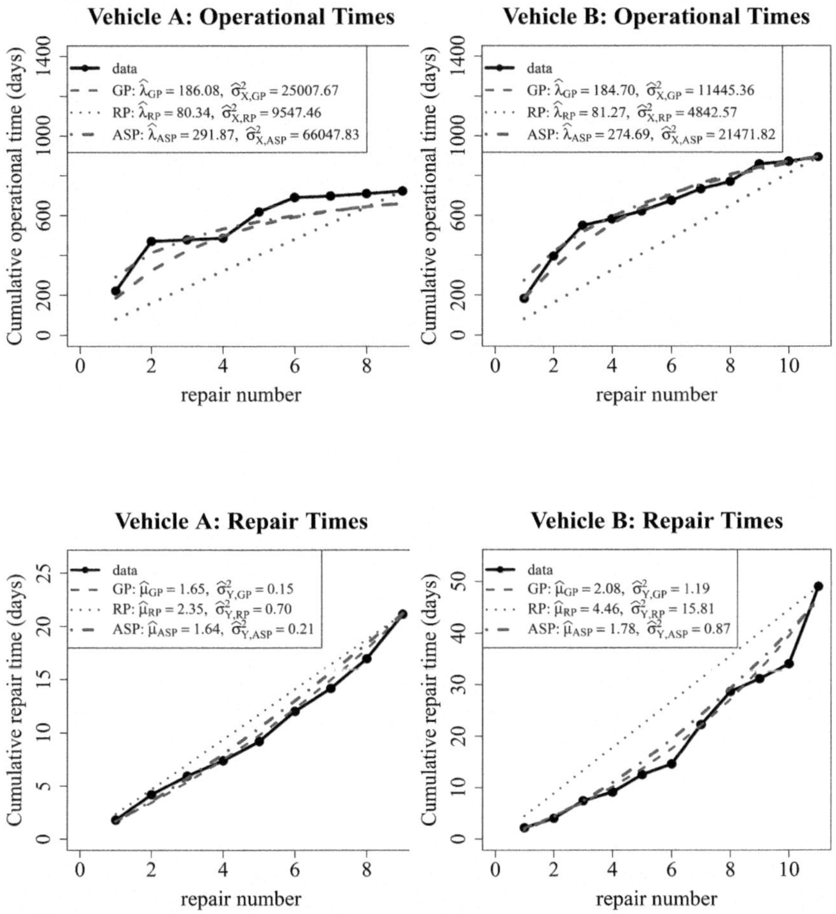

Figure 15.4 Observed/modelled data and fitted values for two vehicles from the warranty claims database.

15.7 CONCLUSION

In this chapter we introduce and study the AAS process and illustrate its possible application in warranty modelling. We focus on two counting processes associated with the AAS process: (1) $N(t)$ - the number of cycles up to time t and (2) $M(t)$ - the number of failures up to time t, for $t > 0$. We propose two numerical approaches for the approximation of their mean and variance functions. For a system which can be successfully modelled by an AAS process with parameters $\{\alpha, F_{X_1}(t);\ \beta, F_{Y_1}(t)\}$ with ratios $\alpha > 0$ and

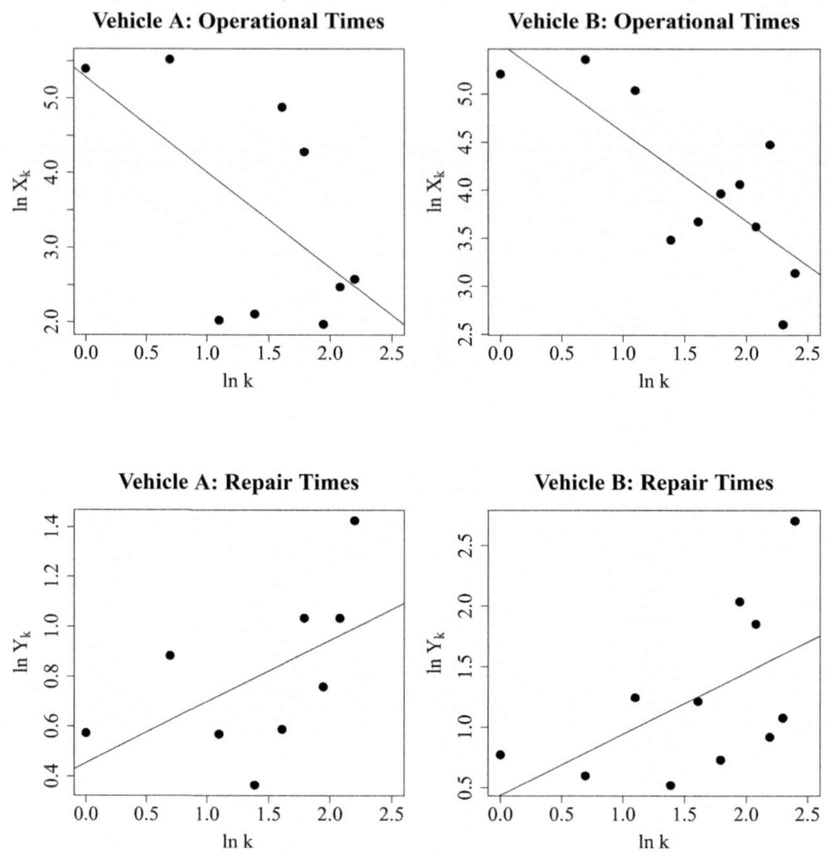

Figure 15.5 Plot to investigate if data are consistent with an α-series process.

$\beta < 0$, these results can be used to compute the mean value and the variance of the warranty cost under different warranty strategies. These results could be particularly useful for designing better warranty strategies as well as assisting the producers in allocating appropriate funds to the warranty reserves related to their products. In addition, we have proposed a procedure to fit the AAS process to warranty data.

When considering the use of the AAS process in modelling real system reliability or warranty data, a natural question immediately arises: when should we use the AAS process and when we should resort to modelling based on the AR process, the AG process or any of the other geometric-like processes (GLP or GL processes), reviewed in [2] and [24]. The answer to this question is not obvious. For a start, we need to have some data related to the failure/repair process of the system that we wish to study. If these data support the idea of an existing monotonic trend within the observations, then it might be appropriate to use a GLP to model it. If a monotonic trend is not observed, then it is better to employ some other stochastic process (e.g., non-homogeneous Poisson process) to model the system.

So, if a monotonic trend is present within the data, the choice between the geometric, the α-series or any of the other GL processes should follow standard statistical approaches/criteria for model selection. These statistical approaches/criteria are well documented in a vast body of literature, (e.g., see [10]).

In this chapter, we have fitted the AAS process to real warranty data from an automotive manufacturer. Although it is a good start in demonstrating the usefulness of the AAS process, an application of the model to a larger dataset will be an interesting next step in our future work. Of course, as usual, finding real failure/repair data is not an easy task, mostly due to commercial confidentiality.

As mentioned above, there are various GLPs that can be used to model ageing systems [2, 24]. Another direction for future work is to extend the results presented in the current chapter to other GLPs, e.g., the extended geometric process [27]. Consideration of alternative warranty policies, as in [26], is another area for future work.

ACKNOWLEDGEMENTS

This research was supported by JSPS KAKENHI Grant-in-Aid for Scientific Research (C), Grant Number: 18K04621.

REFERENCES

1. Arnold, R., Chukova, S., Hayakawa, Y., & Marshall, S. (2019). Warranty cost analysis with an alternating geometric process. *Proceedings of the Institution of Mechanical Engineers, Part O: Journal of Risk and Reliability*, 233(4), 698-715.

2. Arnold, R., Chukova, S., Hayakawa, Y., & Marshall, S. (2020). Geometric-Like Processes: An Overview and Some Reliability Applications. *Reliability Engineering & System Safety*, 201:106990.

3. Arnold, R., Chukova, S., Hayakawa, Y., & Marshall, S. (2021). Mean and variance of an alternating geometric process: An application in warranty cost analysis. *Quality and Reliability Engineering International*, 1-18.

4. Aydoğdu, H., & Altındağ, Ö. (2016). Computation of the mean value and variance functions in geometric process. *Journal of Statistical Computation and Simulation*, 86(5), 986-995.

5. Aydoğdu, H., & Kara, M. (2012). Nonparametric Estimation in α-series Processes. *Computational Statistics & Data Analysis*, 56(1), 190-201.

6. Braun, W. J., Li, W., & Zhao, Y. Q. (2005). Properties of the geometric and related processes. *Naval Research Logistics*, 52(7), 607-616.

7. Braun, W. J., Li, W., & Zhao, Y. Q. (2008). Some theoretical properties of the geometric and α-series processes. *Communications in Statistics - Theory and Methods*, 37(9), 1483-1496.

8. Chukova, S., & Hayakawa, Y. (2004). Warranty cost analysis: Non-zero repair time. *Applied Stochastic Models in Business and Industry*, 20(1), 59-71.

9. Chukova, S., & Hayakawa, Y. (2004). Warranty cost analysis: Renewing warranty with non-zero repair time. *International Journal of Reliability, Quality and Safety Engineering*, 11(02), 93-112.

10. Claeskens, G., & Hjort, N. L. (2008). *Model Selection and Model Averaging*. Cambridge Series in Statistical and Probabilistic Mathematics. Cambridge University Press, Cambridge.

11. Biçer, H. D. (2019). Statistical inference for alpha-series process with the generalized Rayleigh distribution. *Entropy*, 21(5), 451.

12. Gautschi. W. (2012). *Numerical Analysis*, second edition. Springer, New York.

13. Kara, M., Altındağ, Ö., Pekalp, M. H., & Aydoğdu, H. (2019). Parameter estimation in α-series process with lognormal distribution. *Communications in Statistics - Theory and Methods*, 48(20), 4976-4998.

14. Kara, M., Aydoğdu, H., & Şenoğlu, B. (2017). Statistical inference for α-series process with gamma distribution. *Communications in Statistics - Theory and Methods*, 46(13), 6727-6736.

15. Kara, M., Türkşen, Ö., & Aydoğdu, H. (2017) Statistical inference for α-series process with the inverse Gaussian distribution. *Communications in Statistics - Simulation and Computation*, 46(6), 4938-4950.

16. Lai, C. D., & Xie, M. (2003). Concepts and Applications of Stochastic Aging in Reliability. In Pham, H., Eds., *Handbook of Reliability Engineering*, Springer, London, 165-180.

17. Lam, Y. (2007). *The Geometric Process and Its Applications*, World Scientific, Hackensack, NJ.

18. Marshall, S., Arnold, R., Chukova, S., & Hayakawa, Y. (2018). Warranty cost analysis: Increasing warranty repair times. *Applied Stochastic Models in Business and Industry*, 34(4), 544-561.

19. Murthy, D. N. P., Rausand, M., & Østerås, T. (2008). *Product Reliability: Specification and Performance*. Springer Series in Reliability Engineering. Springer-Verlag, London.

20. Nakagawa, T. (2005). *Maintenance Theory of Reliability*. Springer Series in Reliability Engineering. Springer-Verlag, London.

21. Press, W. H., Teukolsky, S. A., Vetterling, W. T., & Flannery, B. P. (2007). *Numerical Recipes 3rd Edition: The Art of Scientific Computing*, third edition. Cambridge University Press, New York, NY, USA.

22. R Core Team. (2020). *R: A Language and Environment for Statistical Computing*. Vienna, Austria.

23. Ross. S. M. (2008). *Stochastic Processes*, second edition. Wiley, New Delhi, India.

24. Wu, D., Peng, R., & Wu, S. (2020). A review of the extensions of the geometric process, applications, and challenges. *Quality and Reliability Engineering International*, 36(2), 436-446.

25. Xie, M. (1989). On the solution of renewal-type integral equations. *Communications in Statistics - Simulation and Computation*, 18(1), 281-293.

26. Yedida, S., Munavar, M. U., & Ranjani, R. (2012). Warranty cost analysis using alternating quasi-renewal processes with a warranty option. *International Journal of Systems Science*, 43(3), 507-517.

27. Zhang, Y. L. & Wang, G. J. (2016). An extended geometric process repair model for a cold standby repairable system with imperfect delayed repair. *International Journal of Systems Science: Operations & Logistics*, 3(3), 163-175.

28. Zuo, K., & Xiao, M. (2020). A repairable multi-state system with a general α-series process and an order-replacement policy. *Communications in Statistics - Theory and Methods*, 0(0), 1-18.

16 Optimum Staggered Testing Strategy for Redundant Safety Instrumented Systems with Different Testing Intervals

Sun-Keun Seo
Dong-A University, South Korea

Won Young Yun
Pusan National University, South Korea

CONTENTS

16.1 INTRODUCTION

IEC (International Electrotechnical Commission) 61508 was designated to announce the requirements for the functional safety of Safety Instrumented Systems (SIS) in 2000 and functional safety of SIS has been required in several industry areas. Based on IEC 61508, which is the basic international standard about functional safety, several international standards customized for functional safety in specific industries have been established, for example, the process industry (IEC 61511, 2002), car industry (ISO 20262, 2011), machinery products, railway vehicles, medical devices, and nuclear power plants.

In IEC 61508, the operating modes of SIS are categorized into low- and high-demand modes. In IEC 61508 (2010 version), if the demand rate is greater than one per year, it is classified as a high-demand operating mode and continuous operation is also included in this mode. Otherwise, the operation is classified as a lower-demand operating mode.

PFD (Average Probability of Failure on Demand) and PFH (Probability of Dangerous Failure per Hour) are used as the system performance measure instead of instantaneous availability, unreliability, and steady state availability in IEC 61508. PFD is similar to un-availability and PFH is similar to rate of occurrence of failure (ROCOF) in repairable systems and we focus on the PFD of redundant safety instrumented systems in this paper.

In general, failures of units related to functional safety result in serious system reliability and safety problems and redundant units are added to minimize the failure probability of safety units on demand. Additionally, the failures of units related to functional safety are usually hidden failures and we should take a proof test (inspection) periodically to find the hidden failures of units related to safety.

The most important issue in the proof test performed for SIS with redundant structure is how to determine testing time points of the units because the testing method with equal testing time points and cycle for all units may be less efficient in finding hidden failures of units and minimize the system PFD. Staggered testing methods in which different testing time points are assigned to different units are better to reduce the system PFD in cases with short testing times (Contini et al. 2013; Liu and Rausand, 2013).

Green (1972) considered staggered testing methods with uniformly divided testing intervals in which the starting points of testing time points of units are uniformly distributed to divide the testing cycles with same time span. Rouvroye and Wiegerinck (2006) derived the PFD under two testing methods by using continuous time Markov chain. Vaurio (1980, 2011) proposed a method for estimating the approximate un-availability under the simultaneous and uniformly divided testing intervals for SIS with M-out-of-N structures. In case of SIS with two identical units, Green (1972) found that the optimal staggered time is a half of the check interval. Part 6 of IEC 61508 (2010) provides approximate formulas to obtain the PFD for SIS with 1, 2, and 3 units in cases with the equal failure rate and the equal proof testing interval for all units. Additionally, the standard recommends testing intervals for staggered testing under the same conditions (same units and same testing intervals). Liu (2014) investigated in detail the optimality of staggered testing in SIS with 1-out-of -2 structure, assuming that the failure rates of units and the testing intervals are different.

Seo and Yun (2021) reviewed the optimality of existing testing schemes and obtain the optimal staggered testing in SIS with 1- and 2-out-of-3 structures but this paper focuses on the optimal staggered testing methods with different grouping and testing intervals for SIS with three units based on the PFD and the expected cost rate.

This paper is organized as follows. Section 16.2 summarizes the optimal staggered testing models with equal testing intervals. In Section 16.3, the optimal staggered testing methods with different testing intervals are obtained to minimize the PFD and Section 16.4 considers a cost model under staggered testing in SIS with three units and obtains the optimal staggered testing models to minimize the expected cost rate and PFD in numerical examples. Finally, Section 16.5 contains the conclusions of the paper.

16.2 PFD OF REDUNDANT SAFETY INSTRUMENTED SYSTEMS WITH 2 AND 3 UNITS

In this section, we summarize the optimal staggered testing methods in SIS with 2 and 3 in the existing studies. In IEC 61508 (2010), the PFD (Average Probability of Failure on Demand) is defined as the average un-availability of safety systems inspected periodically with testing interval τ, and is given as

$$PFD = \frac{MDT(\tau)}{\tau} = \frac{\int_0^\tau q_S(t)\mathrm{d}t}{\tau} \tag{16.1}$$

where $MDF(\tau)$ is the expected down-time during $(0, \tau]$ and $q_S(t)$ is the system unavailability at time t. Exponential distributions are used for failure times of units in SIS and if the failures of the units follow exponential distributions, the un availability of unit i at time t is approximately

$$q_i(t) = 1 - \mathrm{e}^{-\lambda_i t} \approx \lambda_i t, \tag{16.2}$$

when $\lambda_i t$ is close to 0. The approximate un-availability is mainly used to calculate the system PFD in late sections (refer IEC 61508). In this paper, we assume that failed units detected at periodic testing are repaired as good as new and repair times are negligible. We use Equation 16.1 as an optimization criterion to determine the optimal staggered testing strategies and sometimes approximately derive the PFD.

16.2.1 OPTIMAL STAGGERED TESTING IN SIS WITH 1 OUT OF 2 STRUCTURES

Liu (2014) derived the exact formula of the PFD in the safety instrumented system with 1 out of 2 structure and obtained the optimal staggered testing strategies as follows:

I. Equal failure rate and equal testing interval: $c_a^* = 1/2$ (refer Green (1972) for approximate PFD and Rausand and Hoyland (2004) for exact PFD respectively) where c_a, $0 \leq c_a < 1$ is the staggered proportion and for example, $c_a = 1/2$ means that one unit is tested at the middle time points of equal testing period, τ.

II. Equal failure rate and different testing intervals ($\tau_1 = m\tau_2, 1 < m$): $c_a^* = 1/2$ where τ_i: testing interval of unit i, $i = 1, 2$

III. Different failure rates and equal testing interval: $c_a^* \approx 1/2$

IV. Different failure rate and different testing intervals ($\tau_1 = m\tau_2$, $1 < m$): $c_a^* = 1/2$

If we use an approximate PFD in Case (iv), we can know the optimal $c_a^* = 1/2$ as follows: Based on Figure 16.1 (a case with $m = 3$), un-availabilities of units and system are given as

$$q_{1k}(t) = \lambda_1[t - (k-1)\tau], \quad (k-1)\tau \le t < k\tau, \quad k = 1, \dots, m$$

$$q_2(t) = \begin{cases} q_{21}(t) & = \lambda_2(t + \tau - t_a), \quad 0 \le t < t_a \\ q_{22}(t) & = \lambda_2(t - t_a), \quad t_a \le t < m\tau \end{cases} \tag{16.3a}$$

$$q_S(t) = \begin{cases} q_{11}(t)q_{21}(t), & 0 \le t < t_a \\ q_{11}(t)q_{22}(t), & t_a \le t < \tau \\ q_{11}(t)q_{23}(t), & (k-1)\tau \le t < m\tau, \quad k = 2, \dots, m \end{cases} \tag{16.3b}$$

Let the last testing time of unit 1 be 0 and then the last testing time of unit 2 is $t_a - \tau$ (refer to Figure 16.1 (case with $m = 3$)).

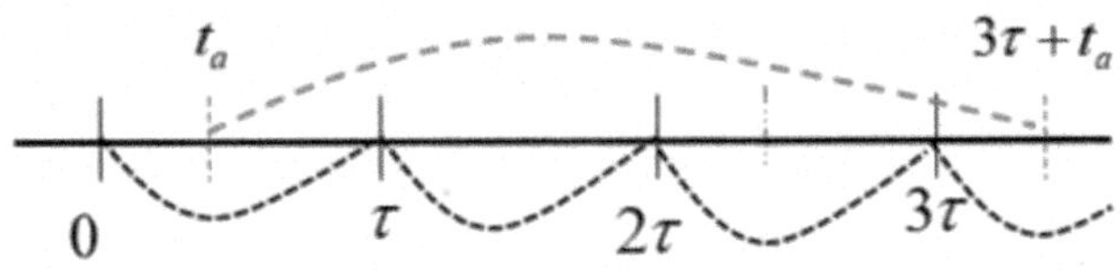

Figure 16.1 Staggered testing scheme in SIS with 1 out of 2 structure ($m = 3$)

Thus, the duration of one renewal cycle is $m\tau$ and let the starting testing time of unit 2 be $t_a = c_a\tau$. Then the PFD is given as

$$PFD = \frac{\int_0^{m\tau} q_S(t)dt}{m\tau} = \frac{\lambda_1\lambda_2\tau^2(6c_a^2 - 6c_a + 3m + 1)}{12} \tag{16.4}$$

By setting the first derivative of Equation 16.4 to 0, we can obtain the optimal $c_a^* = 1/2$ and $PFD^* = \lambda_1\lambda_2\tau^2(6m-1)/24$. Additionally, in Case (iii) with $m = 1$, we can find easily $c_a^* = 1/2$ and $PFD^* = 5\lambda_1\lambda_2\tau^2/24$.

16.2.2 OPTIMAL STAGGERED TESTING IN SIS WITH 1 OUT OF 3 STRUCTURES (EQUAL TESTING INTERVAL)

In this sub-section, we consider safety instrumented systems with 1 out of 3 structure and let the failure rates of three units be λ_1, $\lambda_2 = h_2\lambda_1$, $\lambda_3 = h_3\lambda_1$, $0 \le h_3 \le h_2 \le 1$, respectively. At first, we assume that the testing intervals of three units are equal as τ.

Case with simultaneous testing

At first, we consider the simple testing case in which three units are always tested together and then the PFD is obviously

$$PFD = \frac{h_2 h_3 (\lambda_1 \tau)^3}{4} \tag{16.5}$$

because the PFD of SIS with 1 out of n structure in which testing intervals are equal and n units are tested simultaneously is $PFD = (\lambda\tau)^n/(n+1)$ (cases with equal failure rates) (Rausand and Hoyland, 2004).

Case with three different testing times during one testing period

In this case, three units are tested at different time points during one testing period and we need two staggered proportions c_a, c_b. If failure rates of all units in SIS with 1 out of n structure are equal and a uniformly staggered testing method (the starting time points: $0, \tau/n, 2\tau/n, \ldots, (n-1)\tau/n$) is used, then the PFD is given as $PFD = n!(n+3)(\lambda\tau)^n/[4n^n(n+1)]$ (Green(1972)). Thus, the PFD in case with $n = 3$, $PFD = (\lambda\tau)^3/12$.

When the failure rates of units are different (refer to Figure 16.3 as a similar case), the optimal staggered testing scheme to minimize the PFD is $c_a^* = c_b^* = 1/3$ (testing staggered uniformly). In this case, the restriction, $c_a = c_b$ in Section 16.3 is not required and the optimal value of the PFD is (Seo and Yun, 2021)

$$PFD^* = \frac{h_2 h_3 (\lambda_1 \tau)^3}{12}. \tag{16.6}$$

Case with two different testing times during one testing period

For SIS with 1 out of 3 structure, we can consider two different testing time points during one testing period, and then we need to determine one staggered proportion, c_a, and select two units to test together (grouping method). Then the optimal staggered proportion value and PFD are given (refer to Figure 16.2 and Seo and Yun, 2021).

For the case of $\{1+2, 3\}$:

$$c_a^* = \frac{1}{\sqrt{3}}, \qquad PFD^* = \frac{h_2 h_3 (\lambda_1 \tau)^3 (27 - 8\sqrt{3})}{108} \tag{16.7}$$

where $\{1+2, 3\}$ means a grouping method in which unit 1 and 2 are tested together.

For the case of $\{1, 2+3\}$:

$$c_a^* = 1 - \frac{1}{\sqrt{3}}, \qquad PFD^* = \frac{h_2 h_3 (\lambda_1 \tau)^3 (27 - 8\sqrt{3})}{108}. \tag{16.8}$$

When the case of $\{1+3, 2\}$ is considered, c_a^*, PFD^* are same as the case of $\{1+2, 3\}$.

16.3 STAGGERED TESTING STRATEGIES WITH DIFFERENT TESTING INTERVALS

In this section, we consider the SIS system with 1 out of 3 structure in IEC 61508, and assume that three units have different failure rates, and study various grouping cases. In IEC 61508, the cases with the same failure rates are studied but even though several units perform the same function for safety, units of redundant systems, for example, shut-down valves, are operated under different operation conditions and the failure rates can be different (Liu, 2014). Seo and Yun (2021) studied the optimal staggered testing strategies with equal testing intervals. In this paper, we consider the optimal staggered testing strategies with different testing intervals.

In general, the testing intervals of units with low failure rates should be larger than ones of the units with high failure rates among three units. For simplicity of modelling, we assume that the testing intervals of units with low failure rates are integer multipliers of the interval of the unit with the highest failure rate. Additionally, we also consider various grouping cases with two and three different testing intervals for SIS with 1 out of 3 structure.

In this section, let the failure rates of three units be λ_1, $\lambda_2 = h_2\lambda_1$, $\lambda_3 = h_3\lambda_1$, respectively. When we make two groups with three units and assign two different testing intervals (without loss of generality, $h_1 = 1 \geq h_2 \geq h_3$), the case of $(\tau_1, \tau_2) : \tau_3 = 1 : m$ is better than the case of $\tau_1 : (\tau_2, \tau_3) = 1 : m$ in terms of PFD. Figure 16.2 shows the case of $(\tau_1, \tau_2) : \tau_3 = 1 : m$ $(m = 2)$ and the un-availabilities of unit 2, unit 3, and the system are given (the un-availability of unit 1 is equal to Equation 16.3,

$$q_{2k}(t) = \lambda_2[t - (k-1)\tau], \quad (k-1)\tau \leq t < k\tau, \quad k = 1, \ldots, m$$

$$q_3(t) = \begin{cases} q_{31}(t) = \lambda_3(t + \tau - t_a), & 0 \leq t < t_a \\ q_{32}(t) = \lambda_3(t - t_a), & t_a \leq t < m\tau \end{cases}$$

$$q_S(t) = \begin{cases} q_{11}(t)q_{21}(t)q_{31}(T) = h_2h_3\lambda_1^3t^2(t + m\tau - t_a), & 0 \leq t < t_a \\ q_{11}(t)q_{21}(t)q_{32}(T) = h_2h_3\lambda_1^3t^2(t - t_a), & t_a \leq t < \tau \\ q_{1k}(t)q_{2k}(t)q_{32}(T) = h_2h_3\lambda_1^3(t - k\tau)^2(t - t_a), & k\tau \leq t < (k+1)\tau, \\ & k = 1, \ldots, m-1. \end{cases}$$

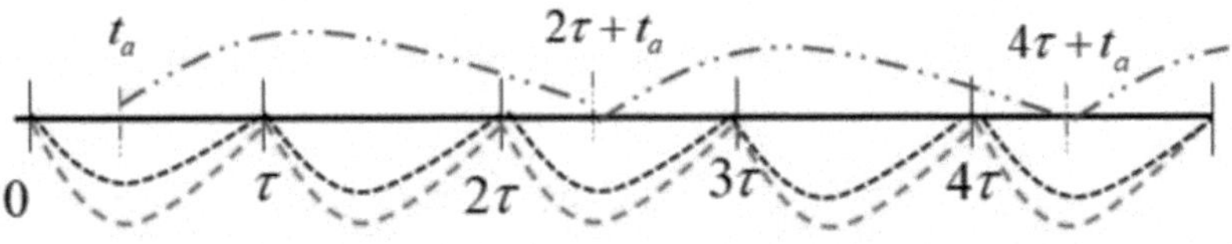

Figure 16.2 Staggered testing scheme structure:2 groups with 1: 2

Let $\tau_1 : (\tau_2, \tau_3) = 1 : m$

$$PFD = \frac{\int_0^{m\tau} q_S(t)\mathrm{d}t}{m\tau} = \frac{h_2 h_3 (\lambda_1 \tau)^3 (4c_a^3 - 4c_a + 2m + 1)}{12}. \tag{16.9}$$

By setting the first derivative of Equation 16.9 to 0, the optimal $c_a^* = 1/\sqrt{3}$ and the optimal value of the PFD is

$$PFD^* = \frac{h_2 h_3 (\lambda_1 \tau)^3 (18m + 9 - 8\sqrt{3})}{108}. \tag{16.10}$$

In the case of $\tau_1 : (\tau_2, \tau_3) = 1 : m$, the optimal $c_a^* = \left(m + 1 - \sqrt{m^2 + 1/3}\right)/2$ and the optimal value is closing to 1/2 from $1 - 1/\sqrt{3}$ as m increases. The PFD is

$$PFD^* = \frac{h_2 h_3 (\lambda_1 \tau)^3 \left[9m^2(m+2) - \sqrt{3}(3m^2+1)^{3/2}\right]}{108}. \tag{16.11}$$

In the case with $m = 1$, Equation 16.11 is equal to Equation 16.10 and eventually it becomes Equation 16.7.

16.3.1 CASES WITH THREE GROUPS AND TWO DIFFERENT TESTING INTERVALS

Case of (1: 1: m)

When we make three groups and assign different testing intervals (without loss of generality, $h_1 = 1 \geq h_2 \geq h_3$), the three testing intervals should satisfy the inequality, $\tau_1 \leq \tau_2 \leq \tau_3$ for minimizing PFD. When we consider two different testing intervals, the case in which $\tau_1 = \tau_2 = \tau$, $\tau_3 = m\tau$, $c_a + c_b < 1$ is investigated in this section. If we apply the different starting testing times to units in the first testing period, we can mitigate and smooth the testing loads over each testing interval. From Figure 16.3 (case with $m = 2$), the un-availabilities of unit 2, unit 3, and the system are given (the un-availability of unit 1 is equal to Equation 16.3a),

$$q_2(t) = \begin{cases} q_{21}(t) = \lambda_2(t + \tau - t_a), & 0 \leq t < t_a \\ q_{22k}(t) = \lambda_2[t - (k-1)\tau - t_a], & (k-1)\tau + t_a \leq t < k\tau + t_a, \\ & k = 1, \ldots, m-1 \\ q_{23}(t) = \lambda_2[t - (m-1)\tau - t_a], & (m-1)\tau + t_a \leq t < m\tau \end{cases} \tag{16.12a}$$

$$q_3(t) = \begin{cases} q_{31}(t) = \lambda_2(t + m\tau - t_a - t_b), & 0 \leq t < t_a + t_b \\ q_{32}(t) = \lambda_3(t - t_a - t_b), & t_a + t_b \leq t < m\tau \end{cases} \tag{16.12b}$$

$$q_S(t) = \begin{cases} q_{11}(t)q_{21}(t)q_{31}(t), & 0 \le t < t_a \\ q_{11}(t)q_{221}(t)q_{31}(t), & t_a \le t < t_a + t_b, \\ q_{11}(t)q_{221}(t)q_{32}(t), & t_a \le t < \tau, \\ q_{1k}(t)q_{22,k-1}(t)q_{32}(t), & (k-1)\tau + t_a \le t < (k-1)\tau + t_a, \\ & k = 2, \ldots, m-1 \\ q_{1k}(t)q_{22k}(t)q_{32}(t), & (k-1)\tau + t_a \le t < k\tau, \\ & k = 2, \ldots, m-1 \\ q_{1m}(t)q_{22,m-1}(t)q_{32}(t), & (m-1)\tau \le t < (m-1)\tau + t_a \\ q_{1m}(t)q_{23}(t)q_{32}(t), & (m-1)\tau \le t < m\tau. \end{cases} \tag{16.12c}$$

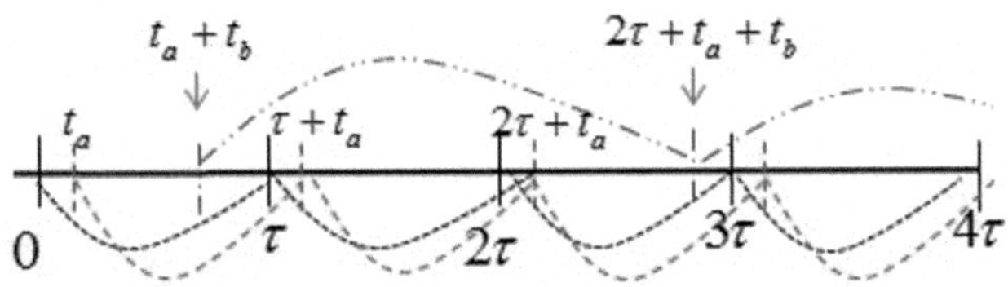

Figure 16.3 Staggered testing scheme: 3 groups with 1 : 1 : 2

It is difficult to find the optimal c_a^*, c_b^* analytically, and let $c_a = c_b$ and we minimize $PDF(m\tau)$. Then, the optimal value of c_a^* is

$$c_a^* = \frac{m+3}{2(m+5)} < \frac{1}{2} \tag{16.13}$$

and the optimal value of PFD is

$$PFD^* = \frac{h_2 h_3 (\lambda_1 \tau)^3 (5m^2 + 26m - 7)}{48(m+5)}. \tag{16.14}$$

In a special case with $m = 1$, the optimal $c_a^* = 1/3$ and the optimal value of PFD is equal to Equation 16.6.

Case of (1: m: m)

In the case with $(1 : m : m)$, un-availabilities of unit 1 and 3 are equal to Equations, 16.3a and 16.12b, respectively. The un-availabilities of unit 2 and the system are given (refer to Figure 16.3 with $m = 2$)

$$q_2(t) = \begin{cases} q_{21}(t) = \lambda_2(t + \tau - t_a), & 0 \le t < t_a \\ q_{22}(t) = \lambda_2(t - t_a), & t_a \le t < m\tau \end{cases}$$

$$q_S(t) = \begin{cases} q_{11}(t)q_{21}(t)q_{31}(t), & 0 \le t < t_a \\ q_{11}(t)q_{22}(t)q_{31}(t), & t_a \le t < t_a + t_b, \\ q_{11}(t)q_{21}(t)q_{32}(t), & t_a + t_b \le t < \tau, \\ q_{1k}(t)q_{21}(t)q_{31}(t), & (k-1)\tau + t_a \le t < k\tau, \quad k = 2, \ldots, m. \end{cases}$$

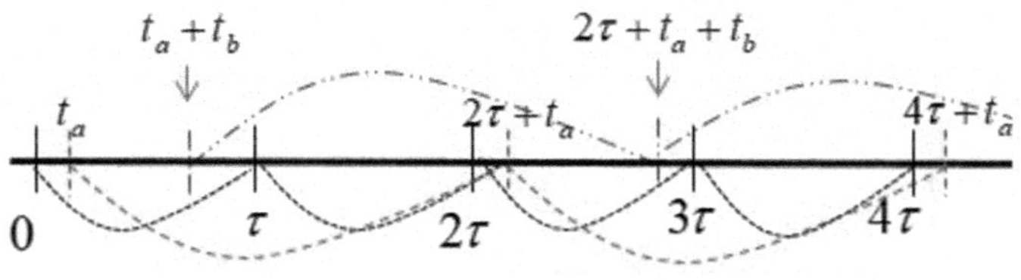

Figure 16.4 Staggered testing scheme: 3 groups with 1 : 2 : 2

It is also difficult to find the optimal c_a^*, c_b^* analytically. Let $c_a = c_b$ and we minimize $PDF(m\tau)$. Then, the optimal value of c_a^* is

$$c_a^* = \frac{3m+1}{4(m+2)}. \tag{16.15}$$

In the case with $m = 2$, $c_a^* = 7/16$ and in the case with $m = 3$, $c_a^* = 1/2$. Thus, as in Figure 16.3, only the case with $m = 2$ guarantees $c_a^* + c_b^* < 1$ and additionally, in the case with $m = 1$ (case with equal testing interval), $c_a^* = 1/3$ from Equation 16.15. In the case with $m = 2$, the optimal value of the PFD is

$$PFD^* = \frac{173h_2h_3(\lambda_1\tau)^3}{384}. \tag{16.16}$$

In the case with $m = 3$, $PFD^* = [9h_2h_3(\lambda_1\tau)^3]/8$ and the optimal value is greater than the optimal value in case with $m = 2$ relatively. The case with $1 : m : m$ reduces the testing numbers but the of PFD is higher than the value in case with $1 : 1 : m$ rule.

16.3.2 CASES WITH THREE DIFFERENT TESTING INTERVALS

In this section, we consider extended cases in which all three units have different testing intervals. When we make three groups and assign different testing intervals (without loss of generality, $h_1 = 1 \ge h_2 \ge h_3$), the case of $\tau_1 \le \tau_2 \le \tau_3$ is good for minimizing the PFD. At first, the case in which $\tau_1 = \tau$, $\tau_2 = 2\tau$, $\tau_3 = 3\tau$, $c_a + b_b < 1$, is studied and then 6τ is the duration of a renewal cycle. From Figure 16.5, we can derive un-availabilities of units and the system is as follows.

$$q_{1k}(t) = \lambda_1[t - (k-1)\tau], \quad (k-1)\tau \le t < k\tau, \quad k = 1, 2, \ldots, 6$$

$$q_2(t) = \begin{cases} q_{21}(t) = \lambda_2(t + 2\tau - t_a), & 0 \le t < t_a \\ q_{22}(t) = \lambda_2(t - t_a), & t_a \le t < 2\tau + t_a \\ q_{23}(t) = \lambda_2(t - 2\tau - t_a), & 2\tau + t_a \le t < 4\tau + t_a \\ q_{24}(t) = \lambda_2(t - 4\tau - t_a), & 4\tau + t_a \le t < 6\tau \end{cases}$$

$$q_3(t) = \begin{cases} q_{31}(t) = \lambda_3(t - 3\tau - t_a), & 0 \le t < t_a + t_b \\ q_{32}(t) = \lambda_2(t - t_a - t_b) & t_a + t_b \le t < 3\tau + t_a + t_b \\ q_{33}(t) = \lambda_2(t - 3\tau - t_a - t_b), & 3\tau + t_a + t_b \le t < 6\tau \end{cases}$$

$$q_S(t) = \begin{cases} q_{11}(t)q_{21}(t)q_{31}(t), & 0 \le t < t_a \\ q_{11}(t)q_{22}(t)q_{31}(t), & t_a \le t < t_a + t_b, \\ q_{11}(t)q_{22}(t)q_{32}(t), & t_a + t_b \le t < \tau, \\ q_{12}(t)q_{22}(t)q_{32}(t), & \tau \le t < 2\tau, \\ q_{13}(t)q_{22}(t)q_{32}(t), & 2\tau \le t < 2\tau + t_a \\ q_{13}(t)q_{23}(t)q_{32}(t), & 2\tau + t_a \le t < 3\tau \\ q_{14}(t)q_{23}(t)q_{32}(t), & 3\tau \le t < 3\tau + t_a + t_b \\ q_{14}(t)q_{23}(t)q_{33}(t), & 3\tau + t_a + t_b \le t < 4\tau \\ q_{15}(t)q_{23}(t)q_{33}(t), & 4\tau \le t < 4\tau + t_a \\ q_{15}(t)q_{24}(t)q_{33}(t), & 4\tau + t_a \le t < 5\tau \\ q_{16}(t)q_{24}(t)q_{33}(t), & 5\tau + t_a \le t < 6\tau \end{cases} \tag{16.17}$$

and

$$PFD = \frac{\int_0^{6\tau} q_S(t)\mathrm{d}t}{6\tau}. \tag{16.18}$$

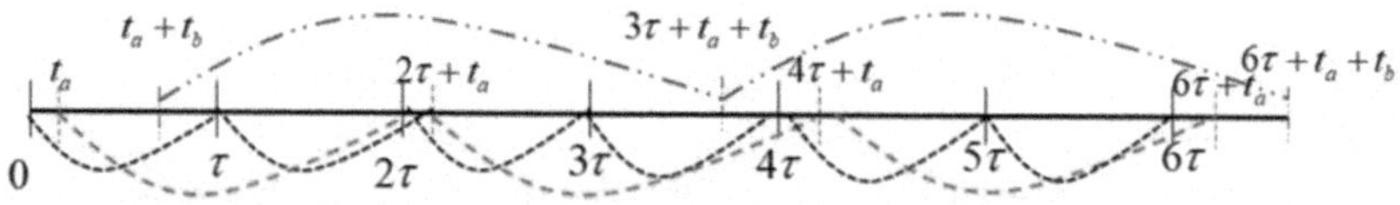

Figure 16.5 Staggered testing scheme: 3 groups with 1 : 2 : 3

Using Equation 16.18, the optimal $c_a^* = 5/18$ under the condition, $c_a = c_b$ and the value of the PFD is

$$PFD = \frac{71}{108} h_2 h_3 (\lambda_1 \tau)^3. \tag{16.19}$$

If we consider another case with $\tau_1 = \tau$, $\tau_2 = 2\tau$, $\tau_3 = 4\tau$, then $c_a^* + c_b^* > 1$ and it is difficult to investigate the general cases in which $\tau_1 < \tau_2 = m_2\tau_1 < \tau_3 = m_3\tau$; $1 < m_2 < m_3$. For the general cases, further studies are required.

16.3.3 COMPARISON BETWEEN DIFFERENT TESTING STRATEGIES

In previous sections, we studied various staggered testing schemes and in this section, we compare the testing methods proposed with different grouping and proportion rules. When the failure rates of units are different, the testing methods with equal testing intervals ($m = 1$) are better to minimize the PFD but it is more reasonable to assign long testing intervals to units with low failure rates because the testing number and costs will be reduced. In this subsection, we compare various staggered testing methods with different numbers of testing intervals and grouping cases for the redundant safety instrumented systems with 1 out of 3 structure. Table 16.1 shows the optimal c_a^*, the relative ratio of PFD to the minimum value, and the average testing numbers for various testing strategies. Thus, Table 16.1 helps us to select the most efficient testing methods. For example, let's compare two methods with $(1:1:1)$ and $(1:1)$ rules. The method with $(1:1:1)$ is used as the basic method for comparison in Table 16.1. The PFD of two methods is $1:1.46$ and the basic one, the method with the $(1:1:1)$ rule is better than the method with the $(1:1)$ rule but the testing numbers of the two methods are $3:2$ and the second one is better than the first one. Thus, we compare the PFD and the testing numbers of various staggered testing strategies and can select the best methods among various methods.

Table 16.1

Comparison of staggered testing schemes with different groups and testing intervals

number of groups	testing intervals	c_a^*	ratio of PFD	average testing number
	$1:1$	$1\sqrt{3} \approx 0.58$	$\frac{27-8\sqrt{3}}{9} \approx 1.460$	2
2: $\{1+2, 3\}$	$1:2$	$1\sqrt{3} \approx 0.58$	$\frac{45-8\sqrt{3}}{9} \approx 3.460$	1.5
	$1:3$	$\frac{1}{\sqrt{3}} \approx 0.58$	$\frac{63-8\sqrt{3}}{9} \approx 5.460$	4/3
	$1:1:1$	$\frac{1}{3}$	1	3
	$1:1:2$	$\frac{5}{14}$	$\frac{65}{28} = 2.321$	2.5
3: $c_a^* = c_b^*$	$1:1:3$	$\frac{3}{8}$	$\frac{29}{8} = 3.625$	$\frac{7}{3}$
	$1:2:2$	$\frac{7}{16}$	$\frac{2076}{384} = 5.406$	2
	$1:2:3$	$\frac{5}{18}$	$\frac{71}{9} = 7.889$	$\frac{11}{6}$

16.4 COST MODELS OF STAGGERED TESTING STRATEGIES

In this section, we propose a cost model for staggered testing problems that includes three cost terms, fixed and variable testing costs and downtime cost. Thus, if we test several units together, the total testing cost is g_a+kg_b in which g_a is the fixed cost and g_b is the variable cost per unit. In order to build the cost model for the staggered testing problems, we use the expected cost rate as an optimization criterion and obtain the expected cost and duration of a renewal cycle. The renewal cycle in staggered testing models is $m'\tau$ and m' is the least common multiple of different testing intervals of units. Additionally, we consider the downtime cost per unit time, g_c and the expected cost rate is given as

$$ECR = \frac{\sum_{j=1}^{l}\left[(g_a + k_j g_b) + g_c\int_{t_{j-1}}^{t_j}(t_j - t)f_{S,j}(t)\mathrm{d}t\right]}{m'\tau} \tag{16.20}$$

where l is the number of tests per cycle, t_i is ith testing time and $t_0 = 0$, $t_l = m'\tau$. $f_{S,j}(t)$ is the probability density function of the system failure time during $[t_{i-1}, t_i]$. Without loss of generality, we can let $g_b = 1$, $r_a = g_a/g_b$, $r_c = g_c/g_a$. Finally, we can find the expected cost rate simplified as follows:

$$ECR = \frac{\sum_{j=1}^{l}\left[(r_a + k_j) + r_c\int_{t_{j-1}}^{t_j} g_{S,j}(t)\mathrm{d}t\right]}{m'\tau} = \frac{r_a l + \sum_{j=1}^{l} k_j}{m'\tau} + r_c PFD \tag{16.21}$$

where $q_{S,j}(t)$ is the system un-availability during $[t_{i-1}, t_i]$ and the following relation is used

$$\int_{t_{j-1}}^{t_j}(t_j - t)f_{S,j}(t)\mathrm{d}t = \int_{t_{j-1}}^{t_j} q_{S,j}(t)\mathrm{d}t - (t_j - t_{j-1})q_{S,j}(t_{j-1}),$$
$$q_{S,j}(t_{j-1}) = 0, \quad j = 1, \ldots, l. \tag{16.22}$$

For example, let's consider the case with two groups ($\{1+2, 3\}$) and $m = 2$, and then the coefficient, $c_a^* = 1/3$. The period of a renewal cycle is $[0, 2\tau]$ and $l = 3$, $t_1 = \tau/\sqrt{3}$, $t_2 = \tau$, $k_1 = 2$, $k_2 = 1$, $k_3 = 2$. The expected cost rate is

$$ECR = \frac{3r_a + 5}{2\tau} + \frac{r_c h_2 h_3 (\lambda_1\tau)^3(45 - 8\sqrt{3})}{108}. \tag{16.23}$$

Thus, we can also obtain the optimal τ as follows.

$$\tau^* = \left[\frac{(270 + 48\sqrt{3})(3r_c + 5)}{611 r_c h_2 h_3 \lambda_1^3}\right]^{1/4}.$$

By a simple example, we want to show how to compare various testing strategies and determine the best one. At first, we consider testing models with equal testing intervals and two or three grouping methods. The values of model parameters, the PFD, and the expected cost rates are as follows;

Case with three groups: $l = 3$, $k_1 = k_2 = k_3 = 1$, $c_a^* = c_b^* = 1/3$,

$$PFD^* = \frac{h_2 h_3 (\lambda_1 \tau)^3}{12}, \qquad ECR_3 = \frac{3r_a + 3}{\tau} + \frac{r_c h_2 h_3 (\lambda_1 \tau)^3}{12}$$

Case with two groups ($\{1+2, 3\}$): $l = 2$, $k_1 = 2$, $k_2 = 1$, $c_a^* = 1/\sqrt{3}$

$$PFD^* = \frac{h_2 h_3 (\lambda_1 \tau)^3 (27 - 8\sqrt{3})}{108}, ECR_2 = \frac{2r_a + 3}{\tau} + \frac{r_c h_2 h_3 (\lambda_1 \tau)^3 (27 - 8\sqrt{3})}{108} \tag{16.24}$$

and the difference between two costs is

$$ECR_3 - ECR_2 = \frac{r_a}{\tau} - \frac{r_c h_2 h_3 (\lambda_1 \tau)^3 (9 - 4\sqrt{3})}{54}.$$

Thus, if

$$r_c h_2 h_3 (\lambda_1)^3 \tau^4 < \frac{54 r_a}{9 - 4\sqrt{3}} \approx 26.06 r_a \tag{16.25}$$

the case with two groups ($\{1+2, 3\}$) is better than the case with three groups.

Numerical examples

In the previous section, we compared various testing strategies with equal testing intervals and now we compare various testing models with different testing intervals. We assume that $h_2 = 1/2$, $h_3 = 1/3$ and the failure rates of three units are different but $\lambda_1 \tau = 0.1$, $r_a = 3$, $r_b = 1$. $h_3 = 1/3$ and m is at most 3. Thus, we consider the following 11 cases and compare their un-availabilities and the expected cost rates.

2 groups($\{1+2, 3\}$) $\rightarrow 1:1,\ \ 1:2,\ \ 1:3$

2 groups($\{1, 2+3\}$) $\rightarrow 1:1,\ \ 1:2,\ \ 1:3$

3 groups($c_a^* + c_b^* < 1, c_a^* = c_b^*$) $\rightarrow 1:1:1,\ \ 1:1:2,\ \ 1:1:3,\ \ 1:2:2,\ \ 1:2:3$

As the basic case, the simultaneous testing method to test units all together is also analyzed. Table 16.2 shows the optimal PFD and the optimal values of the expected cost rate which are dependent on $r_c \lambda_1$. From the PFD point of view, we can know that the PFD of the testing method with three groups and (1 : 1 : 1) is lowest and two methods with two groups and two methods with three groups among 11 cases are better than the simultaneous testing method.

Now let's check the expected cost rates of various models, and at first, some inferior schemes are excluded in comparison, for example, cases with two groups ($\{1+2, 3\} : 1 : 1$) in simultaneous and staggered testing (7 cases with underlines in PFD and refer to Table 16.2). Finally, we have 5 candidates which include the simultaneous testing case and two methods with two groups and (1 : 1) have the same expected cost rate.

The best testing strategy to minimize the expected cost rate among 5 testing methods depends on the values of λ_1, r_c and for example, if $\lambda_1 = 0.01$ and $g_c = 25000g_b$, the testing method with two groups ($\{1+2,3\},\{1,2+3\}$) and 1 : 1 rule is the optimal method to minimize the expected cost rate. The optimal testing interval is 10 and the starting staggered testing is 5.77 or 4.23. In another case with $\lambda_1 = 0.02$, $r_c = 100$, the testing method with two groups($\{1,2+3\}$) and 1 : 3 rule is the optimal one but PFD of the method is highest.

Table 16.2

Summary of a numerical example

Testing schemes			PFD^*	ECR
Simultaneous			0.0000417	$60\lambda_1 + 0.0000417\tau_c$
		1: 1	0.0000203	$90\lambda_1 + 0.0000203\tau_c$
2 groups	$\{1+2,3\}$	1: 2	0.0000481	$70\lambda_1 + 0.0000481\tau_c$
		1: 3	0.0000759	$76.67\lambda_1 + 0.0000759\tau_c$
	$\{1,2+3\}$	1: 1	0.0000203	$90\lambda_1 + 0.0000203\tau_c$
		1: 2	0.0000969	$65\lambda_1 + 0.0000969\tau_c$
		1: 3	0.0002290	$56.67\lambda_1 + 0.0002290\tau_c$
		1: 1: 1	0.0000139	$120\lambda_1 + 0.0000139\tau_c$
		1: 1: 2	0.0000322	$100\lambda_1 + 0.0000322\tau_c$
3 groups		1: 1: 3	0.0000503	$93.33\lambda_1 + 0.0000503\tau_c$
		1: 2: 2	0.0000751	$80\lambda_1 + 0.0000751\tau_c$
		1: 2: 3	0.0001096	$608.3\lambda_1 + 0.0001096\tau_c$

Thus, if the cost terms, failure rates, and parameters of testing intervals are given, then we should consider reasonable values of the coefficient, m, and various testing schemes. Based on the PFD and the expected cost rates, we can determine the best staggered testing method to minimize the expected cost rate.

16.5 CONCLUSIONS

In this paper, we considered staggered testing problems to optimize the testing time points in redundant units of safety instrumented systems. Especially, we studied staggered testing methods to improve the system PFD in which different testing time points are assigned to units. If we assume negligible testing times, the PFD defined in IEC 61508 can be improved by shifting the testing time points during each testing period (staggered testing). In existing papers related to staggered testing, redundant units have the same failure rates, equal testing intervals and same patterns in testing cycles.

Liu (2014) studied the optimality of staggered testing in SIS with 1-out-of-2 structure with different failure rates. Seo and Yun (2021) summarized staggered testing methods for SIS systems with 2 and 3 units. In this paper, we focused on SIS systems with 1 out of 3 structure and staggered testing methods with different testing intervals. The PFD and the expected cost rates of various staggered strategies with different grouping rules are derived and compared. Thus, based on the two optimization criteria, we can select the best staggered strategies. For further studies, we can investigate the effect of repair times to the optimal staggered testing strategies to minimize the PFD and the expected cost rate. Additionally, the complex systems have different safety-related modules, require more complicated checking scenarios for system safety, and more efficient testing strategies should be provided in practice.

ACKNOWLEDGMENT

This research was supported by the Basic Science Research Program through the National Research Foundation of Korea (NRF), which is funded by the Ministry of Education (NRF-2019R1F1A1062353).

REFERENCES

1. Contini, S., Copelli, S., Raboni, M., Torretta, V., Cattaneo, C. S., & Rota, R. (2013). IEC 61508: Effect of Test Policy on the Probability of Failure on Demand of Safety Instrumented Systems. *Chemical Engineering Transactions*, 33, 487-492.

2. Green, A. E., & Bourne, A. J. (1972). *Reliability Technology* , John Wiley & Sons.

3. IEC 61508 (2000). *Functional Safety of Electrical/Electronic/Programmable Electronic (E/E/PE) Safety Related Systems*, first edition. Part 1-7, IEC, Switzerland.

4. IEC 61508 (2010). *Functional Safety of Electrical/Electronic/Programmable Electronic (E/E/PE) Safety Related Systems*, second edition. Part 1-7, IEC, Switzerland.

5. IEC 61511 (2003). IEC 61511 Standard. Functional safety - Safety Instrumented Systems for the Process Industry Sector, Part 1-3, IEC, Switzerland.

6. ISO 26262 (2011). Road Vehicles-Functional Safety, Part 1-10, ISO, Switzerland.

7. Liu, L. (2014). Optimal Staggered Testing Strategies for Heterogeneously Redundant Safety Systems. *Reliability Engineering and System Safety*, 126, 65-71.

8. Liu Y. L., & Rausand, M. (2013). Reliability Effects of Testing Strategies on Safety- Instrumented Systems in Different Demand Modes. *Reliability Engineering and System Safety*, 119, 235-243.

9. Rausand, M., & Hoyland, A. (2004). *System Reliability: Models, Statistical Methods and Applications*, second edition. Wiley.

10. Rouvroye, J. L., & Wiegerinck, J. A. M. (2006). Minimizing Costs while Meeting Safety Requirements: Modeling Deterministic (Imperfect) Staggered Tests Using Standard Markov Models for SIL Calculations, *ISA Transactions*, 45, 611-621.

11. Seo, S. K., & Yun, W. Y. (2021). Optimum Staggered Testing Strategy for 1- and 2-out-of-3 Redundant Safety Instrumented Systems. In Ram, M., & Pham, H. Eds., *Safety and Reliability Modeling and its Applications*, Elsevier, 227-243.

12. Vaurio, J. K. (1980). Availability of Redundant Safety Systems with Common-Mode and Undetected Failures. *Nuclear Engineering and Design* , 58, 415-424.

13. Vaurio, J. K. (2011). Un-availability Equations of k-out-of-n Systems. *Reliability Engineering and System Safety*, 96, 350-352.

17 Modules of Multi-State Systems: Introduction to the Three Modules Theorem

Fumio Ohi
Nagoya Institute of Technology, Japan

CONTENTS

17.1 INTRODUCTION

Practical systems are usually created by hierarchically organizing modules (sub-systems), not by directly connecting elements. In this sense, structural and probabilistic examinations via modules of a system are crucial subjects.

Birnbaum and Esary [3] examined a module of a binary-state system, gave a characterization of it and proved the Three Modules Theorem. The characterization is generalized to the concept of "representative" which is defined in the subsection 17.4.1. The theorem was also proved set-theoretically later by Butterworth [5]. Furthermore, for stochastically independent components, stochastic bounds via a modular decomposition for the reliability of a binary-state system were presented by using the minimal cut and path sets by Bodin [4]. These stochastic bounds are extended to a multi-state system with totally and partially ordered state spaces by Ohi [8, 9], respectively, where the normal property is crucially assumed for structure functions.

For an extension of the Three Modules Theorem to a multi-state case, there are some examinations by Ohi and Nishida [7] and Shinmori, Ohi, Hagihara and Nishida [10]. However, the works are not necessarily satisfying. The theorem was shown to partly hold in the situation of Ohi and Nishida [7]. Shinmori et al. [10], restricting the class of multi-state systems to be EBW (extended Barlow and Wu), proved the theorem under somewhat strong structural conditions.

In this manuscript, summarizing these works about modules in our ordered set theoretical framework, we explore what kinds of conditions are necessary for the Three Modules Theorem to hold for multi-state systems.

In Section 17.2, we give an ordered set theoretical basis for multi-state systems, referring Ohi and Nishida [7] and Ohi [8, 9]. For example, f-order and f-equivalence relations are defined and used for examinations of modules in Section 17.4.

In Section 17.3, we show a definition of a multi-state system and some terminologies, where the state spaces of the system and the components are assumed to be finite totally ordered sets, which may be extended to the case of partially ordered state spaces. Refer to Ohi [9], where the the Three Modules Theorem is not examined.

In Section 17.4, introducing the concept of a representative, a definition and an alternative equivalent one of a module are given for a multi-state system, where a relation among minimal (maximal) state vectors of modules and the system is shown.

In Section 17.5, we introduce a hierarchy of multi-state systems. The ideas of an m-homogeneous system, a Barlow-Wu system (Barlow and Wu [2]), an EBW (extended Barlow-Wu) system and an EEBW (extended EBW) system are presented. For the latter two ideas, refer to Shinmori et al. [10, 11]. Furthermore, we show the Three Modules Theorem for a binary-state system proved by Birnbaum and Esary [3].

In Sections 17.6 and 17.7, in our ordered set theoretical context, we show Shinmori et al. [10]'s examinations about modules of an EEBW systems and the Three Modules Theorem for an EBW system, which seems to be closest to the binary case at the present time. Furthermore, with the work of Ohi and Nishida [7], we give a perspective for the Three Modules Theorem of a multi-state system and suggest that the normal property is necessary for the theorem. However, the situation seems to be complicated.

We note that the ordered sets in this manuscript are assumed to be finite.

17.2 ORDERED SET THEORETICAL PRELIMINARIES

Here, we present general ordered set theoretical examinations of the basic concepts of reliability theory.

For an ordered set $(X, \leqq)$ and a subset $A \subseteqq X$, $MI(A)$ and $MA(A)$ denote respectively the sets of all the minimal and all the maximal elements of A. For elements $a,\ b \in X$, $a < b$ denotes $a \leqq b$ and $a \neq b$. For intervals, we use terminologies, $[a, \rightarrow) = \{x \mid a \leqq x,\ x \in X\}$ and $(\leftarrow, a] = \{x \mid x \leqq a,\ x \in X\}$.

Definition 17.2.1 (Definition of Normal Property) *Generally an increasing mapping f from an ordered set $(X, \leqq)$ to an ordered set $(Y, \leqq)$ is called normal, when for every $y \in Y$,*

$$MI\left(f^{-1}(y)\right) = MI\left(f^{-1}[y, \rightarrow)\right), \tag{17.1}$$

$$MA\left(f^{-1}(y)\right) = MA\left(f^{-1}(\leftarrow, y]\right) \tag{17.2}$$

hold. Shinmori et al. [10] call (17.1) minimally coincident, and similarly (17.2) maximally coincident.

17.2.1 COMPOSITE FUNCTION

Theorem 17.2.1 *Suppose $(X, \leqq)$, $(Y, \leqq)$ and $(Z, \leqq)$ are ordered sets and $f : X \rightarrow Z$, $g : X \rightarrow Y$, $h : Y \rightarrow Z$ are increasing mappings satisfying $f(x) = h(g(x))$ for every $x \in X$, i.e., f is the composite of h and g. Then for every $a \in Z$, we have*

$$MI\left(f^{-1}[a, \rightarrow)\right) \subseteqq \bigcup_{y \in MI(h^{-1}[a, \rightarrow))} MI\left(g^{-1}[y, \rightarrow)\right), \tag{17.3}$$

$$MI\left(f^{-1}(a)\right) \supseteqq \bigcup_{y \in MI(h^{-1}(a))} MI\left(g^{-1}(y)\right), \tag{17.4}$$

$$MA\left(f^{-1}(\leftarrow, a]\right) \subseteqq \bigcup_{y \in MA(h^{-1}(\leftarrow, a])} MA\left(g^{-1}(\leftarrow, y]\right), \tag{17.5}$$

$$MA\left(f^{-1}(a)\right) \supseteqq \bigcup_{y \in MA(h^{-1}(a))} MA\left(g^{-1}(y)\right). \tag{17.6}$$

Furthermore, if f, g and h are normal, then the equalities in place of the above inclusion relations apparently hold.

Proof *We only prove the inclusion relations (17.3) and (17.4), and the others are similarly proved by dual examinations.*

the left-hand side of (17.3) $\subseteqq$ the right-hand side of (17.3)

Suppose x is an element of the left-hand side of (17.3), so to say,

$$x \in MI\left(f^{-1}[a, \rightarrow)\right). \tag{17.7}$$

Since f is the composite function of g and h, we have $f(x) = h(g(x)) \geqq a$ and then the following chain holds.

$$g(x) \in h^{-1}[a, \rightarrow) \quad \Rightarrow \quad \exists b \in MI\left(h^{-1}[a, \rightarrow)\right),\ b \leqq g(x) \quad \Rightarrow \quad x \in g^{-1}[b, \rightarrow)$$

$$\Rightarrow \quad \exists c \in MI\left(g^{-1}[b,\rightarrow)\right), \; c \leqq x \, . \tag{17.8}$$

On the other hand, from $g(c) \geqq b$*, we have* $f(c) = h(g(c)) \geqq h(b) \geqq a$*. Then*

$$c \in f^{-1}[a,\rightarrow). \tag{17.9}$$

Therefore, (17.7), (17.8) and (17.9) imply $x = c$*, and hence* x *is an element of the right-hand side of (17.3).*

the left-hand side of (17.4) $\supseteqq$ *the right-hand side of (17.4)*

Suppose x *is an element of the right-hand side of (17.4), so to say,*

$$\exists b \in MI\left(h^{-1}(a)\right), \quad x \in MI\left(g^{-1}(b)\right). \tag{17.10}$$

Apparently, the following equalities hold.

$$g(x) = b, \quad f(x) = h(g(x)) = h(b) = a.$$

Then we have,

$$\exists c \in MI\left(f^{-1}(a)\right), \quad x \geqq c, \quad f(c) = h(g(c)) = a.$$

Hence, from the above,

$$\begin{aligned} b = g(x) \geqq g(c), \;\; g(c) \in h^{-1}(a), \;\; b \in MI\left(h^{-1}(a)\right) \\ \Longrightarrow \;\; b = g(c), \;\; \textit{equivalently}, \;\; c \in g^{-1}(b), \end{aligned}$$

on the other hand

$$x \geqq c, \quad x \in MI\left(g^{-1}(b)\right),$$

and then $x = c$ *holds, i.e.,* x *is an element of the left-hand side of (17.4), recalling that* $c \in MI\left(f^{-1}(a)\right)$*.*

When the mappings are normal, for example, $MI\left(f^{-1}[a,\rightarrow)\right) = MI\left(f^{-1}(a)\right)$ *holds, and then equalities hold in (17.3)* $\sim$ *(17.6).*

Theorem 17.2.2 *Under the conditions of Theorem 17.2.1, we have for every* $a \in Z$*,*

$$MI\left(h^{-1}(a)\right) \subseteqq \left\{g(x) \;:\; x \in MI\left(f^{-1}(a)\right)\right\}, \tag{17.11}$$

$$MI\left(h^{-1}[a,\rightarrow)\right) \subseteqq \left\{g(x) \;:\; x \in MI\left(f^{-1}[a,\rightarrow)\right)\right\}. \tag{17.12}$$

Proof *Proof of (17.11)* *For* $a \in Z$*, supposing* $y \in MI\left(h^{-1}(a)\right)$*, we have from Theorem 17.2.1 (17.4)*

$$MI\left(g^{-1}(y)\right) \subseteqq MI\left(f^{-1}(a)\right).$$

Then, for $x \in MI\left(g^{-1}(y)\right)$*,* $g(x) = y$*,* $x \in MI\left(f^{-1}(a)\right)$ *and (17.11) hold.*

Proof of (17.12) *Suppose* $a \in Z$ *and* $y \in MI\left(h^{-1}[a,\rightarrow)\right)$. *For* $x \in MI\left(g^{-1}(y)\right)$, *we have* $g(x) = y$, $f(x) = h(g(x)) \geqq a$ *and* $x \in f^{-1}[a,\rightarrow)$, *and then* $\exists w \in MI\left(f^{-1}[a,\rightarrow)\right)$, $w \leqq x$ *holds. Hence*

$$\begin{aligned} g(w) &\leqq g(x) = y \in MI\left(h^{-1}[a,\rightarrow)\right), \\ h(g(w)) &= f(w) \geqq a \Longrightarrow g(w) \in h^{-1}[a,\rightarrow), \end{aligned}$$

from which $g(w) = y$ *holds, since* y *is a minimal element of* $h^{-1}[a,\rightarrow)$. *The above examination is summarised as*

$$\forall y \in MI\left(h^{-1}[a,\rightarrow)\right),\ \exists w \in MI\left(f^{-1}[a,\rightarrow)\right),\ g(w) = y,$$

which shows (17.12).

Theorem 17.2.3 *(i)* f *: normal* $\Longrightarrow h$ *: normal.* *(ii)* g, h *: normal* $\Longrightarrow f$ *: normal.*

Proof *Proof of (i)* *For* $y \in MI\left(h^{-1}[a,\rightarrow)\right)$, *from (17.12), we have*

$$\exists x \in MI\left(f^{-1}[a,\rightarrow)\right),\ g(x) = y.$$

Since f *is normal,* $f(x) = a$ *holds. On the other hand, since* f *is a composite of* g *and* h, $f(x) = h(g(x)) = h(y) = a$.

Proof of (ii) *Since* g *and* h *are normal, we have* $MI\left(h^{-1}[a,\rightarrow)\right) = MI\left(h^{-1}(a)\right)$ $(a \in Z)$ *and* $MI\left(g^{-1}[b,\rightarrow)\right) = MI\left(g^{-1}(b)\right)$ $(b \in Y)$. *Then (17.3) may be rewritten as*

$$MI\left(f^{-1}[a,\rightarrow)\right) \subseteqq \bigcup_{y \in MI(h^{-1}(a))} MI\left(g^{-1}(y)\right).$$

Hence, for $x \in MI\left(f^{-1}[a,\rightarrow)\right)$,

$$\exists b \in MI\left(h^{-1}(a)\right), \quad x \in MI\left(g^{-1}(b)\right).$$

Finally, we have $f(x) = h(g(x)) = h(b) = a$, *which denotes that* f *is normal.*

17.2.2 PRODUCT ORDERED SET

In this subsection, we give some examinations about minimal and maximal elements of some inverse images with respect to increasing mappings defined on a product ordered set.

First, we suppose $(X_1, \leqq)$, $(X_2, \leqq)$ and $(Z, \leqq)$ to be finite ordered sets and f to be an increasing mapping from the product ordered set $X_1 \times X_2$ to Z.

Definition 17.2.2 (Definition of f-order and f-equivalence relation) *For $a, b \in X_1$, f-order ($\overset{f}{\leqq}$) and f-equivalence relation ($\overset{f}{=}$) are defined as the following.*

$$\text{(i) } a \overset{f}{\leqq} b \iff \forall x_2 \in X_2, \quad f(a, x_2) \leqq f(b, x_2).$$
$$\text{(ii) } a \overset{f}{=} b \iff \forall x_2 \in X_2, \quad f(a, x_2) = f(b, x_2).$$

We notice that the relation $\overset{f}{\leqq}$ is a pseudo order on X_1 and is different from the order $\leqq$ defined previously on X_1.

Writing the quotient space of X_1 with respect to the equivalence relation $\overset{f}{=}$ as $\widetilde{X}_1$, we define an order $\overset{f}{\leqq}$ between every equivalent classes Q and R of $\widetilde{X}_1$ as

$$Q \overset{f}{\leqq} R \iff q \in Q, \quad r \in R, \quad q \overset{f}{\leqq} r,$$

of which definition does not depend on the choice of $q \in Q$ and $r \in R$. $(\widetilde{X}_1, \overset{f}{\leqq})$ is easily proved to be an ordered set. We may have a similar discussion on X_2.

We define an increasing mapping $\widetilde{f} : \widetilde{X}_1 \times \widetilde{X}_2 \to Z$ on the product ordered set $(\widetilde{X}_1, \overset{f}{\leqq}) \times (\widetilde{X}_2, \overset{f}{\leqq})$ as

$$Q_1 \in \widetilde{X}_1, \quad Q_2 \in \widetilde{X}_2, \quad \widetilde{f}(Q_1, Q_2) = f(q_1, q_2), \quad q_1 \in Q_1, \quad q_2 \in Q_2,$$

of which definition does not depend on the choice of $q_1 \in Q_1$ and $q_2 \in Q_2$.

We may consider f and $\widetilde{f}$ to be equivalent in the context of reliability theory from what the f-equivalence denotes.

Definition 17.2.3 (Definition of Relevant Property) *X_1 is called f-relevant, when every two different elements of X_1 are not f-equivalent.*

$\widetilde{X}_1$ is apparently $\widetilde{f}$-relevant, which denotes that even if $(X_1, \leqq)$ and $(X_2, \leqq)$ are not f-relevant, we may equivalently transform them to the relevant ones by considering $(\widetilde{X}_1, \overset{f}{\leqq})$, $(\widetilde{X}_2, \overset{f}{\leqq})$ and $\widetilde{f}$.

Theorem 17.2.4 *When X_1 is a totally ordered set, X_1 is f-relevant if and only if the following condition holds.*

$$\forall a \in X_1 \text{ such that } a \neq MI(X_1), \ \exists (x_1, x_2) \in \bigcup_{z \in Z} MI\left(f^{-1}(z)\right), \quad x_1 = a.$$

Proof *When the condition is satisfied, a and the predecessor of a, $a-1$, are apparently not f-equivalent, and then X_1 is f-relevant. If $a \in X_1$ does not satisfy the condition, we have $a \overset{f}{=} a-1$ which denotes that X_1 is not f-relevant.*

Additionally we assume finite ordered sets $(Y_1, \leqq)$, $(Y_2, \leqq)$ and increasing mappings $g_i : X_i \to Y_i$ $(i = 1, 2)$ and $h : Y_1 \times Y_2 \to Z$, which satisfy

$$(x_1, x_2) \in X_1 \times X_2, \quad f(x_1, x_2) = h\left(g_1(x_1), g_2(x_2)\right),$$

where we may consider (g_1, g_2) to be a mapping from $Y_1 \times Y_2$ to Z, and then the mapping f is the composite of (g_1, g_2) and h.

Theorem 17.2.5 *(i) For minimal elements of $g_1^{-1}(b)$ $(b \in Y_1)$, we have the following inclusion relation.*

$$\forall b \in Y_1, \quad MI\left(g_1^{-1}(b)\right) \supseteqq \bigcup_{a \in Z} \left\{x_1 : g_1(x_1) = b, \ (x_1, x_2) \in MI\left(f^{-1}(a)\right)\right\}.$$

(ii) When Y_1 is a totally ordered set and h-relevant, the inverse inclusion relation in the above holds.

Proof *(i) Supposing x_1 to be an element of the right-hand side, we have*

$$\exists a \in Z, \ \exists x_2 \in Y_2 \text{ such that } (x_1, x_2) \in MI\left(f^{-1}(a)\right),$$
$$g(x_1) = b, \quad f(x_1, x_2) = h(g_1(x_1), g_2(x_2)) = h(b, g_2(x_2)) = a,$$

from which the inclusion relation, $x_1 \in MI\left(g_1^{-1}(b)\right)$, clearly holds.

(ii) Suppose Y_1 is h-relevant. From Theorem 17.2.4, for $b \in Y_1$, we have

$$\exists a \in Z, \ \exists y_2 \in G_2, \ (b, y_2) \in MI\left(h^{-1}(a)\right).$$

Furthermore, for $x_1 \in MI\left(g_1^{-1}(b)\right)$ and $x_2 \in MI\left(g_2^{-1}(y_2)\right)$, we have

$$f(x_1, x_2) = h(g_1(x_1), g_2(x_2)) = h(b, y_2) = a.$$

On the other hand, if $(x_1', x_2') \leqq (x_1, x_2)$ and $f(x_1', x_2') = a$, then

$$f(x_1', x_2') = h(g_1(x_1'), g_2(x_2')) = a \implies (g_1(x_1'), g_2(x_2')) \in h^{-1}(a),$$
$$(g_1(x_1'), g_2(x_2')) \leqq (b, y_2), \quad (b, y_2) \in MI\left(h^{-1}(a)\right).$$

From the minimality of (b, y_2),

$$(g_1(x_1'), g_2(x_2')) = (b, y_2) \implies x_1' \in g_1^{-1}(b), \ x_2' \in g_2^{-1}(y_2).$$

Then, from $x_1 \in MI\left(g_1^{-1}(b)\right)$, $x_2 \in MI\left(g_2^{-1}(y_2)\right)$ and $(x_1', x_2') \leqq (x_1, x_2)$, we have $(x_1, x_2) = (x_1', x_2')$, and therefore (x_1, x_2) is a minimal element of $f^{-1}(a)$.

17.3 BASIC CONCEPTS

The concept of a multi-state system is defined as the following Definition 17.3.1, referring to Ohi and Nishida [7].

Definition 17.3.1 (Definition of a System) *A multi-state system is a triplet (Ω_C, S, φ), simply called a system φ, satisfying the following conditions.*

(1) C is a non-empty finite set denoting the set of components, usually assumed to be $\{1, 2, \cdots n\}$, where $i \in C$ is the index of the component and n also denotes the number of components.

(2) Ω_i $(i \in C)$ is a finite totally ordered non-empty set, denoting the state space of the component i. We may assume $\Omega_i = \{0, 1, \cdots N_i\}$, where 0 and N_i denote the failure and the normal states, respectively. The order defined on Ω_i is $0 < 1 < \cdots < N_i$. We assume $N_i \geqq 1$ to avoid a tedious case.

Similarly, S is a non-empty finite totally ordered set, denoting the state space of the system.

(3) $\Omega_C = \prod_{i \in C} \Omega_i$ is the product ordered set of Ω_i $(i \in C)$, i.e., $\Omega_C = \{(x_1, \cdots, x_n) \mid x_i \in \Omega_i \ (i \in C)\}$.

(4) φ, called a structure function, is an increasing surjection from Ω_C to S. The system is simply called a system φ without any confusion.

When all the state spaces of a system are binary, i.e., $\{0, 1\}$, the system is called a binary-state system. A system which is not a binary-state system is usually called a multi-state system, however, in this manuscript, we call uniformly a system and when necessary a binary-state or multi-state system.

We use commonly the symbol $\leqq$ for the orders considered in the sequel. Then for $\mathbf{x} = (x_1, \cdots, x_n)$ and $\mathbf{y} = (y_1, \cdots y_n) \in \Omega_C$, $\mathbf{x} \leqq \mathbf{y}$ denotes $x_i \leqq y_i$ $(i \in C)$, and $\mathbf{x} < \mathbf{y}$ denotes $x_i \leqq y_i$ $(\forall i \in C)$ and $x_j \neq y_j$ $(\exists j \in C)$. Since the structure function $\varphi : \Omega_C \to S$ is increasing, $\varphi(\mathbf{x}) \leqq \varphi(\mathbf{y})$ for $\mathbf{x}$ and $\mathbf{y} \in \Omega_C$ such that $\mathbf{x} \leqq \mathbf{y}$. For the examinations of modules, we define exceptionally an order $\overset{\varphi}{\leqq}$.

For a system (Ω_C, S, φ), by the increasing property of the structure function, we have easily the following equalities for every $s \in S$.

$$\varphi^{-1}[s, \to) = \{\mathbf{x} \in \Omega_C \mid \varphi(\mathbf{x}) \geqq s\} = \bigcup_{\mathbf{m} \in MI(\varphi^{-1}[s,\to))} [\mathbf{m}, \to), \tag{17.13}$$

$$\varphi^{-1}(\leftarrow, s] = \{\mathbf{x} \in \Omega_C \mid \varphi(\mathbf{x}) \leqq s\} = \bigcup_{\mathbf{M} \in MA(\varphi^{-1}(\leftarrow,s])} (\leftarrow, \mathbf{M}]. \tag{17.14}$$

When the system is a binary-state system, $MI\left(\varphi^{-1}[1, \to)\right) = MI\left(\varphi^{-1}(1)\right)$ and $MA\left(\varphi^{-1}(\leftarrow, 0]\right) = MA\left(\varphi^{-1}(0)\right)$ are the sets of the well-known minimal path and minimal cut vectors, respectively.

Generally, a system is not necessarily normal. However, a binary-state system is clearly normal. Normal property is used in Ohi [8, 9] to examine stochastic bounds for a stochastic performance via a modular decomposition, where a practical meaning of the normal property is also presented.

For $A \subseteqq C$, we write $\Omega_A = \prod_{i \in A} \Omega_i$ for the product ordered set of Ω_i ($i \in A$), and an element of Ω_A as $\mathbf{x}_A$, which is, dropping off A, simply written as $\mathbf{x}$, when there is no confusion. Following this rule, an element of Ω_i is written as $x_{\{i\}}$, which is, however, written as x_i simply. An element of Ω_A is called a state vector denoting a combination of states of the components of A.

For $A \subseteqq B \subseteqq C$ and $\mathbf{x}_B \in \Omega_B$, $\mathbf{x}_A$ is the projection of $\mathbf{x}_B$ to Ω_A. Precisely we should write as $\mathbf{x}_{\{B\}\{A\}}$, however, the above style causes no confusion. For example, when $B = \{1, 3, 4, 5, 6\}$ and $A = \{1, 3, 6\}$, for $\mathbf{x}_B = (x_1, x_3, x_4, x_5, x_6) \in \Omega_B$, we have $\mathbf{x}_A = (x_1, x_3, x_6)$, i.e., which is a state vector consisting of the states of the components of A in $\mathbf{x}_B$.

For $B \subseteqq C$, suppose $\{A_1, \cdots, A_m\}$ is a finite non-empty partition of B, i.e., $A_k \neq \phi$ $(1 \leqq k \leqq m)$, $A_k \cap A_l = \phi$ $(k \neq l)$, $\cup_{k=1}^m A_k = B$. $(\mathbf{x}_{A_1}, \cdots, \mathbf{x}_{A_m})$ is an element of Ω_B, of which projection to Ω_{A_k} is $\mathbf{x}_{A_k}$.

Following Definition 17.2.3, the component i is said to be relevant, if the state space Ω_i is φ-relevant. When all the components are relevant, the system is called coherent. If the component i is not relevant, taking the quotient space $\widetilde{\Omega}_i$ of Ω_i with respect to φ-equivalent relation, we may have a relevant system considered to be equivalent to the system φ. Then, the relevant condition is not restricting. For a binary-state system, the above relevant property is the well-known relevant property in Barlow and Proschan [1].

17.4 A MODULE OF A SYSTEM

In this section, referring to Ohi and Nishida [7], and Ohi [8, 9], we present basic concepts relating to a module of a multi-state system.

17.4.1 DEFINITION AND BASIC PROPERTIES

Definition 17.4.1 (Definition of a Module) *For a system (Ω_C, S, φ), a finite partition $\{A_1, \cdots, A_m\}$ of C is called a modular decomposition of the system, when there exist systems $(\Omega_{A_i}, S_i, \chi_i)$ $(i = 1, \cdots, m)$ and $(\prod_{i=1}^m S_i, \psi, S)$ such that*

$$\forall \mathbf{x} \in \Omega_C,\ \varphi(\mathbf{x}) = \psi\left(\chi_1\left(\mathbf{x}_{A_1}\right), \cdots, \chi_m\left(\mathbf{x}_{A_m}\right)\right). \tag{17.15}$$

Each component is trivially a kind of system with the identity mapping on the state space of the component. A subset A of C is called a module, when $\{A, \{k\} : k \in C \backslash A\}$ is a modular decomposition and in this case we may write (17.15) as

$$\forall \mathbf{x} \in \Omega_C,\ \varphi(\mathbf{x}) = \psi\left(\chi\left(\mathbf{x}_A\right), \mathbf{x}_{C \backslash A}\right), \tag{17.16}$$

where χ and ψ are structure functions of systems (Ω_A, S_A, χ) and $(S_A \times \Omega_{C \backslash A}, S, \psi)$, respectively.

When the systems χ_i $(i = 1, \cdots, m)$ and ψ are relevant (normal), the modular decomposition is called a relevant (normal) modular decomposition.

Suppose $A \subsetneqq C$ is a module of a system (Ω_C, S, φ). Since S_A is a totally ordered set, for every $\mathbf{x},\ \mathbf{y} \in \Omega_A$, $\mathbf{x} \overset{\varphi}{\leqq} \mathbf{y}$ or $\mathbf{y} \overset{\varphi}{\leqq} \mathbf{x}$ holds. Then, $(\Omega_A, \overset{\varphi}{\leqq})$ is a pseudo totally ordered set. On the other hand, if $(\Omega_A, \overset{\varphi}{\leqq})$ is a pseudo totally ordered set, then A is a module of the system φ, i.e., S_A, χ and ψ satisfying (17.16) are easily constructed. Then we have the following theorem about a necessary and sufficient condition for $A \subsetneqq C$ to be a module.

Theorem 17.4.1 *$A \subsetneqq C$ is a module of a system (Ω_C, S, φ) if and only if $(\Omega_A, \overset{\varphi}{\leqq})$ is a pseudo totally ordered set, or equivalently $(\widetilde{\Omega}_A, \overset{\varphi}{\leqq})$ is a totally ordered set.*

For a totally ordered set $(\widetilde{\Omega}_A, \overset{\varphi}{\leqq})$, supposing

$$\widetilde{\Omega}_A = \{Q_1, Q_2, \cdots, Q_q\}, \quad Q_1 \overset{\varphi}{<} Q_2 \overset{\varphi}{<} \cdots \overset{\varphi}{<} Q_q,$$

and $\mathbf{a}_i \in Q_i$ $(i = 1, 2, \cdots, q)$ to be a representative of Q_i, we have

$$(i)\ \mathbf{a}_1 \overset{\varphi}{<} \mathbf{a}_2 \overset{\varphi}{<} \cdots \overset{\varphi}{<} \mathbf{a}_q, \qquad (ii)\ \forall \mathbf{x} \in \Omega_A,\ \exists \mathbf{a}_j,\ \ \mathbf{x} \overset{\varphi}{=} \mathbf{a}_j. \tag{17.17}$$

On the contrary, if there exist $\mathbf{a}_1, \cdots, \mathbf{a}_q \in \Omega_A$ satisfying the two relations of (17.17), then we have the quotient space $\widetilde{\Omega}_A$ with respect to $\overset{\varphi}{=}$, where $Q_j = \{\mathbf{x}_A | \mathbf{x}_A \overset{\varphi}{=} \mathbf{a}_j\}$, and the ordered set $(\widetilde{\Omega}_A, \overset{\varphi}{\leqq})$ is a totally ordered set, i.e., A is a module. Such a set of state vectors $\{\mathbf{a}_1, \cdots, \mathbf{a}_q\}$ is called a representative of A in this thesis. Then we have alternatively a necessary and sufficient condition for $A \subsetneqq C$ to be a module.

Theorem 17.4.2 *$A \subsetneqq C$ is a module if and only if there exists a representative of A.*

Theorem 17.4.3 *Suppose (Ω_C, S, φ) is a normal system and has a normal modular decomposition $\{A_1, \cdots, A_m\}$. Then we have the following equalities. For $s \in S$,*

$$MI\left(\varphi^{-1}[s, \rightarrow)\right) = \bigcup_{(t_1,\ \cdots,\ t_m) \in MI(\psi^{-1}[s, \rightarrow))} \prod_{i=1}^{m} MI\left(\chi^{-1}[t_i, \rightarrow)\right), \tag{17.18}$$

$$MA\left(\varphi^{-1}(\leftarrow, s]\right) = \bigcup_{(t_1,\ \cdots,\ t_m) \in MA(\psi^{-1}(\leftarrow, s]))} \prod_{i=1}^{m} MA\left(\chi^{-1}(\leftarrow, t_i]\right). \tag{17.19}$$

Proof *From the assumption that χ_i $(i = 1, \cdots, m)$ are normal, it is clear that $(\chi_1, \cdots, \chi_m) : \prod_{i=1}^{m} \Omega_{A_i} \rightarrow \prod_{i=1}^{m} S_i$ is normal. And then, this theorem holds by Theorem 17.2.1.*

In our ordered set theoretical context, the proof of Theorem 17.4.3 is simpler than Ohi [8].

For a module $A \subseteqq C$ of a system (Ω_C, S, φ) we have the following theorem for relevant properties of φ, χ and ψ, of which proof is easy and then omitted.

Theorem 17.4.4 (Consistency of Relevance among φ, χ and ψ) *(i) φ : relevant $\Longrightarrow \chi$: relevant. (ii) χ and ψ : relevant $\Longrightarrow \varphi$: relevant.*

17.5 HIERARCHY OF MULTI-STATE SYSTEMS

In this section we classify the systems in the sense of Definition 17.3.1 into several hierarchical classes.

17.5.1 HOMOGENEOUS SYSTEM

Definition 17.5.1 *A system (Ω_C, S, φ) is called an m-homogeneous system, when the following conditions are satisfied.*

(i) The cardinal numbers of the state spaces are the same, i.e., the state spaces may be assumed to be

$$\Omega_i = S = \{0, 1, \cdots, m\}, \ i \in C.$$

(ii) When the levels of the components' states are the same s, then so is the state of the system, i.e.,

$$0 \leqq \forall s \leqq m, \ \varphi(s, \cdots, s) = s.$$

For $A \subseteqq C$, when the states of the components of A are the same state s, the state vector $\underbrace{(s, \cdots, s)}_{|A|}$ is written as $\mathbf{s}_A$, where $|A|$ is the cardinal number of the set A.

Definition 17.5.2 *A modular decomposition $\{A_1, \cdots, A_k\}$ of a system (Ω_C, S, φ) is called m-homogeneous when the systems $(\Omega_{A_i}, S_i, \chi_i)$ $(i = 1, \cdots, k)$ and $\left(\prod_{i=1}^k S_i, S, \psi\right)$ are m-homogeneous. In this case the system φ is necessarily m-homogeneous.*

$A \subseteqq C$ is called an m-homogeneous module of the system φ, when A is an element of an m-modular decomposition.

The next theorem is about a representative of an m-homogeneous module and is clear.

Theorem 17.5.1 *For a system (Ω_C, S, φ), $A \subseteqq C$ is an m-homogeneous module if and only if $\{\mathbf{0}_A, \mathbf{1}_A, \cdots, \mathbf{m}_A\}$ is the representative of A.*

From Theorem 17.5.1, m-homogeneous module A has a representative $\{\mathbf{0}_A, \mathbf{1}_A, \cdots, \mathbf{m}_A\}$, and for $s \neq t$, $\mathbf{s}_A \overset{\varphi}{\neq} \mathbf{t}_A$ holds. Then $\varphi(\mathbf{s}_A, \mathbf{z}) \neq \varphi(\mathbf{t}_A, \mathbf{z})$ $(\exists \mathbf{z} \in \Omega_{C \setminus A})$. Since $\chi(\mathbf{s}_A) = s$ and $\chi(\mathbf{t}_A) = t$, we have $\psi(s, \mathbf{z}) \neq \psi(t, \mathbf{z})$. Therefore ψ is relevant.

Then, with Theorem 17.4.4, we have the following theorem for the consistency of the relevant property among the systems φ, χ and ψ.

Theorem 17.5.2 *Suppose $A \subsetneqq C$ is an m-homogeneous module of an m-homogeneous system (Ω_C, S, φ). φ is relevant if and only if χ and ψ are relevant.*

Definition 17.5.3 *For a system (Ω_C, S, φ) which consists of n components;*

(i) The system φ is called a Barlow-Wu (BW) system, when the following condition is satisfied. There exists a family $\{P_i\}_{i=1}^p$ of subsets of C such that

$$P_i \backslash P_j \neq \phi \ (i \neq j), \quad \bigcup_{i=1}^{p} P_i = C,$$

$$\forall \mathbf{x} \in \Omega_C, \quad \varphi(\mathbf{x}) = \max_{1 \leqq i \leqq p} \min_{j \in P_i} x_j.$$

(ii) The system φ is called an Extended BW (EBW) system, when the following condition is satisfied.

$$\forall s \in S, \quad MI\left(\varphi^{-1}[s, \rightarrow)\right) \subseteqq \{0, s\}^n.$$

(iii) The system φ is called an Extended EBW (EEBW) system, when the following condition is satisfied.

$$\forall s \in S, \quad MI\left(\varphi^{-1}(s)\right) \subseteqq \{0, s\}^n.$$

A Barlow-Wu system is an idea of a multi-state system given by Barlow and Wu [2] in early studies of multi-state systems. This idea utilizes the minimal path sets of a binary-state system and the system is considered to have minimal path sets common to all the states of the system. A Barlow-Wu system is an m-homogeneous and normal system. An extended Barlow-Wu system assumes a family of minimal path sets differently for each state of the system. An EEBW system is not necessarily normal, but is m-homogeneous. The concepts of EBW and EEBW systems are given by Shinmori et al. [10, 11].

The following inclusion relations among the classes of Definition 17.5.3 is clear.

binary-state system $\Longrightarrow$ Barlow-Wu system $\Longrightarrow$ EBW system
$\Longrightarrow$ EEBW system $\Longrightarrow$ m-homogeneous system
$\Longrightarrow$ system with totally ordered state spaces (Definition 17.3.1)
$\Longrightarrow$ system with partially ordered state spaces (Ohi [9])

Similar to binary-state systems, stochastic evaluation of multi-state systems, as stochastic bounds for the reliability of the systems, are based on the families $\left\{MI\left(\varphi^{-1}(s)\right)\right\}_{s \in S}$ and $\left\{MI\left(\varphi^{-1}[s, \rightarrow)\right)\right\}_{s \in S}$. Refer to Ohi [8, 9].

17.5.2 THREE MODULES THEOREM OF BINARY-STATE SYSTEMS

In this subsection, we present the Three Modules Theorem of a binary-state system by Birnbaum and Esary [3].

When $A \subsetneqq C$ is a module of a binary-state system (Ω_C, S, φ), we assume that systems (Ω_A, S_A, χ) and $\left(S_A \times \Omega_{C \setminus A}, S, \psi\right)$ are binary-state systems. Then, recalling Theorem 17.5.1, we have the next theorem for A to be a module of the binary-state system.

Theorem 17.5.3 *$A \subsetneqq C$ is a module of a binary-state system (Ω_C, S, φ) if and only if*

$$\forall \mathbf{x} \in \Omega_A, \quad \mathbf{x} \overset{\varphi}{=} \mathbf{0}_A \quad or \quad \mathbf{x} \overset{\varphi}{=} \mathbf{1}_A. \tag{17.20}$$

Using Theorem 17.5.3, we have the following novel Three Modules Theorem, which was proved first by Birnbaum and Esary [3], and later proved set-theoretically by Butterworth [5].

Theorem 17.5.4 (Three Modules Theorem) *Suppose (Ω_C, S, φ) is a binary-state system. When A_1, A_2 and A_3 are mutually disjoint nonempty subsets of C, and $A_1 \cup A_2$ and $A_2 \cup A_3$ are modules of the system, then the following holds.*

(i) Each of A_1, A_2 and A_3 is a module of the system.

(ii) $A_1 \cup A_2 \cup A_3$ is a module of the system.

(iii) The modules $(\Omega_{A_1}, \{0,1\}, \chi_1)$, $(\Omega_{A_2}, \{0,1\}, \chi_2)$ and $(\Omega_{A_3}, \{0,1\}, \chi_3)$ are constructed in parallel or series in the system φ.

The Three Modules Theorem has not been proved even for a multi-state system of Definition 17.3.1 under appropriate conditions.

17.6 EEBW SYSTEM

In this Section 17.6, restricting the class of systems to EEBW, we introduce the examinations by Shinmori et al. [10] in our ordered set theoretical context. These examinations lead us to an if and only if condition for a subset of components to be a module of an EEBW system and the three modules theorem for EBW systems, which is proved under restricted structural conditions. These examinations suggests a point for the theorem to hold for a multi-state system.

Lemma 17.6.1 *For an EEBW system (Ω_C, S, φ) and nonempty subset $A \subsetneqq C$:*

(i) We have the following.

$$\forall s, \forall t \in S \ (s \neq t), \ \forall \mathbf{x} \in MI\left(\varphi^{-1}(s)\right), \ \forall \mathbf{y} \in MI\left(\varphi^{-1}(t)\right),$$

$$\mathbf{x}_A \neq \mathbf{0}_A \ and \ \mathbf{y}_A \neq \mathbf{0}_A \implies \mathbf{x}_A \overset{\varphi}{\neq} \mathbf{y}_A$$

(ii) When the system φ is coherent, $s \neq t$ implies $\mathbf{s}_A \overset{\varphi}{\neq} \mathbf{t}_A$.

Proof *(i) If* $\mathbf{x}_A \overset{\varphi}{=} \mathbf{y}_A$, $\varphi\left(\mathbf{x}_A, \mathbf{y}_{C\setminus A}\right) = \varphi(\mathbf{y}) = t$ *holds, then*

$$\exists \mathbf{w} \in MI\left(\varphi^{-1}(t)\right), \quad \mathbf{w} \leqq \left(\mathbf{x}_A, \mathbf{y}_{C\setminus A}\right).$$

By the pattern of minimal state vectors of the EEBW system, $\mathbf{x}_A \in \{0, s\}^{|A|}$, $\mathbf{w}_A \in \{0, t\}^{|A|}$ *and furthermore* $s < t$, *then* $\mathbf{w}_A \leqq \mathbf{x}_A$ *implies* $\mathbf{w}_A = \mathbf{0}_A$. *Therefore we have* $\mathbf{w} < \mathbf{y}$, *which contradicts both* $\mathbf{w}$ *and* $\mathbf{y}$ *belonging to* $MI\left(\varphi^{-1}(t)\right)$. *We notice that this examination is based on the assumptions* $\mathbf{x}_A \neq \mathbf{0}_A$ *and* $\mathbf{y}_A \neq \mathbf{0}_A$.

(ii) From the pattern of minimal state vectors of a EEBW system, $\varphi(\mathbf{r}) = r\ (\forall r \in S)$ *and*

$$\forall r \in S,\ \forall \mathbf{x} \in MI\left(\varphi^{-1}(r)\right), \quad \mathbf{x} \leqq \left\{ \begin{array}{c} \left(\mathbf{x}_A, \mathbf{r}_{C\setminus A}\right) \\ \left(\mathbf{r}_A, \mathbf{x}_{C\setminus A}\right) \end{array} \right\} \leqq \mathbf{r}.$$

We have

$$\forall r \in S,\ \forall \mathbf{x} \in MI\left(\varphi^{-1}(r)\right), \quad \varphi(\mathbf{x}) = \varphi(\mathbf{x}_A, \mathbf{r}_{C\setminus A}) = \varphi(\mathbf{r}_A, \mathbf{x}_{C\setminus A}) = \varphi(\mathbf{r}) = r.$$

Hence, since the system φ *is coherent, we have*

$$\forall r \in S,\ \exists \mathbf{x} \in MI\left(\varphi^{-1}(r)\right), \quad \mathbf{x}_A \neq \mathbf{0}_A,\ \varphi\left(\mathbf{r}_A, \mathbf{x}_{C\setminus A}\right) = r.$$

Supposing $\mathbf{s} \overset{\varphi}{=} \mathbf{t}$ *and* $s < t$, *we have from the above,*

$$\exists \mathbf{x} \in MI\left(\varphi^{-1}(t)\right), \quad \varphi\left(\mathbf{t}_A, \mathbf{x}_{C\setminus A}\right) = \varphi\left(\mathbf{s}_A, \mathbf{x}_{C\setminus A}\right) = t,$$

of which the last equality is impossible from the pattern of the minimal state vectors.

The next theorem states that an m-homogeneous module of a coherent EEBW system is necessarily an EEBW module, i.e., χ and ψ are EEBW.

And, along with Theorem 17.5.2, system φ is coherent EEBW if and only if an m-homogeneous module A is coherent EEBW.

Theorem 17.6.1 *When* A *is an* m*-homogeneous module of a coherent EEBW system* (Ω_C, S, φ), *for every* $s\ (0 < s \leqq m)$ *we have*

$$MI\left(\chi^{-1}(s)\right) = \left\{\mathbf{x}_A \mid \mathbf{x}_A \neq \mathbf{0}_A,\ \mathbf{x} \in MI\left(\varphi^{-1}(s)\right)\right\}, \tag{17.21}$$

$$MI\left(\psi^{-1}(s)\right) = \left\{\left(\chi\left(\mathbf{x}_A\right), \mathbf{x}_{C\setminus A}\right) \mid \mathbf{x} \in MI\left(\varphi^{-1}(s)\right)\right\}. \tag{17.22}$$

Proof *Proof of (17.21) Since* φ *is coherent, then for every* $s\ (0 < s \leqq m)$ *the set on the right-hand side of (17.21) is non-empty.*

Since A *is an* m*-homogeneous module,* ψ *is relevant from Theorem 17.5.2, and then by Theorem 17.2.5 (ii), we have*

$$\forall s\ (0 \leqq s \leqq m), \quad MI\left(\chi^{-1}(s)\right) = \bigcup_{t=0}^{m} \left\{\ \mathbf{x}_A \mid \chi(\mathbf{x}_A) = s,\ \mathbf{x} \in MI\left(\varphi^{-1}(t)\right)\right\}.$$

Furthermore, by Lemma 17.6.1 (i),

$$\begin{aligned}&\forall s(0 < s \leqq m), \quad \exists\, t(0 < t \leqq m),\\ &MI\left(\chi^{-1}(s)\right) = \left\{\mathbf{x}_A \mid \mathbf{x}_A \neq \mathbf{0}_A,\ \mathbf{x} \in MI\left(\varphi^{-1}(t)\right)\right\} \qquad (17.23)\end{aligned}$$

and $MI\left(\chi^{-1}(s)\right) \subseteqq \{0,\ t\}^{|A|}$ *by the pattern of minimal state vectors of* φ.

If $s < t$, *we have a contradiction with* $\chi\left(\mathbf{s}_A\right) = s$.

If $t < s$, *from (17.23), we have*

$$\forall \mathbf{x}_A \in MI\left(\chi^{-1}(s)\right), \ s = \chi(\mathbf{x}_A) \leqq \chi(\mathbf{t}_A) = t,$$

which is a contradiction.

Therefore, (17.21) is proved to hold.

Proof of (17.22) *By Theorem 17.2.2 (17.11) and (17.21), (17.22) is apparent and the proof is omitted here.*

Theorem 17.6.2 *When* (Ω_C, S, φ) *is a coherent EEBW system and* $A \subseteqq C$ *is an* m*-homogeneous module of the system* φ, *necessarily a coherent EEBW module, we have the following.*

$$\begin{aligned}&\forall s\ (0 < s \leqq m),\ \forall \mathbf{x},\ \forall \mathbf{y} \in MI\left(\varphi^{-1}(s)\right),\\ &\qquad \mathbf{x}_A \neq \mathbf{0}_A,\ \mathbf{y}_A \neq \mathbf{0}_A \implies \left(\mathbf{x}_A, \mathbf{y}_{C\setminus A}\right) \in MI\left(\varphi^{-1}(s)\right). \qquad (17.24)\end{aligned}$$

Proof *From Theorem 17.6.1, for* $\mathbf{x},\ \mathbf{y} \in MI\left(\varphi^{-1}(s)\right)$ *such that* $\mathbf{x}_A \neq \mathbf{0}_A,\ \mathbf{y}_A \neq \mathbf{0}_A$,

$$\begin{aligned}&\mathbf{x}_A \in MI\left(\chi^{-1}(s)\right), \quad \mathbf{y}_A \in MI\left(\chi^{-1}(s)\right),\\ &\chi(\mathbf{x}_A) = s, \quad \chi(\mathbf{y}_A) = s,\\ &\left(\chi(\mathbf{x}_A), \mathbf{x}_{C\setminus A}\right) \in MI\left(\psi^{-1}(s)\right), \quad \left(\chi(\mathbf{y}_A), \mathbf{y}_{C\setminus A}\right) \in MI\left(\psi^{-1}(s)\right).\end{aligned}$$

Noting that

$$\varphi\left(\mathbf{x}_A, \mathbf{y}_{C\setminus A}\right) = \psi\left(\chi(\mathbf{x}_A), \mathbf{y}_{C\setminus A}\right) = \psi\left(s, \mathbf{y}_{C\setminus A}\right) = \psi\left(\chi(\mathbf{y}_A), \mathbf{y}_{C\setminus A}\right) = \varphi(\mathbf{y}) = s,$$

we have

$$\exists\, \mathbf{z} \in MI\left(\varphi^{-1}(s)\right), \quad \mathbf{z} \leqq \left(\mathbf{x}_A, \mathbf{y}_{C\setminus A}\right).$$

For the proof of this theorem, it is sufficient to show that the equality holds in the above inequality.

If $\mathbf{z}_A = \mathbf{0}_A$, *we have* $\mathbf{z} < \mathbf{y}$, *which contradicts the minimality of* $\mathbf{y}$. *Then* $\mathbf{z}_A \neq \mathbf{0}_A$ *holds. By the previous Theorem 17.6.1,* $\mathbf{z}_A \in MI\left(\chi^{-1}(s)\right)$. *Since* $\mathbf{x}_A \in MI\left(\chi^{-1}(s)\right)$ *and* $\mathbf{z}_A \leqq \mathbf{x}_A$ *hold, then we have* $\mathbf{z}_A = \mathbf{x}_A$.

Again from the previous theorem, we have

$$\begin{aligned}&\left(\chi(\mathbf{z}_A), \mathbf{z}_{C\setminus A}\right) \in MI\left(\psi^{-1}(s)\right), \quad \chi(\mathbf{z}_A) = s,\\ &\left(\chi(\mathbf{x}_A), \mathbf{y}_{C\setminus A}\right) \in MI\left(\psi^{-1}(s)\right), \quad \chi(\mathbf{x}_A) = s.\end{aligned}$$

Along with $(s, \mathbf{z}_{C\setminus A}) \leqq (s, \mathbf{y}_{C\setminus A})$, $\mathbf{z}_{C\setminus A} = \mathbf{y}_{C\setminus A}$ *holds.*

Therefore, $\mathbf{z} = (\mathbf{x}_A, \mathbf{y}_{C\setminus A})$ *follows.*

Lemma 17.6.2 *Suppose (Ω_C, S, φ) is a coherent EEBW system and A is a nonempty subset of C. When elements $\mathbf{x} \in \Omega_C$ and $\mathbf{z} \in \Omega_{C\setminus A}$ satisfy the following conditions (1) and (2);*

$$\begin{aligned}
(1)\quad & \exists s\ (0 < s \leqq m),\ \exists \mathbf{a} \in MI\left(\varphi^{-1}(s)\right) \text{ such that } \mathbf{a}_A \neq \mathbf{0}_A, \\
& \qquad \forall r\ (s < r \leqq m),\ \forall \mathbf{y} \in MI\left(\varphi^{-1}(r)\right) \text{ such that } \mathbf{y}_A \neq \mathbf{0}_A, \\
& \qquad\qquad \mathbf{a}_A \leqq \mathbf{x}_A, \quad \mathbf{y}_A \nleqq \mathbf{x}_A, \\
(2)\quad & \varphi(\mathbf{s}_A, \mathbf{z}) = t,
\end{aligned}$$

then we have

$$\begin{aligned}
(i)\quad & s < t \implies \varphi(\mathbf{x}_A, \mathbf{z}) = \varphi(\mathbf{s}_A, \mathbf{z}) = t, \\
(ii)\quad & t \leqq s \implies \varphi(\mathbf{x}_A, \mathbf{z}) \leqq \varphi(\mathbf{s}_A, \mathbf{z}) = t.
\end{aligned}$$

Proof *Proof of (i), the case of $s < t$*

From $\varphi(\mathbf{s}_A, \mathbf{z}) = t$,

$$\exists \mathbf{w} \in MI\left(\varphi^{-1}(t)\right), \ \mathbf{w} \leqq (\mathbf{s}_A, \mathbf{z}).$$

By the pattern of the minimal state vectors of EEBW and $s < t$, we have $\mathbf{w}_A = \mathbf{0}_A$. Then, $\mathbf{w} \leqq (\mathbf{x}_A, \mathbf{z})$ holds and hence $\varphi(\mathbf{x}_A, \mathbf{z}) \geqq t$ follows.

If $\varphi(\mathbf{x}_A, \mathbf{z}) = r$ and $r > t$,

$$\exists \mathbf{y} \in MI\left(\varphi^{-1}(r)\right), \ \mathbf{y} \leqq (\mathbf{x}_A, \mathbf{z})$$

and then $\mathbf{y}_A \leqq \mathbf{x}_A$. Because of the condition on s, $\mathbf{y}_A = \mathbf{0}_A$ and then $\mathbf{y} \leqq (\mathbf{s}_A, \mathbf{z})$ follows. Therefore, we have $\varphi(\mathbf{y}) = r \leqq \varphi(\mathbf{s}_A, \mathbf{z}) = t$, which contradicts $t < r$.

Proof of (ii), the case of $t \leqq s$

- *If $t < \varphi(\mathbf{x}_A, \mathbf{z}) \leqq s$, setting $\varphi(\mathbf{x}_A, \mathbf{z}) = r$ $(t < r \leqq s)$,*

$$\exists \mathbf{y} \in MI\left(\varphi^{-1}(r)\right), \ \mathbf{y} \leqq (\mathbf{x}_A, \mathbf{z}),$$

where we notice $\mathbf{y} \in \{0, r\}^n$. Then, since $r < s$ and $\mathbf{y}_A \leqq \mathbf{s}_A$, we have $\mathbf{y} \leqq (\mathbf{s}_A, \mathbf{z})$ and $t < \varphi(\mathbf{y}) = r \leqq \varphi(\mathbf{s}_A, \mathbf{z}) = t$, which is a contradiction.

- *If $s < \varphi(\mathbf{x}_A, \mathbf{z})$, setting $\varphi(\mathbf{x}_A, \mathbf{z}) = r$ $(s < r)$,*

$$\exists \mathbf{y} \in MI\left(\varphi^{-1}(r)\right), \ \mathbf{y} \leqq (\mathbf{x}_A, \mathbf{z}).$$

By the condition on s, noticing that $s < r$, $\mathbf{y}_A = \mathbf{0}_A$ follows. Then $\mathbf{y} \leqq (\mathbf{s}_A, \mathbf{z})$ implies $r = \varphi(\mathbf{y}) < \varphi(\mathbf{s}_A, \mathbf{z}) = t$, which contradicts $t \leqq s < r$.

Theorem 17.6.3 *Suppose (Ω_C, S, φ) is a coherent EEBW system. A nonempty subset $A \subseteqq C$ is an m-homogeneous module if the following two conditions hold.*

$$(1)\quad \forall s (0 < s \leqq m),\ \forall\ \mathbf{x},\ \forall\ \mathbf{y} \in MI\left(\varphi^{-1}(s)\right),$$

$$\mathbf{x}_A \neq \mathbf{0}_A,\ \mathbf{y}_A \neq \mathbf{0}_A \implies (\mathbf{x}_A, \mathbf{y}_{C\setminus A}) \in MI\left(\varphi^{-1}(s)\right),$$

(2) $\forall s,\ \forall t\ (0 < t < s \leqq m),\ \forall\ \mathbf{x} \in MI\left(\varphi^{-1}(s)\right),$

$$\mathbf{x}_A \neq \mathbf{0}_A, \implies \exists\ \mathbf{y} \in MI\left(\varphi^{-1}(t)\right) \text{ such that } \mathbf{y}_A \neq \mathbf{0}_A,\ \mathbf{y}_A < \mathbf{x}_A.$$

Proof *It is sufficient to prove*

$$\forall\ \mathbf{x}_A \in \Omega_A,\ \exists\ s\ (0 \leqq s \leqq m),\ \ \mathbf{x}_A \stackrel{\varphi}{=} \mathbf{s}_A$$

when (1) and (2) are satisfied. In this proof, when $\mathbf{x} \in \Omega_C$ *satisfies the conditions of Lemma 17.6.2, we show from (1) and (2) of this theorem that the following equality holds by (ii) of Lemma 17.6.2,*

$$\text{when } t \leqq s,\ \ \varphi(\mathbf{x}_A, \mathbf{z}) = \varphi(\mathbf{s}_A, \mathbf{z}).$$

In the cases other than the conditions of Lemma 17.6.2, it is easily shown that $\mathbf{x}_A \stackrel{\varphi}{=} \mathbf{0}_A$ *or* $\mathbf{x}_A \stackrel{\varphi}{=} \mathbf{m}_A$ *holds.*

(I) Proof in the case of $t = s$ *Since* $\varphi(\mathbf{s}_A, \mathbf{z}) = s$, *we have*

$$\exists \mathbf{y} \in MI\left(\varphi^{-1}(s)\right),\ \ \mathbf{y} \leqq (\mathbf{s}_A, \mathbf{z}).$$

(I-i) In the case of $\mathbf{y}_A = \mathbf{0}_A$ *From* $\mathbf{y} \leqq (\mathbf{x}_A, \mathbf{z})$, $\varphi(\mathbf{y}) = s \leqq \varphi(\mathbf{x}_A, \mathbf{z})$ *holds. On the other hand, from Lemma 17.6.2, we have* $\varphi(\mathbf{x}_A, \mathbf{z}) \leqq \varphi(\mathbf{s}_A, \mathbf{z}) = s$. *Then,* $\varphi(\mathbf{x}_A, \mathbf{z}) = \varphi(\mathbf{s}_A, \mathbf{z}) = s$ *follows.*

(I-ii) In the case of $\mathbf{y}_A \neq \mathbf{0}_A$ *From the conditions of Lemma 17.6.2,*

$$\exists \mathbf{a} \in MI\left(\varphi^{-1}(s)\right),\ \ \mathbf{a}_A \leqq \mathbf{x}_A,\ \mathbf{a}_A \neq \mathbf{0}_A.$$

For the minimal state vectors $\mathbf{y}$ *and* $\mathbf{a}$ *of* $MI\left(\varphi^{-1}(s)\right)$ *satisfying respectively* $\mathbf{y}_A \neq \mathbf{0}_A$ *and* $\mathbf{a}_A \neq \mathbf{0}_A$, *by the assumptions of this theorem, we have* $(\mathbf{a}_A, \mathbf{y}_{C\setminus A}) \in MI\left(\varphi^{-1}(s)\right)$ *and then,* $(\mathbf{a}_A, \mathbf{y}_{C\setminus A}) \leqq (\mathbf{x}_A, \mathbf{z})$ *holds. Therefore,*

$$s = \varphi(\mathbf{a}_A, \mathbf{y}_{C\setminus A}) \leqq \varphi(\mathbf{x}_A, \mathbf{z}) \leqq \varphi(\mathbf{s}_A, \mathbf{z}) = s,$$

and $\varphi(\mathbf{x}_A, \mathbf{z}) = \varphi(\mathbf{s}_A, \mathbf{z})$ *holds.*

(II) Proof in the case of $t < s$ *Since* $\varphi(\mathbf{s}_A, \mathbf{z}) = t$, *we have*

$$\exists\ \mathbf{y}^t \in MI\left(\varphi^{-1}(t)\right),\ \ \mathbf{y}^t \leqq (\mathbf{s}_A, \mathbf{z}).$$

(II-i) In the case of $\mathbf{y}^t_A = \mathbf{0}_A$ *Since* $\mathbf{y}^t = (\mathbf{0}_A, \mathbf{y}^t_{C\setminus A}) \leqq (\mathbf{x}_A, \mathbf{z})$, *from Lemma 17.6.2, we have*

$$t = \varphi(\mathbf{y}^t) \leqq \varphi(\mathbf{x}_A, \mathbf{z}) \leqq \varphi(\mathbf{s}_A, \mathbf{z}) = t,$$

and then $\varphi(\mathbf{x}_A, \mathbf{z}) = \varphi(\mathbf{s}_A, \mathbf{z}) = t$ *holds.*

(II-ii) In the case of $\mathbf{y}^t_A \neq \mathbf{0}_A$ *From the assumption of Lemma 17.6.2,*

$$\exists \mathbf{x}^s \in MI\left(\varphi^{-1}(s)\right),\ \ \mathbf{x}^s_A \neq \mathbf{0}_A,\ \ \mathbf{x}^s_A \leqq \mathbf{x}_A.$$

From the condition (2) of this theorem,

$$\exists \mathbf{x}^t \in MI\left(\varphi^{-1}(t)\right), \quad \mathbf{x}_A^t \neq \mathbf{0}_A, \quad \mathbf{x}_A^t < \mathbf{x}_A^s.$$

And for minimal state vectors $\mathbf{x}^t$ *and* $\mathbf{y}^t$ *in* $MI\left(\varphi^{-1}(t)\right)$*, from assumption (1) of this theorem,*

$$\left(\mathbf{x}_A^t, \mathbf{y}_{C\setminus A}^t\right) \in MI\left(\varphi^{-1}(t)\right).$$

Thus, we have

$$\left(\mathbf{x}_A^t, \mathbf{y}_{C\setminus A}^t\right) < \left(\mathbf{x}_A^s, \mathbf{y}_{C\setminus A}^t\right) \leqq (\mathbf{x}_A, \mathbf{z}),$$

$$t = \varphi\left(\mathbf{x}_A^t, \mathbf{y}_{C\setminus A}^t\right) \leqq \varphi\left(\mathbf{x}_A^s, \mathbf{y}_{C\setminus A}^t\right) \leqq \varphi\left(\mathbf{x}_A, \mathbf{z}\right).$$

Lemma 17.6.2 implies $\varphi(\mathbf{x}_A, \mathbf{z}) \leqq \varphi(\mathbf{s}_A, \mathbf{z}) = t$*, and then, with* $t \leqq \varphi(\mathbf{x}_A, \mathbf{z})$ *from the above, we have* $\varphi(\mathbf{x}_A, \mathbf{z}) = \varphi(\mathbf{s}_A, \mathbf{z})$*.*

It is easily shown that for an EEBW system, the system is normal if and only if it is EBW. Then for an EBW system, a necessary and sufficient condition for a subset $A \subseteqq C$ to be a module is that the conditions of (1) and (2) of Theorem 17.6.3 hold.

Corollary 17.6.1 *For a coherent EBW system, a nonempty subset* $A \subseteqq C$ *is a module if and only if the conditions (1) and (2) of Theorem 17.6.3 hold.*

Proof *The necessity is from Theorem 17.6.2, the sufficiency is from Theorem 17.6.3.*

The conditions of Theorem 17.6.3 are sufficient for a subset of C to be a module, when the system is EEBW. However, Corollary 17.6.1 tells us that for an EBW system, the conditions are necessary and sufficient.

17.7 INTRODUCTION TO THREE MODULES THEOREM FOR MULTI-STATE SYSTEMS

Following the necessary and sufficient conditions of Corollary 17.6.1 for the EBW case, Shinmori et al. [10] have proved the next theorem with some additional structural assumptions to Theorem 17.6.3.

Theorem 17.7.1 *Suppose* (Ω_C, S, φ) *is a coherent EBW system. When nonempty subsets* A_1*,* A_2 *and* A_3 *of* C *are mutually exclusive,* $A_1 \cup A_2$ *and* $A_2 \cup A_3$ *are modules, and either one of the following two conditions is satisfied. Writing* $A = A_1 \cup A_2 \cup A_3$*,*

$$\begin{aligned}
&(1) \quad \forall s \ (0 < s \leqq m),\ \forall \mathbf{x} \in MI\left(\varphi^{-1}(s)\right), \\
&\qquad \mathbf{x}_A \neq \mathbf{0}_A \implies \forall\ A_i\ (i = 1, 2, 3), \quad \mathbf{x}_{A_i} \neq \mathbf{0}_{A_i},
\end{aligned}$$

(2) $\forall s\ (0 < s \leqq m),\ \forall \mathbf{x} \in MI\left(\varphi^{-1}(s)\right),$

$$\mathbf{x}_A \neq \mathbf{0}_A \implies \exists A_i(i=1,2,3),\ \forall A_j(j \neq i),\ \mathbf{x}_{A_i} \neq \mathbf{0}_{A_i},\ \mathbf{x}_{A_j} = \mathbf{0}_{A_j}.$$

Then we have the following:
(i) A_1, A_2 and A_3 are modules.
(ii) $A_1 \cup A_3$ is a module.
(iii) $A_1 \cup A_2 \cup A_3$ is a module.

The conditions (1) and (2) of Theorem 17.7.1 denote series and parallel systems composed of A_1, A_2 and A_3, respectively. For the binary case, as shown in Theorem 17.5.4, this structural property is proved to be a result and not assumptions.

On the other hand, Ohi and Nishida [7] have defined another concept of relevant property as the following. For a system (Ω_C, S, φ), the component $i \in C$ is called relevant when the following conditions hold ;

$$\forall r,\ \forall s\ \in S\ (r \neq s),\ \exists k,\ \exists l \in \Omega_i,\ \exists \mathbf{x} \in \Omega_{C\setminus\{i\}},\ \ \varphi(k_i, \mathbf{x}) = r,\ \ \varphi(l_i, \mathbf{x}) = s,$$

for which we do not have generally an inclusion relation with the relevant property of this manuscript. Under this alternative relevancy, Ohi and Nishida [7] have proved the following theorem.

Theorem 17.7.2 *Let (Ω_C, S, φ) be a coherent system, i.e., φ is increasing and every component is relevant of Ohi and Nishida [7]. When A_1, A_2 and A_3 are nonempty mutually exclusive subsets of C and $A_1 \cup A_2$, $A_2 \cup A_3$ are modules, then $A_1 \cup A_2 \cup A_3$, A_1 and A_3 are modules.*

We notice that A_2 is not proved to be a module in the above theorem. However, this theorem does not need a structural condition among A_1, A_2 and A_3 as (1) and (2) of Theorem 17.7.1.

The perspective over the Three Modules Theorem is summarized in the following table, which includes the case that the structural properties are not assumed. The situation about the theorem is shown to be complicated.

	structural property	A_1	A_2
Binary-state system	derived	a module	a module
Shinmori et al. [10]	assumed	a module	a module
Shinmori et al. [10]	not assumed	not necessarily	not necessarily
Ohi et al. [7]	not assumed	a module	not necessarily

	A_3	$A_1 \cup A_2 \cup A_3$	$A_1 \cup A_3$
Binary-state system	a module	a module	a module
Shinmori et al. [10]	a module	a module	a module
Shinmori et al. [10]	not necessarily	a module	not necessarily
Ohi et al. [7]	a module	a module	not necessarily

Comparing the two cases that the structural property is assumed and that the property is not assumed, we may observe the drastic difference in the variety of the results, from which some structural conditions seem to be needed for the theorem in a general case of multi-state.

On the other hand, the condition (1) of Theorem 17.6.3 denotes a preservation of minimal property under crossing every two minimal state vectors, which is also given in Theorem 17.4.3 under the assumption of normal property. (2) of Theorem 17.6.3 implies a kind of normal property. From the observation, we may imagine that preservation property under the crossing operation and the normal property would play a crucial role for the proof of Three Modules Theorem.

Anyway, the Three Modules Theorem is not proved under appropriate conditions and in a full form as the binary-state case.

17.8 CONCLUDING REMARKS

In this thesis, we have redefined the basic notions inherent in reliability theory as relevant property, normal property, module, and so on, in a more generalized ordered set theoretical situation, and proved an inclusion relation among the minimal elements for a composite function $f = h \circ g$. This inclusion relation becomes an equality one, when f, g and h are normal, and plays a crucial role for stochastically evaluating a system's reliability via modular decomposition. With these examinations as a general mathematical framework, we have examined the concept of a module, and have given a necessary and sufficient condition for a set of components to be a module, i.e., the condition is that the set of the components has a representative. We think that this condition may work for evaluating the reliability of a network.

Referring to the works by Shinmori et al. [10, 11], introducing a new class of systems called m-homogeneous in this thesis, the class of systems with totally ordered state spaces are classified into hierarchical classes and it is also shown how the if and only if condition for the modularity is modified along with the hierarchy of multi-state systems.

Finally, we present the Three Modules Theorem for EBW systems by Shinmori et al. [10] and comparing it with the works by Ohi [7] and the binary-state case, examinations of what conditions are required for the Three Modules Theorem to hold in a general setting are given and it is suggested that the normal property would play an important role.

REFERENCES

1. R.E. Barlow and F. Proschan. *Statistical Theory of Reliability and Life Testing.* Holt, Reihart and Winston, Inc., 1975.

2. R.E. Barlow and A.S. Wu. Coherent systems with multi-state components. *Mathematics of Operations Research*, volume 3, pages 275-281, 1978.

3. Z.W. Birnbaum and J.D. Esary. Modules of coherent binary systems. *A Publication of the Society for Industrial and Applied Mathematics*, volume 13, pages 444-462, 1965.

4. L.D.Bodin. Approximation to system reliability using a modular decomposition. *Technometrics*, volume 12, pages 335-344, 1970.

5. R.W. Butterworth. A set theoretic treatment of coherent system. *A Publication of the Society for Industrial and Applied Mathematics*, volume 15, pages 675-688, 1978.

6. J.D. Esary, A.W. Marshall and F. Proschan. Some reliability applications of the hazard transform. *A Publication of the Society for Industrial and Applied Mathematics*, volume 18, pages 849-860, 1970.

7. F. Ohi and T. Nishida. Generalized multistate coherent systems. *J. Japan. Statist. Soc. Journal of the Japan Statistical Society*, volume 13, pages 65-181, 1983.

8. F. Ohi. Steady-state bounds for multi-state systems' reliability via modular decompositions. *Applied Stochastic Models in Business and Industry*, Wiley Online Library, volume 31, pages 307-324, 2015.

9. F. Ohi. Stochastic evaluation methods of a multi-state system via a modular decomposition. *Journal of Computational Science*, volume 17, pages 156-169, 2016.

10. S. Shinmori, F. Ohi, H. Hagihara and T. Nishida. Modules for two classes of multi-state systems. *The Transactions of the IEICE*, volume 72, pages 600-608, 1989.

11. S. Shinmori, H. Hagihara, F. Ohi and T. Nishida, On an extension of Barlow-Wu systems - basic properties, *Journal of the Operations Research Society of Japan*, volume 32, pages 159-172, 1989.

18 A Postponed Repair Model for a Mission-Based System Based on a Three-Stage Failure Process

Jinting Wang
Central University of Finance and Economics, China

Nan Yang
Beijing Jiaotong University, China

CONTENTS

18.1 INTRODUCTION

Preventive maintenance (PM) including inspection, condition monitoring and preventive repair or replacement of defective components is common activity to prevent system's failures [1]. The purpose of inspections is to reveal the system's working state, and thereby to carry out necessary maintenance actions if needed. Periodic inspections, as one of the most common applied inspection strategies because of its implementation simplicity, are widely reviewed in literature (see [2, 3, 4]). Some systems in offices and industries, however, successively execute missions or computer processes and the duration of missions or computer processes is a random variable. For such systems, it would be impossible or impractical to maintain them in a strictly periodic fashion. A representative example is that a database system has to complete non-overlapping missions [5]. For such a system, inspections are generally performed at the completion of each mission to make the best of the idle periods between missions so as to prevent losses of production [6]. Note that more inspection policies than one are recommended in many cases due to the insufficiency of only one type of inspection in terms of the maintenance cost [7, 8, 9].

Christer [10] first proposed a delay time concept to illustrate the necessity of inspection activities and then applied it to preventive maintenance problems. The delay time concept considers the failure process as a two-stage process: the normal stage, which is from new to the initial point of an identifiable defect; and the delay time stage, which is from that defect point to failure. A number of case studies have been reported using the delay time concept with actual applications in industry, see Aven et al. [11], Jones et al. [12] and Wang et al. [13]. The delay-time-based models can be divided into two categories [14]: a component tracking model and a complex system model, where the former refers to a single-component subject to a single failure mode (see [11, 12]) and the latter refers to a system with many components and failure modes (see [13]). For more delay-time-based models of single-component systems and complex systems, see Wang [15]. In this paper, we focus on a single-component model. Such a binary definition of the system's state before failure, however, may be restrictive because it's more likely that the system could undergo several defective states before it completely fails. For example, a three-colour scheme was used to quantify the plant state before failure into green (normal), yellow (need attention), and red (need immediate attention) [16]. Therefore the delay time concept was extended into a three-stage failure process where the traditional delay time was divided into another two stages named minor and major defective stages. Recently, Wang et al. [17, 18] considered the models where four states i.e., normal state, minor defective state, major defective state and failure are assumed. However, it is noted that the above mentioned papers [16, 17, 18] have not addressed the random inspection issue. This motivates our study.

Most preventive maintenance models assume the instantaneous execution of replacement or repair once the defective state is revealed. In contrast, we relax this assumption and allow repair to be postponed for the purpose of cost-saving. This has been applied to the traditional two-stage failure process [19] and the extended three-stage failure process [20]. However, in these papers the decisions about postponed repair are associated with periodic inspections but not with random inspections. In this paper, whether to postpone repair of the major defective system or not depends mainly on the outcome of random inspections. To be specific, if the system is major defective at a random inspection, repair is postponed when the time to the following periodic

Table 18.1
Summary of the most-related research

Articles	Periodic inspection	Random inspection	Failure process	Postpone repair (or replacement)
Wang et al. [1]	Yes	No	a two-stage failure process	No
Yang et al. [8]	Yes	Yes	a two-stage failure process	No
Yang et al. [9]	Yes	Yes	a two-stage failure process	Yes
Wang [17]	Yes	No	a three-stage failure process	No
Wang et al. [18]	Yes	No	a three-stage failure process	Yes
Wang et al. [20]	Yes	No	a three-stage failure process	Yes
Our Work	Yes	Yes	a three-stage failure process	Yes

inspection is less than a given threshold, otherwise repair is immediate. On the other hand, repair is immediate when the system is major defective at a periodic inspection.

Our contributions lie in the following aspects: We consider a three-stage failure process subject to two types of inspections, i.e., periodic and random inspections, which has not been studied in the literature. In such a new maintenance model, the repair action under certain conditions could be delayed. It can reduce the maintenance cost because of enabling maintenance resources to be prepared properly in advance, avoiding excessive maintenance and prolonging the system's expected renewal cycle length. We summarize the major assumptions used in the above-mentioned most-related research articles in Table 18.1 to make it easier for the readers to understand the contributions of our model.

18.2 NOTATIONS AND ASSUMPTIONS

Throughout the paper some notations and assumptions for model-building purposes are presented in Table 18.2. Other notations will be defined when they are needed.

We consider a single-component system subject to a single failure mode. It has the following assumptions.

(1) The system executes missions successively with random durations that follow an independent identical distribution. Mission process: A renewal process with a renewal function $M(t) \equiv \sum_{n=1}^{\infty} H^{(n)}(t)$ and $M_0(t) \equiv \sum_{n=0}^{\infty} H^{(n)}(t)$ where $H^{(0)}(t) \equiv 1$ for $t \geq 1$.

(2) Failure has the following independent three stages, namely, Normal stage with $U(x)$, Minor defective stage $V(y)$, Major defective stage $W(z)$. Throughout of the paper we assume that random variables X, Y, Z have the respective distributions $U(x)$, $V(y)$, $W(z)$, and tau is a random variable of the nth completed mission time, tau_{n+1} is a random variable of the $(n+1)$th

Table 18.2
Notation

Notation	Explanation
Z_n	the duration of the n_{th} mission, $n = 1, 2, \cdots$
$H(t)$	cumulative distribution function (cdf) of Z_n
$H^{(n)}(t)$	n-fold Stieltjes convolution of $H(t)$
$h(t)$	probability density function (pdf) of Z_n
$W(t)$	the expected number of missions successively completed by t
X	random variable representing the duration of the normal state
$U_X(x)$	cumulative distribution function (cdf) of X
$u_X(x)$	probability density function (pdf) of X
Y	random variable representing the duration of the minor defective state
$V_Y(y)$	cumulative distribution function (cdf) of Y
$v_Y(y)$	probability density function (pdf) of Y
Z	random variable representing the duration of the major defective state
$W_Z(z)$	cumulative distribution function (cdf) of Z
$w_Z(z)$	probability density function (pdf) of Z
T	the interval of the periodic inspection
t_i	threshold for postponing repair of a major defective system identified at a random inspection in $(iT, (i+1)T), i = 0, 1, 2, \cdots$
τ	the duration of the n completed missions
ε	the duration of the subsequent $(n+1)$-th mission
C_{ip}	cost of a periodic inspection
C_{ir}	cost of a random inspection
C_{dp}^1	cost of an immediately PM of a minor defective system identified at a periodic inspection
C_{dr}^1	cost of an immediately PM of a minor defective system identified at a random inspection
C_{dp}^2	cost of an immediately PM of a major defective system identified at a periodic inspection
C_{dr}^2	cost of an immediately PM of a major defective system identified at a random inspection
C_{dl}	cost of a postponed repair of a major defective system identified at a random inspection
C_f	cost of an immediate failure replacement

mission time. Then, we have

$$\Pr\{\tau_n \le X, \tau_{n+1} > X + Y + Z - \tau_n | X = x, Y = y, Z = z\}$$
$$= \int_0^x [\int_{x+y+z-\tau}^{\infty} dH(\epsilon)] dH^{(n)}(\tau) = \int_0^x \bar{H}(x + y + z - \tau) dH^{(n)}(\tau). \qquad (*)$$

(3) Failure of the system is self-announcing, whereas the defective states can only be revealed by inspections. Both random and periodic inspections are instantaneous and perfect, i.e., they always reveal the states of the system. Periodic inspection epochs are $iT (i = 1, 2, \ldots)$.

(4) If the system is found to be in a minor defective stage, it is repaired immediately. If the system is found major defective at a periodic inspection, it is repaired immediately. If the system is major defective at a random inspection, repair will be postponed if the time to the following periodic inspection is less than a predetermined threshold, otherwise repair will be immediate. After repair, the system becomes new, i.e., the age of the system is reset to 0. Finally, if the system fails, it is replaced immediately.

18.3 COST MODEL UNDER THE PROPOSED POLICY

18.3.1 EXPECTED NUMBER OF MISSIONS SUCCESSIVELY COMPLETED BY T

Considering that a random inspection is instantaneous, the probability that there are exactly n, $n = 1, 2, \cdots$, missions successively completed by t is

$$P_n(t) = H^{(n)}(t) - H^{(n+1)}(t).$$

Then, the expected number of missions accomplished by t is

$$\begin{aligned} W(t) &= \sum_{n=0}^{\infty} nP_n(t) = \sum_{n=1}^{\infty} nP_n(t) = \sum_{n=1}^{\infty} n\left(H^{(n)}(t) - H^{(n+1)}(t)\right) \\ &= \sum_{n=1}^{\infty} H^{(n)}(t). \end{aligned}$$

By differentiating $W(t)$ with respect to t , we have

$$w(t) = dW(t)/dt = \sum_{n=1}^{\infty} h^{(n)}(t),$$

where $h^{(n)}(t)$ is the derivative of $H^{(n)}(t)$ with respect to t.

18.3.2 THREE RENEWAL CASES AND THE CORRESPONDING OCCURRENCE PROBABILITIES

Under the proposed policy, the mission-based system may be renewed at a failure, a random inspection and a periodic inspection. For these three different renewal cases, the corresponding occurrence probabilities are formulated as follows.

18.3.2.1 A Failure Renewal

According to the assumptions, there are two possible failure scenarios.
(1) Scenario 1

The system fails in $(iT, (i+1)T)$, $i = 0, 1, 2, \cdots$, before any defect is found (see Fig. 18.1). The probability of such a renewal is given by

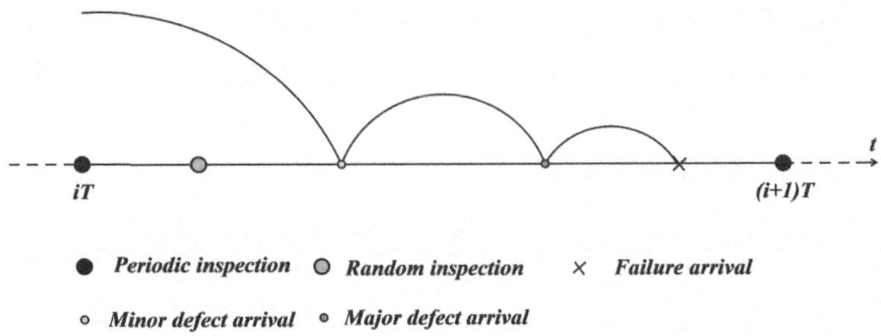

Figure 18.1 Scenario 1 of the failure renewal

$$
\begin{aligned}
&P_{f_1}(i,T) \\
&= \Pr\left\{\begin{matrix} iT < X < (i+1)T, 0 < Y < (i+1)T - X, 0 < Z < (i+1)T - X - Y, \\ 0 \le \tau < X, \varepsilon > X + Y + Z - \tau \end{matrix}\right\} \\
&= \int_{iT}^{(i+1)T} \int_0^{(i+1)T-x} \int_0^{(i+1)T-x-y} \int_{x+y+z}^{\infty} h(\varepsilon)\, u_X(x) v_Y(y)\, w_Z(z)\, d\varepsilon dz dy dx \\
&\quad + \sum_{n=1}^{\infty} \int_{iT}^{(i+1)T} \int_0^{(i+1)T-x} \int_0^{(i+1)T-x-y} \int_{x+y+z-\tau}^{\infty} \int_0^{x} h^{(n)}(\tau)\, h(\varepsilon) \\
&\qquad \times u_X(x)\, v_Y(y)\, w_Z(z)\, d\tau d\varepsilon dz dy dx \\
&= \int_{iT}^{(i+1)T} \int_0^{(i+1)T-x} \int_0^{(i+1)T-x-y} \overline{H}(x+y+z)\, u_X(x)\, v_Y(y)\, w_Z(z)\, dz dy dx \\
&\quad + \sum_{n=1}^{\infty} \int_{iT}^{(i+1)T} \int_0^{(i+1)T-x} \int_0^{(i+1)T-x-y} \int_0^{x} h^{(n)}(\tau)\, \overline{H}(x+y+z-\tau) \\
&\qquad \times u_X(x)\, v_Y(y)\, w_Z(z)\, d\tau dz dy dx.
\end{aligned}
$$

By equation (*) in Assumption (2), the above result can be rewritten as

$$
\int_{iT}^{(i+1)T} \left(\int_0^{(i+1)T-x} \left\{ \int_0^{(i+1)T-x-y} [\overline{H}(x+y+z-\tau) dM_0(\tau)] dZ(z) \right\} dV(y) \right) dU(x), \tag{18.1}
$$

where note that $M_0(t) \equiv \sum_{n=0}^{\infty} H^{(n)}(t)$.

The probability density function (pdf) of failure at $iT+\varphi$ with $0 < \varphi < T$ is

$$
\begin{aligned}
g_{f_1}(i,T,\varphi) = &\int_{iT}^{iT+\varphi} \int_0^{iT+\varphi-x} \overline{H}(iT+\varphi) u_X(x) v_Y(y) w_Z(iT+\varphi-x-y)\, dy dx \\
&+ \sum_{n=1}^{\infty} \int_{iT}^{iT+\varphi} \int_0^{iT+\varphi-x} \int_0^{x} h^{(n)}(\tau) \overline{H}(iT+\varphi-\tau) \\
&\times u_X(x) v_Y(y) w_Z(iT+\varphi-x-y)\, d\tau dy dx.
\end{aligned}
$$

(2) Scenario 2

In this scenario, a major defect is first found at a random inspection in $((i+1)T - t_i, (i+1)T)$, $0 < t_i < T$. Since the time to $(i+1)T$ is smaller than the postponement threshold, repair is postponed to $(i+1)T$. However, a failure occurs during this postponement period, and then the system is replaced immediately (see Fig. 18.2). The probability of such a renewal is given by

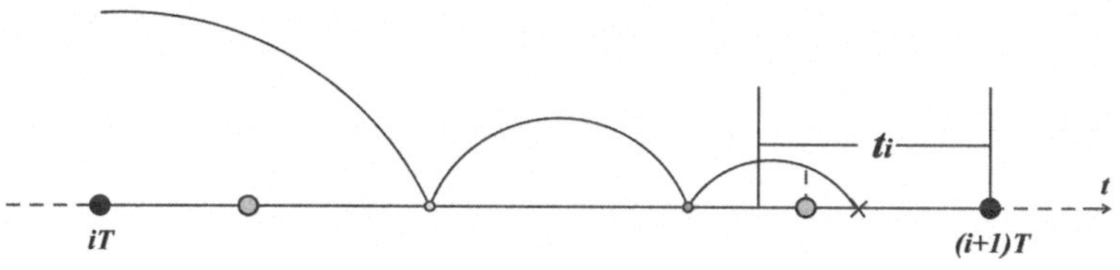

Figure 18.2 Scenario 2 of the failure renewal

$$\begin{aligned}
&P_{f2}(i,T,t_i)\\
&= \Pr\Big\{0 \le \tau < (i+1)T, \max(0,(i+1)T - t_i - \tau) < \varepsilon < (i+1)T - \tau, \max(iT,\tau) <\\
&\qquad X < \tau + \varepsilon, 0 < Y < \tau + \varepsilon - X, \tau + \varepsilon - X - Y < Z < (i+1)T - X - Y\Big\}\\
&= \int_{(i+1)T-t_i}^{(i+1)T} \int_{iT}^{\varepsilon} \int_0^{\varepsilon - x} h(\varepsilon) u_X(x) v_Y(y) \left\{W_Z((i+1)T - x - y) - W_Z(\varepsilon - x - y)\right\} dy dx d\varepsilon\\
&\quad + \sum_{n=1}^{\infty} \int_0^{(i+1)T} \int_{\max(0,(i+1)T-t_i-\tau)}^{(i+1)T-\tau} \int_{\max(iT,\tau)}^{\tau+\varepsilon} \int_0^{\tau+\varepsilon-x} h^{(n)}(\tau) h(\varepsilon)\\
&\qquad\qquad \times u_X(x) v_Y(y) W_Z((i+1)T - x - y)\, dy dx d\varepsilon d\tau\\
&\quad - \sum_{n=1}^{\infty} \int_0^{(i+1)T} \int_{\max(0,(i+1)T-t_i-\tau)}^{(i+1)T-\tau} \int_{\max(iT,\tau)}^{\tau+\varepsilon} \int_0^{\tau+\varepsilon-x} h^{(n)}(\tau) h(\varepsilon)\\
&\qquad\qquad \times u_X(x) v_Y(y) W_Z(\varepsilon + \tau - x - y)\, dy dx d\varepsilon d\tau. \qquad (18.2)
\end{aligned}$$

The pdf of failure at $iT + \varphi$ with $T - t_i < \varphi < T$ is

$$\begin{aligned}
g_{f2}(i,T,t_i,\varphi) &= \int_{(i+1)T-t_i}^{iT+\varphi} \int_{iT}^{\varepsilon} \int_0^{\varepsilon-x} h(\varepsilon) u_X(x) v_Y(y) w_Z(iT + \varphi - x - y)\, dy dx d\varepsilon\\
&+ \sum_{n=1}^{\infty} \int_0^{iT+\varphi} \int_{\max(0,(i+1)T-t_i-\tau)}^{iT+\varphi-\tau} \int_{\max(iT,\tau)}^{\tau+\varepsilon} \int_0^{\tau+\varepsilon-x} h^{(n)}(\tau) h(\varepsilon)\\
&\qquad \times u_X(x) v_Y(y) w_Z(iT + \varphi - x - y)\, dy dx d\varepsilon d\tau.
\end{aligned}$$

The expected number of random inspections due to a failure renewal $E(N_{ir}^f)$ is given by

$$E\left(N_{ir}^{f}\right)=\sum_{i=0}^{\infty}\sum_{n=1}^{\infty}\int_{iT}^{(i+1)T}\int_{0}^{(i+1)T-x}\int_{0}^{(i+1)T-x-y}\int_{x+y+z-\tau}^{\infty}\int_{0}^{x}nh^{(n)}\left(\tau\right)h\left(\varepsilon\right)\times u_X\left(x\right)v_Y\left(y\right)w_Z\left(z\right)d\tau d\varepsilon dzdydx$$
$$+\sum_{i=0}^{\infty}\int_{(i+1)T-t_i}^{(i+1)T}\int_{iT}^{\varepsilon}\int_{0}^{\varepsilon-x}h\left(\varepsilon\right)u_X\left(x\right)v_Y\left(y\right)\times\left\{W_Z\left(\left(i+1\right)T-x-y\right)-W_Z\left(\varepsilon-x-y\right)\right\}dydxd\varepsilon$$
$$+\sum_{i=0}^{\infty}\sum_{n=1}^{\infty}\int_{0}^{(i+1)T}\int_{\max(0,(i+1)T-t_i-\tau)}^{(i+1)T-\tau}\int_{\max(iT,\tau)}^{\tau+\varepsilon}\int_{0}^{\tau+\varepsilon-x}(n+1)\,h^{(n)}(\tau)h(\varepsilon)\times u_X(x)v_Y(y)W_Z((i+1)T-x-y)dydxd\varepsilon d\tau$$
$$-\sum_{i=0}^{\infty}\sum_{n=1}^{\infty}\int_{0}^{(i+1)T}\int_{\max(0,(i+1)T-t_i-\tau)}^{(i+1)T-\tau}\int_{\max(iT,\tau)}^{\tau+\varepsilon}\int_{0}^{\tau+\varepsilon-x}(n+1)h^{(n)}\left(\tau\right)h\left(\varepsilon\right)\times u_X\left(x\right)v_Y\left(y\right)W_Z\left(\varepsilon+\tau-x-y\right)dydxd\varepsilon d\tau. \tag{18.3}$$

18.3.2.2 A Random Inspection Renewal

There are two possible scenarios for a random inspection renewal, as shown in Fig. 18.3 and Fig. 18.4.

(1) Scenario 1

A minor defect is first found by the random inspection within $(iT,(i+1)T)\,, i=0,1,2,\cdots$, and then the system is preventatively repaired immediately (see Fig. 18.3). The probability is

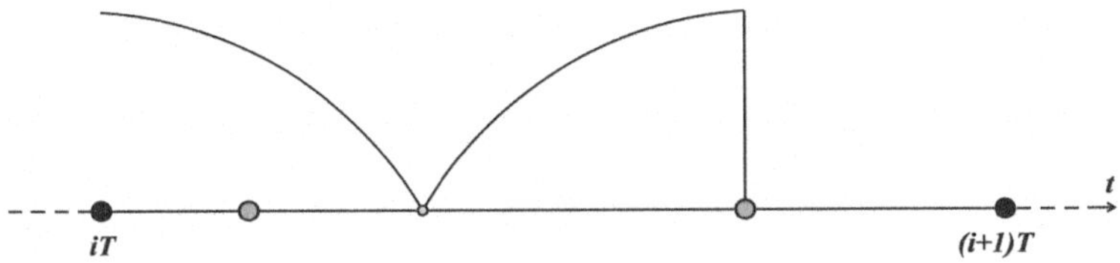

Figure 18.3 Scenario 1 of the random inspection renewal

$$\begin{aligned}&P_{dr}^{1}\left(i,T\right)\\&=\Pr\left\{iT<X<\left(i+1\right)T,0\leq\tau<X,Y>\tau+\varepsilon-X,X-\tau<\varepsilon<\left(i+1\right)T-\tau\right\}\\&=\int_{iT}^{(i+1)T}\int_{\varepsilon\text{-}x}^{\infty}\int_{x}^{(i+1)T}h\left(\varepsilon\right)u_X\left(x\right)v_Y\left(y\right)d\varepsilon dydx\end{aligned}$$

$$+\sum_{n=1}^{\infty}\int_{iT}^{(i+1)T}\int_{\tau+\varepsilon-x}^{\infty}\int_{x-\tau}^{(i+1)T-\tau}\int_{0}^{x} h^{(n)}(\tau)\,h(\varepsilon)\,u_X(x)\,v_Y(y)\,d\tau d\varepsilon dy dx$$

$$=\int_{iT}^{(i+1)T}\int_{x}^{(i+1)T} h(\varepsilon)\,u_X(x)\,\overline{V_Y}(\varepsilon-x)\,d\varepsilon dx$$

$$+\sum_{n=1}^{\infty}\int_{iT}^{(i+1)T}\int_{x-\tau}^{(i+1)T-\tau}\int_{0}^{x} h^{(n)}(\tau)\,h(\varepsilon)\,u_X(x)\,\overline{V_Y}(\tau+\varepsilon-x)\,d\tau d\varepsilon dx. \tag{18.4}$$

Then, the pdf of repair of the minor defective system at $iT+\varphi$ with $0<\varphi<T$ is given by

$$g_{dr}^{1}(i,T,\varphi)=\int_{iT}^{iT+\varphi} h(iT+\varphi)\,u_X(x)\,\overline{V_Y}(iT+\varphi-x)\,dx$$

$$+\sum_{n=1}^{\infty}\int_{iT}^{iT+\varphi}\int_{0}^{x} h^{(n)}(\tau)\,h(iT+\varphi-\tau)\,u_X(x)\,\overline{V_Y}(iT+\varphi-x)\,d\tau dx.$$

(2) Scenario 2

A major defect is first found at a random inspection in $(iT,(i+1)T-t_i)$. Since the time to $(i+1)T$ is larger than the postponement threshold, the system is repaired immediately (See Fig. 18.4). The probability of such a renewal is

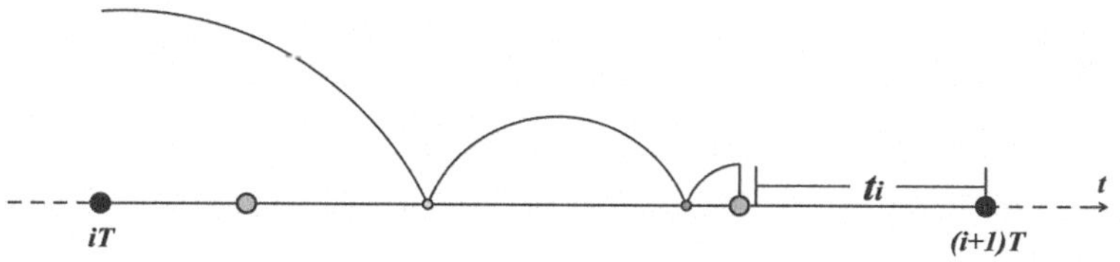

Figure 18.4 Scenario 2 of the random inspection renewal

$$P_{dr}^{2}(i,T,t_i)$$

$$=\Pr\left\{\begin{matrix} iT<X<(i+1)T-t_i, 0<Y<(i+1)T-t_i-X, Z>\tau+\varepsilon-X-Y, \\ 0\le\tau<X, X+Y-\tau<\varepsilon<(i+1)T-t_i-\tau \end{matrix}\right\}$$

$$=\int_{iT}^{(i+1)T-t_i}\int_{0}^{(i+1)T-t_i-x}\int_{x+y}^{(i+1)T-t_i} h(\varepsilon)u_X(x)\,v_Y(y)\,\overline{W_Z}(\varepsilon-x-y)\,d\varepsilon dy dx$$

$$+\sum_{n=1}^{\infty}\int_{iT}^{(i+1)T-t_i}\int_{0}^{(i+1)T-t_i-x}\int_{x+y-\tau}^{(i+1)T-t_i-\tau}\int_{0}^{x} h^{(n)}(\tau)\,h(\varepsilon)$$

$$\times u_X(x)\,v_Y(y)\,\overline{W_Z}(\tau+\varepsilon-x-y)\,d\tau d\varepsilon dy dx. \tag{18.5}$$

The pdf of repair of the major defective system at $iT+\varphi$ with $0<\varphi<T-t_i$

is

$$g_{dr}^{2}(i,T,t_i,\varphi) = \int_{iT}^{iT+\varphi}\int_{0}^{iT+\varphi-x} h(iT+\varphi)\,u_X(x)\,v_Y(y)\,\overline{W_Z}(iT+\varphi-x-y)dydx + \sum_{n=1}^{\infty}\int_{iT}^{iT+\varphi}\int_{0}^{iT+\varphi-x}\int_{0}^{x} h^{(n)}(\tau)\,h(iT+\varphi-\tau)\,u_X(x)\,v_Y(y) \times \overline{W_Z}(iT+\varphi-x-y)\,d\tau dydx.$$

As we did for Eq. (18.3), the expected number of random inspections due to a random inspection renewal $E\left(N_{ir}^{dr}\right)$ is

$$E\left(N_{ir}^{dr}\right) = \sum_{i=0}^{\infty}\int_{iT}^{(i+1)T}\int_{\varepsilon\text{-}x}^{\infty}\int_{x}^{(i+1)T} h(\varepsilon)\,u_X(x)\,v_Y(y)\,d\varepsilon dydx + \sum_{i=0}^{\infty}\sum_{n=1}^{\infty}\int_{iT}^{(i+1)T}\int_{\tau+\varepsilon-x}^{\infty}\int_{x-\tau}^{(i+1)T-\tau}\int_{0}^{x}(n+1)\,h^{(n)}(\tau)\,h(\varepsilon)\,u_X(x)\,v_Y(y)\,d\tau d\varepsilon dydx + \sum_{i=0}^{\infty}\int_{iT}^{(i+1)T-t_i}\int_{0}^{(i+1)T-t_i-x}\int_{x+y}^{(i+1)T-t_i} h(\varepsilon)u_X(x)\,v_Y(y)\,\overline{W_Z}(\varepsilon-x-y)\,d\varepsilon dydx + \sum_{i=0}^{\infty}\sum_{n=1}^{\infty}\int_{iT}^{(i+1)T-t_i}\int_{0}^{(i+1)T-t_i-x}\int_{x+y-\tau}^{(i+1)T-t_i-\tau}\int_{0}^{x}(n+1)h^{(n)}(\tau)\,h(\varepsilon) \times u_X(x)v_Y(y)\,\overline{W_Z}(\tau+\varepsilon-x-y)d\tau d\varepsilon dydx. \tag{18.6}$$

18.3.2.3 A Periodic Inspection Renewal

There are three possible scenarios for a periodic inspection renewal, i.e., two immediate periodic inspection renewals and a postponed repair renewal.

(1) Scenario 1

A minor defect is first found by the periodic inspection at $(i+1)T$, and then the system is repaired immediately (see Fig. 18.5). The probability is

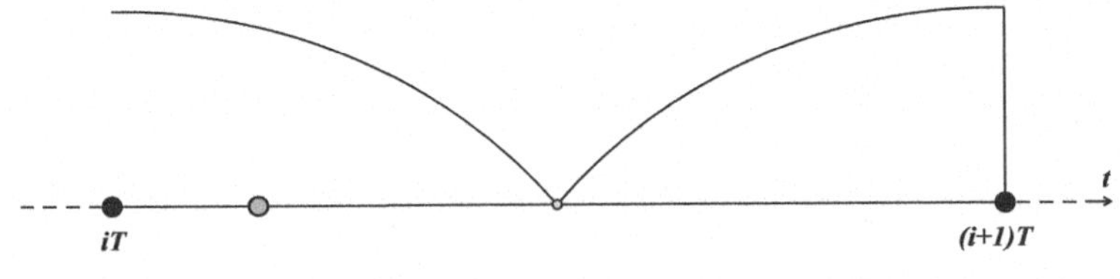

Figure 18.5 Scenario 1 of the periodic inspection renewal

$$P_{dp}^{1}(i,T) = \Pr\{iT < X < (i+1)T, Y > (i+1)T - X, 0 \le \tau < X, \varepsilon > (i+1)T - \tau\}$$
$$= \int_{iT}^{(i+1)T}\overline{H}((i+1)T)\,u_X(x)\,\overline{V_Y}((i+1)T-x)\,dx$$

$$+\sum_{n=1}^{\infty}\int_{iT}^{(i+1)T}\int_{0}^{x} h^{(n)}(\tau)\overline{H}((i+1)T-\tau)\,u_X(x)\,\overline{V_Y}((i+1)T-x)\,d\tau dx. \quad (18.7)$$

(2) Scenario 2

A major defect is first found by the periodic inspection at $(i+1)T$, and then the system is repaired immediately (see Fig. 18.6). The probability is

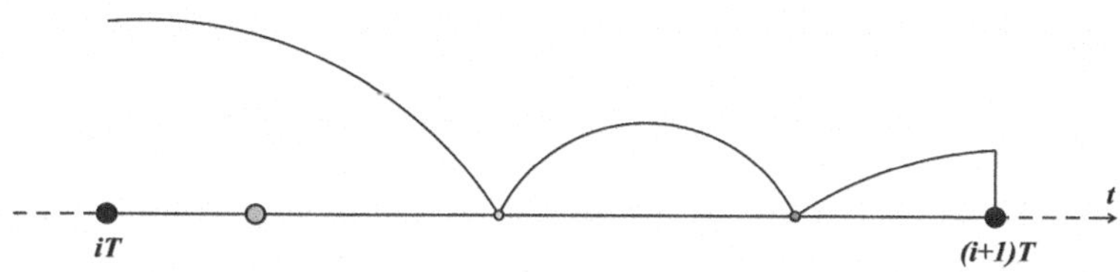

Figure 18.6 Scenario 2 of the random inspection renewal

$$\begin{aligned}
&P_{dp}^{2}(i,T)\\
&=\Pr\Big\{iT<X<(i+1)T, 0<Y<(i+1)T-X, Z>(i+1)T-X-Y,\\
&\qquad\qquad 0\le\tau<X, \varepsilon>(i+1)T-\tau\Big\}\\
&=\int_{iT}^{(i+1)T}\int_{0}^{(i+1)T-x}\int_{(i+1)T-x-y}^{\infty}\int_{(i+1)T}^{\infty} h(\varepsilon)\,u_X(x)\,v_Y(y)\,w_Z(z)\,d\varepsilon dz dy dx\\
&\quad+\sum_{n=1}^{\infty}\int_{iT}^{(i+1)T}\int_{0}^{(i+1)T-x}\int_{(i+1)T-x-y}^{\infty}\int_{(i+1)T-\tau}^{\infty}\int_{0}^{x} h^{(n)}(\tau)\,h(\varepsilon)\\
&\qquad\times u_X(x)\,v_Y(y)\,w_Z(z)\,d\tau d\varepsilon dz dy dx. \quad (18.8)
\end{aligned}$$

(3) Scenario 3

A major defect is first found at a random inspection in $((i+1)T-t_i,(i+1)T), 0<t_i<T$. Then, repair of the major defective system is postponed to $(i+1)T$, before which no failure occurs (see Fig. 18.7). The probability is

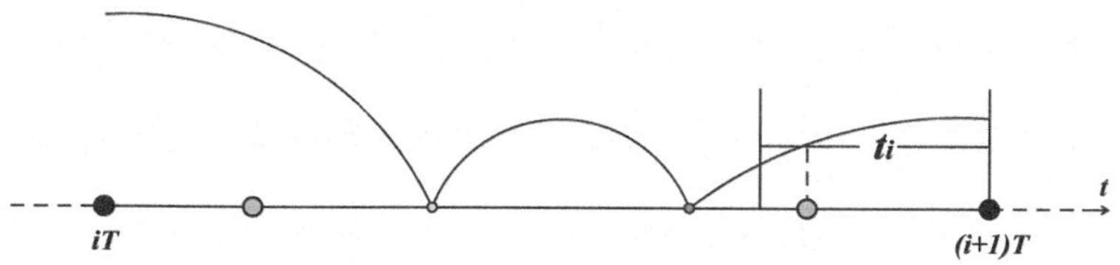

Figure 18.7 Scenario 3 of the random inspection renewal

$$
\begin{aligned}
&P_{dl}(i,T,t_i)\\
&= \Pr\left\{\begin{matrix}\max(iT,\tau)<X<\tau+\varepsilon, 0<Y<\tau+\varepsilon-X, Z>(i+1)T-X-Y,\\ 0\le\tau<(i+1)T, \max(0,(i+1)T-t_i-\tau)<\varepsilon<(i+1)T-\tau\end{matrix}\right\}\\
&= \int_{(i+1)T-t_i}^{(i+1)T}\int_0^{\varepsilon\text{-}x}\int_{iT}^{\varepsilon} h(\varepsilon)u_X(x)v_Y(y)\overline{W_Z}((i+1)T-x-y)\,dxdyd\varepsilon\\
&\quad+\sum_{n=1}^{\infty}\int_0^{(i+1)T}\int_{\max(0,(i+1)T-t_i-\tau)}^{(i+1)T-\tau}\int_0^{\tau+\varepsilon\text{-}x}\int_{\max(iT,\tau)}^{\tau+\varepsilon} h^{(n)}(\tau)h(\varepsilon)\\
&\qquad\times u_X(x)v_Y(y)\overline{W_Z}((i+1)T-x-y)\,dxdyd\varepsilon d\tau. \quad (18.9)
\end{aligned}
$$

Similar to Eqs. (18.3) and (18.6), the expected number of random inspections due to a periodic inspection renewal $E(N_{ir}^{dp})$ is given by

$$
\begin{aligned}
E\left(N_{ir}^{dp}\right) &= \sum_{i=0}^{\infty}\sum_{n=1}^{\infty}\int_{iT}^{(i+1)T}\int_0^x nh^{(n)}(\tau)\overline{H}((i+1)T-\tau)u_X(x)\overline{V_Y}((i+1)T-x)d\tau dx\\
&+\sum_{i=0}^{\infty}\sum_{n=1}^{\infty}\int_{iT}^{(i+1)T}\int_0^{(i+1)T-x}\int_{(i+1)T-x-y}^{\infty}\int_{(i+1)T-\tau}^{\infty}\int_0^x nh^{(n)}(\tau)h(\varepsilon)\\
&\qquad\times u_X(x)v_Y(y)w_Z(z)\,d\tau d\varepsilon dzdydx\\
&+\sum_{i=0}^{\infty}\int_{(i+1)T-t_i}^{(i+1)T}\int_0^{\varepsilon\text{-}x}\int_{iT}^{\varepsilon} h(\varepsilon)u_X(x)v_Y(y)\overline{W_Z}((i+1)T-x-y)\,dxdyd\varepsilon\\
&+\sum_{i=0}^{\infty}\sum_{n=1}^{\infty}\int_0^{(i+1)T}\int_{\max(0,(i+1)T-t_i-\tau)}^{(i+1)T-\tau}\int_0^{\tau+\varepsilon\text{-}x}\int_{\max(iT,\tau)}^{\tau+\varepsilon}(n+1)h^{(n)}(\tau)h(\varepsilon)\\
&\qquad\times u_X(x)v_Y(y)\overline{W_Z}((i+1)T-x-y)dxdyd\varepsilon d\tau. \quad (18.10)
\end{aligned}
$$

18.3.3 THE EXPECTED RENEWAL CYCLE COST

Using Eqs. (18.1), (18.2) and (18.3), the expected cost due to a failure renewal is given by

$$
\begin{aligned}
E(C_F(T,t_i)) &= C_f\sum_{i=0}^{\infty}\{P_{f1}(i,T)+P_{f2}(i,T,t_i)\}\\
&\quad+C_{ip}\sum_{i=0}^{\infty}\{iP_{f_1}(i,T)+iP_{f2}(i,T,t_i)\}+C_{ir}E\left(N_{ir}^f\right).
\end{aligned}
$$

Using Eqs. (18.4), (18.5) and (18.6), the expected cost due to a random inspection renewal is given by

$$
\begin{aligned}
E(C_{DR}(T,t_i)) &= C_{dr}^1\sum_{i=0}^{\infty}P_{dr}^1(i,T)+C_{dr}^2\sum_{i=0}^{\infty}P_{dr}^2(i,T,t_i)\\
&\quad+C_{ip}\sum_{i=0}^{\infty}\left\{iP_{dr}^1(i,T)+iP_{dr}^2(i,T,t_i)\right\}+C_{ir}E\left(N_{ir}^{dr}\right).
\end{aligned}
$$

Using Eqs. (18.7), (18.8), (18.9) and (18.10), the expected cost due to a periodic inspection renewal is given by

$$E\left(C_{DP}\left(T,t_i\right)\right) = C_{dp}^1 \sum_{i=0}^{\infty} P_{dp}^1\left(i,T\right) + C_{dp}^2 \sum_{i=0}^{\infty} P_{dp}^2\left(i,T\right) + C_{dl} \sum_{i=0}^{\infty} P_{dl}\left(i,T,t_i\right)$$
$$+ C_{ip} \sum_{i=0}^{\infty} \left\{\left(i+1\right) P_{dp}^1\left(i,T\right) + \left(i+1\right) P_{dp}^2\left(i,T\right) + iP_{dl}\left(i,T,t_i\right)\right\}$$
$$+ C_{ir} E\left(N_{ir}^{dp}\right).$$

To sum up, the expected renewal cycle cost is

$$E\left(C\left(T,t_i\right)\right) = E\left(C_F\left(T,t_i\right)\right) + E\left(C_{DR}\left(T,t_i\right)\right) + E\left(C_{DP}\left(T,t_i\right)\right).$$

18.3.4 THE EXPECTED RENEWAL CYCLE LENGTH

The expected renewal cycle length is

$$E\left(L\left(T,t_i\right)\right) = E\left(L_F\left(T,t_i\right)\right) + E\left(L_{DR}\left(T,t_i\right)\right) + E\left(L_{DP}\left(T,t_i\right)\right)$$
$$= \sum_{i=0}^{\infty} \int_0^T \left(iT+\varphi\right) g_{f_1}\left(i,T,\varphi\right) d\varphi + \sum_{i=0}^{\infty} \int_{T-t_i}^T \left(iT+\varphi\right) g_{f_2}\left(i,T,t_i,\varphi\right) d\varphi$$
$$+ \sum_{i=0}^{\infty} \int_0^T \left(iT+\varphi\right) g_{dr}^1\left(i,T,\varphi\right) d\varphi + \sum_{i=0}^{\infty} \int_0^{T-t_i} \left(iT+\varphi\right) g_{dr}^2\left(i,T,t_i,\varphi\right) d\varphi$$
$$+ \sum_{i=0}^{\infty} \left(i+1\right) T \left(P_{dp}^1\left(i,T\right) + P_{dp}^2\left(i,T\right) + P_{dl}\left(i,T,t_i\right)\right).$$

In virtue of the renewal-reward theorem, the long-run expected cost per unit time under the proposed maintenance policy can be formulated as follows,

$$\overline{C}\left(T,t_i\right) = \frac{E\left(C\left(T,t_i\right)\right)}{E\left(L\left(T,t_i\right)\right)}.$$

The aim is to find the optimal interval T and postponement threshold t_i that minimize $\overline{C}\left(T,t_i\right)$.

18.4 THREE MAINTENANCE POLICIES

For comparison, we propose three additional maintenance policies, i.e., a pure periodic inspection policy, a pure random inspection policy and a periodic and random inspection policy without postponed repair for the mission-based system.

- **Pure periodic inspection policy**: The system only undergoes periodic inspections, and repair or replacement is always immediate.

Table 18.3
Distribution functions and their parameters

Normal stage	Minor defective stage	Major defective stage
$U_X(x) = 1 - e^{-\eta_1 x}$	$V_Y(y) = 1 - e^{-\eta_2 y}$	$W_Z(z) = 1 - e^{-\eta_3 z}$
$\eta_1 = 0.063$	$\eta_2 = 0.104$	$\eta_3 = 0.166$

Table 18.4
Cost parameters

C_f	C_{dp}^1	C_{dr}^1	C_{dp}^2	C_{dr}^2	C_{dl}	C_{ip}	C_{ir}
1300	150	150	200	200	100	50	10

- **Pure random inspection policy**: The system is only inspected at the completion of each mission, and repair or replacement is always immediate once the system is defective or fails.

- **Periodic and random inspection policy without postponed repair**: For this policy, both random and periodic inspections are scheduled, and repair or replacement is always immediate.

The renewal probabilities for these three policies can be formulated in a similar way to the policy proposed in Section 18.3, which are omitted here due to space limitation.

18.5 NUMERICAL EXAMPLES

We assume the three stages of the system follow three exponential distributions. The distribution functions are given in Table 18.3, and the cost parameters are shown in Table 18.4. From the parameters in Table 18.3, the expected length of every stage can be obtained, $EX = 15.87$, $EY = 9.62$ and $EZ = 6.02$. This is consistent with the situation in practical industry that the three stages in the deterioration process shorten gradually. We also suppose that $Z_n, n = 1, 2, \cdots$ follow an exponential distribution and the distribution function is $H(t) = 1 - e^{-\lambda t}$ (see Nakagawa et al. [21] and Zhao et al. [22]), then

$$h^{(n)}(t) = \frac{\lambda(\lambda t)^{n-1} e^{-\lambda t}}{(n-1)!}.$$

Under a pure periodic inspection policy, it should be noted that when $T \to \infty$, there is actually no preventive action, and replacements are carried out only when the system fails. Therefore, $\overline{C}^{(1)}(\infty)$ is a positive and finite constant, i.e., $\overline{C}^{(1)}(\infty) \in (0, \infty)$. Additionally, the continuity of $\overline{C}^{(1)}(T)$ holds. Then there exits an optimal solution T^* that minimizes the objective function $\overline{C}^{(1)}(T)$ and $\overline{C}^{(1)}(T^*)$ is finite. From Figure 18.8, we can see that the minimal expected cost per unit time is obtained at $T^* = 7$.

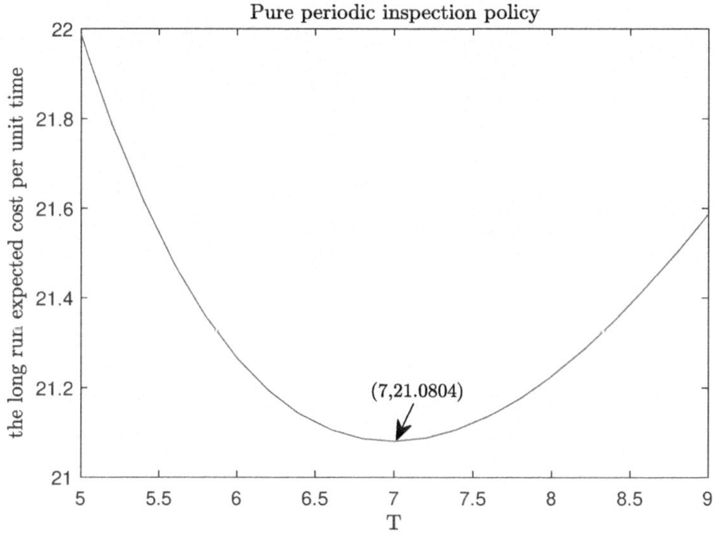

Figure 18.8 The long-run expected cost per unit time in terms of T under a pure periodic inspection policy

Figure 18.9 shows that the long-run expected cost per unit time in terms of T under periodic and random inspection policy without postponed repair have the same tendency as those in a pure periodic inspection policy.

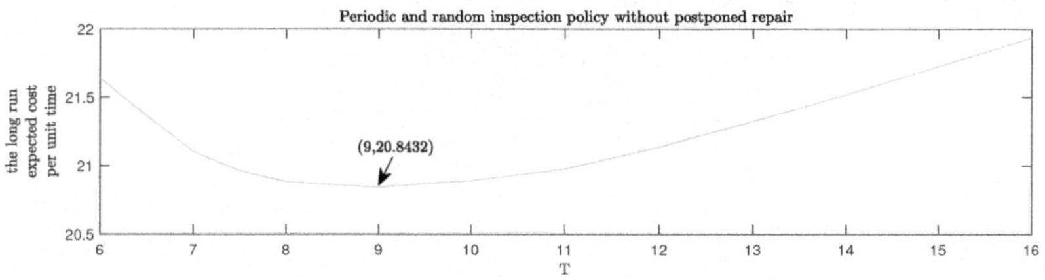

Figure 18.9 The long-run expected cost per unit time in terms of T under periodic and random inspection policy without postponed repair

For the assumed parameter settings, we obtain the long-run expected cost per unit time with different periodic inspection interval T and postponement threshold t_i under the proposed policy, as shown in Figure 18.10 (or Table 18.5). The search for the optimal T^* and $t_i{}^*$ is carried out by enumeration.

Table 18.5

The long-run expected cost per unit time in terms of T and t_i under the proposed policy

t_i	T					
	9	10	11	12	13	14
1	21.2577	20.9926	20.8833	20.8759	21.6569	22.9372
2	20.6985	20.3413	20.1418	20.0478	20.6821	21.7824
3	20.4091	19.9798	19.7083	19.5441	20.0696	21.0377
4	20.2916	19.8097	19.4830	19.2629	19.7076	20.5778
5	20.2757	19.7614	19.3965	19.1348	19.5211	20.3205
6	20.3082	19.7823	19.3970	**19.1086**	19.4564	20.2077
7	20.3495	19.8312	19.4435	19.1442	19.4719	20.1951
8	20.3751	19.8779	19.5038	19.2090	19.5339	20.2472

From Table 18.4, we can uncover the optimal T^* which minimized the long-run expected cost per unit time for a fixed t_i. Then we can uncover the optimal ${t_i}^*$ which minimized the long-run expected cost per unit time for a fixed T. Therefore $\overline{C}\,(T, t_i)$ is jointly convex in $(T,\ t_i)$, then there exists an optimal $(T^*,\ {t_i}^*)$ which minimizes $\overline{C}\,(T, t_i)$.

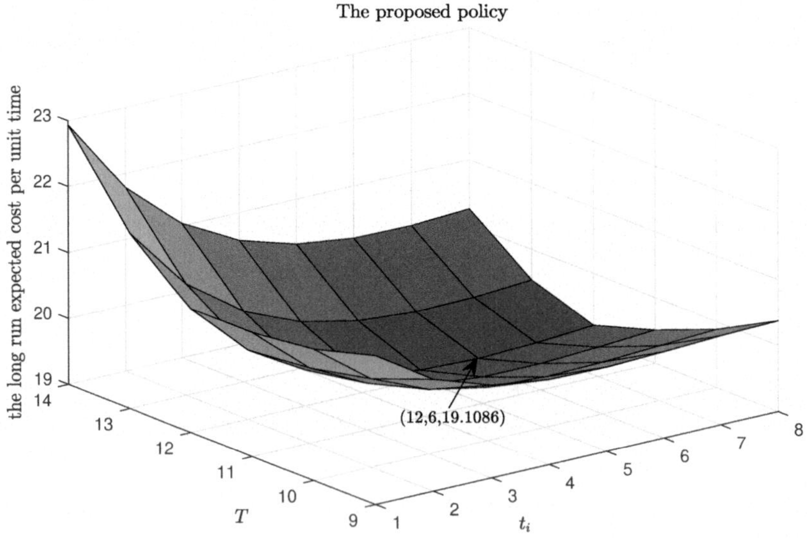

Figure 18.10 The long-run expected cost per unit time in terms of T and t_i under the proposed policy

Table 18.6

A comparison between the proposed policy and another three policies in terms of the optimal decision variables (PPI: Pure periodic inspection; PRI: Pure random inspection policy; PRI-WP: Periodic and random inspection without postponed repair)

	The Proposed Policy	PPI policy	PRI policy	PPI-WPR policy
Optimal decision variables	$(T^* = 12, t_i^* = 6)$	$T^* = 7$	–	$T^* = 9$
Optimal long-run expected cost per unit time	19.1086	21.0804	21.7334	20.8432

Table 18.6 gives a comparison between the proposed policy and another three policies in terms of the optimal long-run expected cost per unit time and the corresponding decision variables. Compared with these three policies, the cost savings achieved by the proposed policy are 10.32%, 13.74%, and 9.08%, respectively. However, the outcome depends mainly on the specified cost parameters.

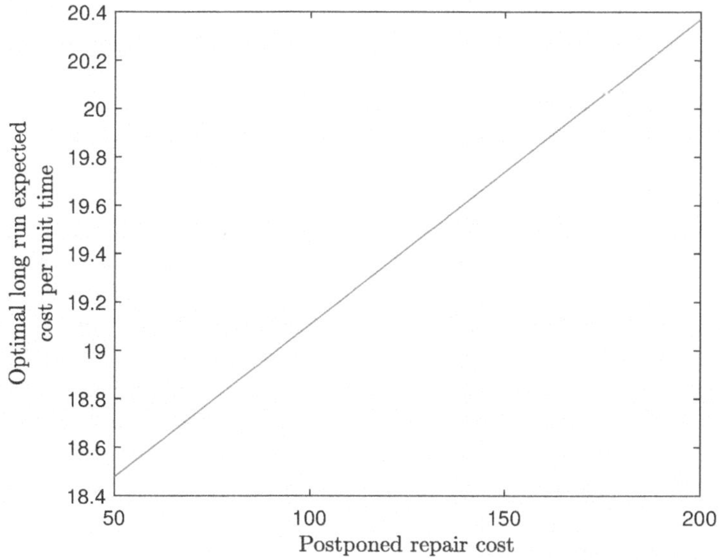

Figure 18.11 The optimal long-run expected cost per unit time versus the postponed repair cost

It is worthwhile to investigate the sensitivity of the optimal long-run expected cost per unit time with respect to the postponed repair cost. Consider the range $50 \leq C_{dl} \leq 200$, and keep the other cost parameters unchanged as in Table 18.4. Figure 18.11 shows that the optimal long-run expected cost per unit time increases with the increase of C_{dl}. In addition, when the postponed repair cost approaches the immediate repair cost, the difference between the proposed policy and a periodic and random inspection policy without postponed repair is negligible in terms of the optimal long-run expected cost per unit time.

18.6 CONCLUSIONS AND FURTHER RESEARCH

In this paper, a combined periodic and random inspection policy was proposed for a mission-based system based on a three-stage failure process. If the system is major defective at a random inspection, repair will be postponed if the time to the following periodic inspection is less than a predetermined threshold, otherwise repair will be immediate. If the system is minor defective at a periodic inspection, it is repaired immediately. The renewal probabilities of the system have been presented and then the long-run expected cost per unit time were derived accordingly using the renewal-reward theorem. Three additional policies, i.e., a pure periodic inspection policy, a pure random inspection policy, and a periodic and random inspection policy without postponed repair were proposed for comparison purpose. The numerical example showed that when both the random inspection cost and the postponed repair cost are low, the proposed policy is more cost-effective compared with other three policies.

There are several further research topics worthy of investigating in the future. (1) If the system is found to be in a minor defective stage, the system may not be necessarily replaced or repaired immediately. (2) Future research could be linking maintenance with spare part provision. If a delayed repair is carried out, the ordering quantity of spare parts will be affected. (3) The combined random and periodic inspection policy can be adapted to multi-component or multi-failure mode systems, which are usually observed in industry. (4) Imperfect maintenance (including imperfect inspections and imperfect repairs) for a system subject to random and periodic inspection is worth further studying. (5) Many industrial systems (e.g., transportation devices, manufacturing systems, energy generation systems and oil pipeline networks) suffer from inevitable failures due to complex degradation processes and environmental conditions such as random shocks. A replacement policy should be proposed to deal with both internal deterioration and external shock damages for a system subject to random and periodic inspection.

ACKNOWLEDGMENTS

We acknowledge support from the National Natural Science Foundation of China (Grant nos. 71871008, 71571014) and support from the Emerging Interdisciplinary Project of Central University of Finance and Economics (grant no. 21XXJC010).

REFERENCES

1. Wang, W., & Wang, H. (2015). Preventive replacement for systems with condition monitoring and additional manual inspections. *European Journal of Operational Research*, 247(2), 459-471.

2. Taghipour, S., Banjevic, D., & Jardine, A. K. S. (2010). Periodic inspection optimization model for a complex repairable system. *Reliability Engineering and System Safety*, 2010, 95(9), 944-952.

3. Golmakani, H. R., & Moakedi, H. (2012). Periodic inspection optimization model for a two-component repairable system with failure interaction. *Computers and Industrial Engineering*, 63(3), 540-545.

4. Golmakani, H. R., & Moakedi, H. (2012). Periodic inspection optimization model for a multi-component repairable system with failure interaction. *The International Journal of Advanced Manufacturing Technology*, 61(1-4), 295-302.

5. Zhao, X., Qian, C., Nakamura, S., & Nakagwa, T. (2012). Random Inspection Policies for a Database System. *Computer Science & Service System (CSSS), 2012 International Conference on.* IEEE, 191-194.

6. Pinedo, M. L. (2016). *Scheduling: Theory, Algorithms, and Systems.* Springer.

7. Taghipour, S., & Banjevic, D. (2012). Optimal inspection of a complex system subject to periodic and opportunistic inspections and preventive replacements. *European Journal of Operational Research*, 220(3), 649-660.

8. Yang, L., Ma, X., & Zhao, Y. (2016). Random and periodic inspection of mission-based systems with a defective stage. *Reliability and Maintainability Symposium (RAMS), 2016 Annual.* IEEE, 1-6.

9. Yang, L., Ma, X., Zhai, Q., & Zhao, Y. (2016). A delay time model for a mission-based system subject to periodic and random inspection and postponed replacement. *Reliability Engineering and System Safety*, 150, 96-104.

10. Christer, A. H. (1976). Innovative decision making, *Proceedings of the NATO Conference on the Role and Effectiveness of Theories of Decision in Practice.* Hodder and Stoughton, 368-377.

11. Aven, T., & Castro, I. T. (2009). A delay-time model with safety constraint. *Reliability Engineering and System Safety*, 94(2), 261-267.

12. Jones, B., Jenkinson, I., & Wang, J. (2009). Methodology of using delay time analysis for a manufacturing industry. *Reliability Engineering and System Safety*, 94(1), 111-124.

13. Wang, W., Banjevic, D., & Pecht, M. A. (2010). Multi-component and multi-failure mode inspection model based on the delay time concept. *Reliability Engineering and System Safety*, 95(8), 912-920.

14. Wang, W. (2008). Delay time modelling. *Complex System Maintenance Handbook.* Springer, London, 345-370.

15. Wang, W. (2012). An overview of the recent advances in delay-time-based maintenance modelling. *Reliability Engineering and System Safety*, 106, 165-178.

16. Wang, W., Scarf, P. A., & Smith, M. A. J. (2000). On the application of a model of condition-based maintenance. *Journal of the Operational Research Society*, 51(11), 1218-1227.

17. Wang, W. (2011). An inspection model based on a three-stage failure process. *Reliability Engineering and System Safety*, 96(7), 838-848.

18. Wang, H., Wang, W., & Peng, R. (2017). A two-phase inspection model for a single component system with three-stage degradation. *Reliability Engineering and System Safety*, 158, 31-40.

19. van Oosterom, C. D., Elwany, A. H., Çelebi, D., & van Houtum, G. J. J. A. N. (2014). Optimal policies for a delay time model with postponed replacement. *European Journal of Operational Research*, 232(1), 186-197.

20. Wang, W., Zhao, F., & Peng, R. (2014). A preventive maintenance model with a two-level inspection policy based on a three-stage failure process. *Reliability Engineering and System Safety*, 121, 207-220.

21. Nakagawa, T., Mizutani, S., & Chen, M. A. (2010). Summary of periodic and random inspection policies. *Reliability Engineering and System Safety*, 95(8), 906-911.

22. Zhao, X., & Nakagawa, T. (2015). Optimal periodic and random inspections with first, last and overtime policies. *International Journal of Systems Science*, 46(9), 1648-1660.